Gas Separation Membranes

Ahmad Fauzi Ismail • Kailash Chandra Khulbe
Takeshi Matsuura

Gas Separation Membranes

Polymeric and Inorganic

Springer

Ahmad Fauzi Ismail
Advanced Membrane Technology Research Center (AMTEC)
Universiti Teknologi Malaysia
Johor Bahru, Malaysia

Takeshi Matsuura
Department of Chemical & Biological Engineering
University of Ottawa
Ottawa, ON, Canada

Kailash Chandra Khulbe
Industrial Membrane Research Laboratory
Department of Chemical & Biological Engineering
University of Ottawa
Ottawa, ON, Canada

ISBN 978-3-319-34327-3 ISBN 978-3-319-01095-3 (eBook)
DOI 10.1007/978-3-319-01095-3

Springer Cham Heidelberg New York Dordrecht London

Softcover reprint of the hardcover 1st edition 2015

Printed on acid-free paper

Springer International Publishing AG Switzerland is part of Springer Science+Business Media (www.springer.com)

Preface

Over the last decade, there have been growing interests in the advancement and applications of membrane-based gas separation technology to tackle various global issues and challenges. Tremendous progress has been made in the development of gas separation membranes based on both inorganic and polymeric materials for applications in a wide range of industrial sectors. Interestingly, new polymers and copolymers as well as advanced materials such as metal organic frameworks (MOF) and composite materials such as polymer inclusion membranes (PIM) have been introduced. Tremendous efforts have also been made to design and fabricate asymmetric membranes based on these emerging materials. Inorganic materials, particularly those in the nano-sized range, have been incorporated in polymeric membranes to prepare mixed matrix membranes (MMMs) with higher selectivity and permeability to surpass the Robeson upper boundary limits.

In this contribution, the authors give a contemporary and comprehensive review of the progress made in the field of gas separation membranes and membrane gas separation processes. This book covers innovative designs and inventions of new materials for polymeric and inorganic membrane preparation. It also emphasizes the recent advances made for the characterizations of membranes (atomic force microscopy (AFM), Fourier transform infrared (FTIR) spectroscopy, X-ray diffraction, electron spin resonance (ESR), and positron annihilation spectroscopy). Recent progresses in membrane module and system design are also included. Likely future R&D directions relevant to the development of gas separation membranes and gas separation processes are also highlighted. Since such a contribution is unprecedented, it is the main intention of the authors to promote this book as a small dictionary to cover a broad range of subjects related to gas separation membranes which include material, theory, preparation, characterization, membrane module, and system design and applications.

The book consists of the following chapters:

Chapter 1 briefly introduces the membrane gas separation processes. The process is outlined and compared with other gas separation processes to highlight the advantages and limitations of membrane gas separation processes.

Chapter 2 deals with the fundamental theory of gas transport through the membrane. Transport equations for porous and nonporous membranes are provided and discussed. Gas transport through rubbery and glassy polymers is distinguished and the transport models for the mixed matrix membrane are outlined.

Chapter 3 delivers the most important information where the membranes developed from different materials, including organic and inorganic, and the mixture of organic and inorganic materials in the composite MMMs are summarized. Carbon-based membranes are also outlined in this chapter. Developments and preparations of MOFs and membranes based on inorganic materials such as zeolites, silica, and metals are also described in this chapter.

Chapter 4 summarizes methods to prepare membrane materials and membranes. Membrane preparation method based on conventional phase inversion technique, hollow fiber spinning, and membrane surface modification by coating and other methods are presented thoroughly in this chapter.

Chapter 5 is a relatively short chapter. Various membrane modules, such as plate and frame, spiral wound, tubular, capillary, and hollow fiber, are briefly outlined and their advantages and disadvantages are shown. The concept of membrane contactors is also included in this chapter.

Chapter 6 includes various applications of membrane gas separation processes. The chapter also deals with membrane separation processes that are related to membrane gas separation, such as pervaporation and membrane distillation. The hybrid processes of membrane gas separation and other separation processes are also outlined in this chapter.

Finally, Chapter 7 summarizes membrane characterization methods which include macroscopic and spectrophotometric methods. Advanced characterization methods such as positron annihilation are also included in this chapter.

The authors believe that readers in universities, research institutions, and industry who are engaged in research on membranes for gas separation processes will benefit from the scientific knowledge contained in this book. It is therefore the authors' wish and ultimate goal to render general yet significant contributions to the further development of membrane science and technology particularly in membrane-based gas separation.

Johor Bahru, Malaysia A.F. Ismail
Ottawa, ON, Canada K.C. Khulbe
Ottawa, ON, Canada T. Matsuura

Contents

Chapter 1
Introduction

1.1 Membrane Separation Processes

Membrane separation technology is based on the interaction of specific gases with the membrane material by a physical or chemical interaction. Membrane processes are considered to be visible and effective technologies for the separation of gaseous mixtures at the industrial scale due to their high efficiency, simple operation, and low cost. Membrane processes encompass a wide range of applications in fluid separation and are now considered a new and emerging separation technology for industrial applications. For several important separation processes, membrane technology has now reached its initial stage of maturity.

As a general definition, a membrane is regarded as a selective barrier between two phases. A membrane is usually a solid film, and occasionally a fluid film, of a small but observable thickness. It is characterized by flux and selectivity properties that provide functional transport across the barrier. The driving force for transport across the membrane is the chemical potential gradient, and the physical structure of the membrane determines the flux. The difference in flux between individual penetrant components determines membrane selectivity. In other words, membrane separation is a technology which selectively separates (fractionates) materials via pores and/or minute gaps in the molecular arrangement of a continuous structure. Membrane separations are often classified by pore size and by the separation driving force. Such classifications include: Microfiltration (MF), Ultrafiltration (UF), Ion-Exchange (IE), and Reverse Osmosis (RO).

A.F. Ismail et al., *Gas Separation Membranes*,
DOI 10.1007/978-3-319-01095-3_1

1.2 Membrane-Based Gas Separation

Membrane-based gas separation can refer to any of a number of techniques used to separate gases, either to give multiple products or to purify a single product. A gas conforms to the shape of a container in which it is held and acquires a uniform density inside the container, even in the presence of *gravity* and regardless of the amount of substance in the container. If not confined to a container, gases, also known as vapors, will disperse into *space*. The *atoms* or *molecules* of matter in a gaseous state move freely among each other, and are, in most instances, packed more loosely than the molecules of the same substance in the *solid* or *liquid* state.

The most important characteristics of gases include:

1. *High compressibility*: External forces can compress a gas and decrease its volume; removing the external force allows the gas volume to increase.
2. *Thermal expandability*: When a gas sample is heated, its volume increases, and when it is cooled its volume decreases.
3. *Low viscosity*: Gases flow much easier than liquids.
4. *Low density*: Gas densities are on the order of grams per liter, whereas liquids and solids are grams per cubic cm—1,000 times greater.
5. *Infinite miscibility*: Gases mix in any proportion, such as in air, and in any combination.

Important variables that describe a gas are pressure (P), temperature (T), and volume (V).

In gas mixtures, the gas behavior depends on the number, not the identity, of gas molecules. The ideal gas equation applies to each gas individually and to the mixture as a whole. Beside these properties, all molecules in gas mixtures behave exactly the same way.

Membrane technology has received significant attention from various industrial sectors and academics in their research because the technology offers the most relevant means of reducing environmental problems and costs. The gas separation membrane field is highly competitive both between companies developing membrane technologies and with other gas separation technology developers. Benny D. Freeman [1] says "Membranes will have a large role to play in important environmental and energy-related processes such as the cost-effective purification of hydrogen and methane."

Membranes are employed in:

- the separation of *nitrogen* or *oxygen* from air;
- separation of *hydrogen* from gases like *nitrogen* and *methane*;
- recovery of hydrogen from product streams of *ammonia* plants;
- recovery of hydrogen in *oil refinery* processes;
- separation of methane from the other components of *biogas*;
- *enrichment of air by oxygen* for medical or metallurgical purposes;
- enrichment of ullage by nitrogen in *inerting systems* designed to prevent fuel tank explosions;
- removal of *water vapor* from *natural gas* and other gases;

- removal of CO_2 from natural gas;
- removal of H_2S from natural gas;
- removal of *volatile* organic liquids (VOL) from air of exhaust streams.

Inorganic membranes are increasingly being explored to separate gas mixtures. Such membranes usually have much higher gas fluxes as compared to polymeric membranes. Furthermore, inorganic membranes are well-known to be more thermally and chemically stable than polymeric membranes.

1.2.1 Historical Background

In the middle of the eighteenth century, Nollet [2] discovered that a pig's bladder passes preferentially ethanol when it was brought into contact on one side with a water–ethanol mixture and on the other side with pure water. This was the first recorded study of membrane phenomena and the discovery of osmosis phenomena. Graham [3] systematically studied on mass transport in semipermeable membranes during diffusion of gases through different media and reported that natural rubber exhibits different permeabilities to different gases. Fick [4] interpreted diffusion in liquids as a function of concentration gradients and derived the Law of Mass Diffusion. Table 1.1 shows briefly the main events related to the development of gas separation membranes.

1.2.2 Scientific and Commercial Development of Membrane Processes

After the invention of a cellulose acetate high-flux asymmetric membrane by Loeb and Sourirajan for RO in 1962, and its application for gas separation that followed, intense development activities were started to produce high-performance membranes economically. Although gas separation membranes have been known for more than a century, only during the last 30 years have membranes been used on an industrial scale for gas separation (GS). Membrane-based GS has grown exponentially since the first industrial application of Prism membranes by Permea (Monsanto) for hydrogen separation from the purge gas stream of ammonia plants. The use of synthetic membranes commercially by industrial gas suppliers, petroleum producers, chemical companies, and refineries began in the early 1980s. Figure 1.1 shows the major events of industrial membrane gas separation that have occurred in the last 30 years.

The largest membrane plant for natural gas processing (CO_2–CH_4 separation) installed in Pakistan in 1995 with spiral wound modules is included in Fig. 1.1, which is a clear example of the easy scale-up of membrane technology [50]. Table 1.2 shows the commercial applications and current major suppliers of membrane gas separation.

Table 1.1 Events in development of gas separation membranes

Scientist (year)	Events
Nollet (1752) [2]	Discovered that a pig's bladder passes preferentially ethanol when it was brought in contact on one side with a water–ethanol mixture and on the other side with pure water. This was the first recorded study of membrane phenomena and the discovery of osmosis phenomena
Graham (1829) [3]	Performed the first recorded experiment on the transport of gases and vapors in polymeric membranes
Fick (1855) [4]	Proposed a quantitative description of material transport through boundary layers
Graham (1866) [3]	Systematically studied on mass transport in semipermeable membranes during diffusion of gases through different media and reported that the natural rubber exhibits different permeabilities to different gases. Graham's Law of Mass Diffusion was proposed
Lord Rayleigh (1900) [5]	Determined relative permeabilities of oxygen, argon, and nitrogen in rubber
Benchold (1907) [6]	Prepared nitrocellulose membranes with graded pore size structure. Defined the relationship between bubble point and temperature, surface tension, and pore radius
Knudsen (1908) [7]	Defined Knudsen diffusion
Shakespear (1917–1920) [8–10]	Found temperature dependency of gas permeability that is independent of partial pressure difference across membranes
Daynes (1920) [11]	Developed time lag method to determine diffusion and solubility coefficient
Barree (1939) [12]	Applied Arrhenius equation for permeabilities and diffusivities
Barrer and Strachan (1955) [13]	Studied the diffusion and the adsorption of permanent gases through compressed carbon powders
Loeb and Sourirajan (1962, 1964) [14, 15]	Developed RO membrane based on cellulose acetate, which provided high fluxes at moderate hydrostatic pressures. Found dried RO membrane can be used for gas separation
Vieth and Sladek (1965) [16]	Proposed models for sorption and diffusion in glassy polymers
Stern et al. (1969) [17]	First to systematically study the transport of gases in high polymers at elevated temperatures
Cynara and Separex Company (1982–1983) [18]	Developed cellulose acetate membranes for the separation of CO_2
Henis and Tripodi (1980) (Monsanto, Inc.) [19]	The first major product Monsanto Prism A® membrane for hydrogen separation
Gies (1986) [20]	All-silica zeolite deca-dodecasil 3R (DD3R)
Paul and Kemp (1973) [21]	First reported MMMs for gas separation
Permea PRISM membrane (1980) [22]	First commercialized gas separation membrane
Kulprathipanja et al. (1988) [23, 24]	Mixed matrix systems of polymer/adsorbent might yield superior separation performance to that of pure polymeric system
Robeson (1991) [25]	Proposed upper bound between gas permeability and selectivity
Iijima (1991) [26]	Discovery of carbon nanotubes (CNTs)

(continued)

Table 1.1 (continued)

Scientist (year)	Events
Suda and Haraya (1997) [27]	Prepared carbon molecular sieve (CMS) membrane, prepared from pyrolization of polyimide, and permeabilities of different gases were studied ($H_2 > He > CO_2 > O_2 > N_2$)
McKeown (1998) [28]	Polymers of intrinsic microporosity (PIMS)
Yang et al. (1999) [29]	Proposed gas separation by zeolite membranes on the basis of different adsorption properties
Caro et al. (2000) [30]	Proposed gas separation by zeolite membranes on the basis of differences in the molecular size and shape
Mahajan and Koros (2002) [31–33]	Application of 4A zeolite in polymers for MMMs membrane preparation
Skoulidas et al. (2002) [34]	Made simulations for both self- and transport diffusivities of light gases such as H_2 and CH_4 in carbon nanotubes and zeolite
Ackerman et al. (2003) [35]	Made simulations for Ar and Ne transport through CNTs
Hinds et al. (2004) [36]	Tried to incorporate aligned CNTs into the polymer matrix and proposed the potential of the nanotubes' inner cores to act as a channel for gas transport
McKeown et al. (2005) [37]	PIMS are excellent performers for gas separation
Chen and Sholl (2006) [38]	Predicted selectivity and flux of CH_4–H_2 separation using single-walled carbon nanotubes as membranes
Gonzo et al. (2006) [39]	Applied Maxwell equation to the performance of mixed matrix membranes (MMMs)
Cong et al. (2007) [40]	Used $BPPO_{dp}$ membranes using both pristine single-wall CNTs (SWNTs) and multi-wall CNTs (MWNTs). Composite membranes increased in CO_2 permeability compared to the corresponding pure-polymer membrane
Husain and Koros (2007) [41]	Increased hydrophobicity of the zeolite surface by capping surface hydroxyls with hydrophobic organic chains via Grignard-type reactions (MMMs preparation)
Himeno et al. (2007) [42]	α-alumina surface coated with DDR zeolite for gas separation
Bergh et al. (2008) [43]	Separation and permeation characteristics of a DDR zeolite membrane (gas separation)
Yoo, Lai, and Jeong group (2009) [44]	The first MOF membranes were reported
Li et al. (2010) [45]	SAPO-34 zeolite membranes for CO_2–CH_4 separation
Aroon et al. (2010) [46]	Fabricated PI/Raw-MWCNTs and PI/Chitosan functionalized MWCNTs (PI/C-f-MWCNTs) MMMs membranes for gas permeation
Mustafa et al. (2010) [47]	Fabricated MMMs consisting of functionalized carbon nanotubes (CNTs) and polyethersulfone (PES) for biogas purification. The PES-modified carbon nanotubes membranes had increased permeability and CO_2–CH_4 selectivity
Betard et al. (2012) [48]	Metal–organic framework (MOF) membrane by stepwise deposition of reactants
Li et al. [49]	Reported the first mixed matrix composite membrane made of commercially available poly(amide-b-ethylene oxide) (Pebax®1657, Arkema)

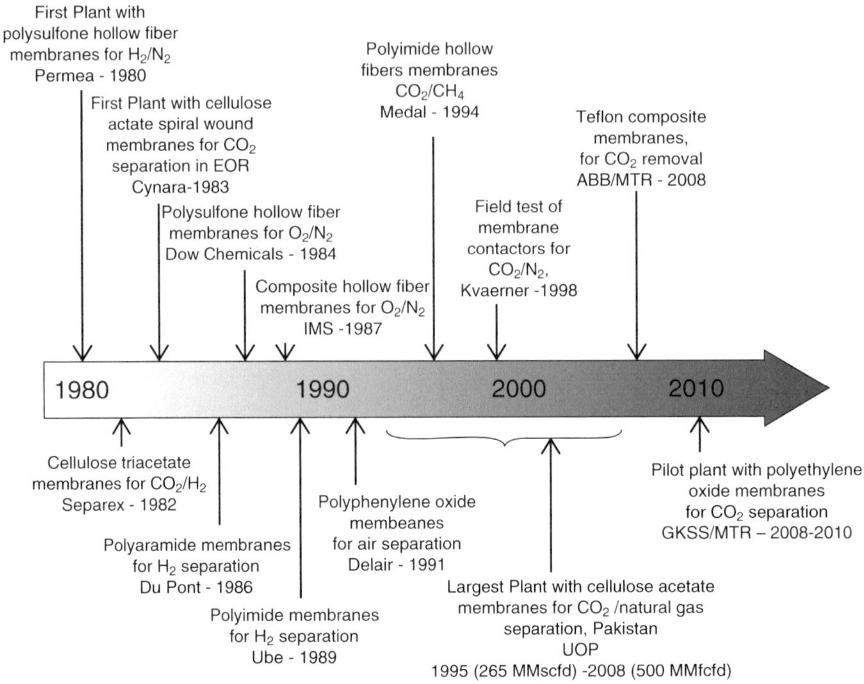

Fig. 1.1 Milestones in the industrial application of membrane gas separation systems

Table 1.2 Commercial applications and current major suppliers of membrane gas separation

Gas separation	Application	Supplier
O_2–N_2	Nitrogen generation Oxygen enrichment	Permea (Air Products) Generon (IGS), IMS (Praxair), Medal (Air Liquid), Parker Gas Separation, Ube
H_2–hydrocarbons	Refinery hydrogen recovery	Air Products, Air Liquid Praxair, Ube
H_2–CO_2	Syngas ratio adjustment	As above
H_2–N_2	Ammonia Purge gas	As above
CO_2–CH_4	Acid gas treatment enhanced oil recovery landfill gas upgrading	Cynara (NATCO), Kvaerner, Air Products, Ube, UOP (Separex)
H_2S–hydrocarbon	Sour gas treating	As above
H_2O–hydrocarbon	Natural gas dehydration	Kvaerner, Air Products
H_2O–air	Air dehydration	Air Products, Parker Balxston Ultratroc, Praxair
Hydrocarbons–air	Pollution control hydrocarbon recovery	Borsig, MTR, GMT, NKK
Hydrocarbons from process streams	Organic solvent recovery Monomer recovery	Borsig, MTR, GMT, SIHI

1.3 Advantages of Membrane Processes

There are many significant advantages of using membranes for industrial processes. In 2002, an ad hoc committee at the International Conference on Membranes and Membrane Processes (ICOM 2002) prepared a report on membrane technology perspectives and needs. The following advantages to membrane processes are listed.

1. Do not involve phase changes or chemical additives.
2. Simple in concept and operation.
3. Modular and easy to scale up.
4. Greater efficiency for raw materials use and potential for recycling of by-products.
5. Equipment size may be decreased.

Furthermore another advantage for membrane devices for gas separation is that they usually operate under continuous steady-state conditions.

Baldus and Tillmman [51] described simple rules for gas separation by membranes to be favorable:

1. When moderate purity recovery is sufficient.
2. When the components to be separated are a considerable amount.
3. When the feed gas is available at the necessary pressure, or when the residue stream is needed at high pressure.
4. When the feed gas contains no substances harmful to the membrane.
5. When a membrane with sufficient selectivity is available.

The choice of membrane material for GS applications is based on specific physical and chemical properties, since a material should be tailored in an advanced way to separate particular gas mixtures. Membrane material should be robust (i.e., long-term and stable) for GS process.

The GS properties of membranes depend upon [52]:

1. Material (permeability, separation factors).
2. Membrane structure and thickness (permeance).
3. Membrane configuration (e.g., flat, hollow fiber).
4. Module and system design.

There are, however, some practical problems or weaknesses with membranes [52]. Membrane performance generally decreases with time. This decrease can be caused by concentration polarization and fouling. Concentration polarization occurs because of limited permeation of certain species. These species will become higher in concentration directly adjacent to the membrane, reducing permeate transport. The magnitude of this effect depends on the type of species used and the flow setup. However, concentration polarization is not a very severe problem for gas separation membranes. Fouling is due to the adsorbed species to the membrane surface as well as inside the pores. This fouling limits or even blocks the permeation of the gas species. Notable fouling species of gas separation membranes are sulfur-containing compounds such as H_2S and SO_2. To mitigate the effects of fouling, membranes can be cleaned by heating and purging with non-adsorbing gases. Small particles can best be removed from the feed flow using a filter.

Membrane deterioration may also be caused by compaction, i.e., a reduction in pore size due to pressurization. This phenomenon occurs with polymeric membranes and is usually irreversible. Most often the pore size does not return to its original value when pressure is decreased.

Other practical considerations come into play when making choices to get the optimal membrane system design. One of them is the effect of the thermal stresses on structural integrity. If temperature variation occurs, several parts of the system may experience different degrees of expansion. If there is no room to accommodate these differences in expansion, the system can be seriously damaged. Beside this, the pressure drop over a membrane unit (not the membrane itself) is directly proportional to the module length. To reduce the pressure drop, it would be advantageous to apply shorter modules; however, shorter modules will require more seals.

Another issue to be considered is the ease in starting up and shutting down the unit. A system design needs to accommodate these effects.

References

1. Freemantle M (2005) Membrane for gas separation. Chem Eng News 83(40):49–57
2. Nollet JA (1752) Recherches sur les causes du bouillonnement de liquids. Hist Acad R Sci: 57, Paris Annee MDCCXLVIII
3. Graham T (1866) On the absorption and dialytic separation of gases by Colloid Septa. Philos Trans R Soc Lond 32:399–439
4. Fick A (1855) Über Diffusion. Pogg Ann Phys Chem 94:59
5. Rayleigh L (1900) On the passage of Argon through thin films of India rubber. Philos Mag 49:220
6. Benchold H (1907) Ultrafiltration. Biochem Z 6:379
7. Knudsen M (1908) The laws of molecular flow and the internal viscous streaming of gases through tubes. Ann Phys 28:75
8. Shakespear GA (1918) On the permeability of films and of proofed fabrics. Adv. Comm. for Aeronautics Report T1164
9. Shakespear GA (1917) Adv. Comm. for Aeronautics Report No. 317 and 516
10. Shakespear GA, Daynes HA (1920) The theory of the Kathrometer. Proc R Soc 97A:273
11. Daynes HA (1920) The process of diffusion through a rubber membrane. Proc R Soc Lond Ser A 97(685):286–307
12. Barreer RM, Rideal EK (1939) Permeation, diffusion and solution of gases in organic polymers. Trans Farad Soc 35:628–643
13. Barrer RM, Strachan E (1955) Sorption, and surface diffusion in microporous carbon cylinders. Proc R Soc Lond Ser A 231(1184):52–74
14. Loeb S, Sourirajan S (1962) Seawater demineralization by means of a semipermeable membrane. In: Gould R (ed) Advances in chemistry, ACS Series No. 38. American Chemical Society, Washington, DC, pp 117–132
15. Loeb S, Sourirajan S (1964) High flow porous membranes for separating water from saline solutions. US Patent 3,133,132, 1964

16. Vieth WR, Sladek KJ (1965) A model for diffusion in a glassy polymer. J Colloid Sci 20(9):1014–1033
17. Stern SA, Mullhaupt JT, Garies PJ (1969) The effect of pressure on the permeation of gases and vapors through polyethylene. Usefulness of the corresponding states principle. AIChE J 15(1):64–73
18. Baker RW (2002) Future directions of membrane gas separation technology. Ind Eng Chem Res 41:1393–1411
19. Henis JMS, Tripodi MK (1980) A novel approach to gas separation using composite hollow fiber membranes. Sep Sci Technol 15:1059–1068
20. Gies H (1986) Studies on clatherasis: VI. Crystal structure of decadodecasil 3R: the missing link between zeolites and clathrasils. Z Kristallogr 175:93–104
21. Paul DR, Kemp DR (1973) The diffusion time lag in polymer membranes containing adsorptive fillers. J Polym Phys 41:79–93
22. Longterm Outlook to 2030 (2010) Natural gas demand and supply. The European Union of the Natural Gas Industry
23. Kulprathipanja S, Neuzil RW, Li NN (1988) Separation of fluids by means of mixed matrixembranes. US Patent 4,740,219, 1988
24. Kulprathipanja S, Neuzil RW, Li NN (1992) Separation of gases by means of mixed matrix membranes. US Patent 5,127,925, 1992
25. Robeson LM (1991) Correlation of separation factor versus permeability of polymeric membrane. J Membr Sci 62:165–185
26. Iijima S (1991) Helical microtubes of graphite carbon. Nature 354:56–58
27. Suda H, Haraya K (1997) Gas permeation through micropores of carbon molecular sieve membranes derived from Kapton polyimide. J Phys Chem B 101:3988–3994
28. McKeown NB (1998) Phthalocyanine materials: synthesis, structure and function. CUP, Cambridge, UK
29. Yang M, Crittenden BD, Perera SP, Moueddeb H, Dalmon JA (1999) The hindering effect of adsorbed components on the permeation of a non-adsorbing component through a microporous silicate membrane: the principle Barrier theory. J Membr Sci 156:1–9
30. Caro J, Noack M, Kolsch P, Schafer R (2000) Zeolite membranes—state of their development and perspective. Microporous Mesoporous Mater 38:3–24
31. Mahajan R, Koros WJ (2002) Mixed matrix membrane materials with glassy polymers. Part 1. Polym Eng Sci 42:1420–1431
32. Mahajan R, Koros WJ (2002) Mixed matrix membrane materials with glassy polymers. Part 2. Polym Eng Sci 42:1432–1441
33. Mahajan R, Koros W (2000) Factors controlling successful formation of mixed matrix gas separation materials. Ind Eng Chem Res 39:2692–2696
34. Skoulidas AI, Ackerman DM, Johnson JK, Sholl DS (2002) Rapid transport in carbon nanotubes. Phys Rev Lett 89(18):185901
35. Ackerman DM, Skoulidas AI, Sholl DS, Johnson JK (2003) Diffusivities of Ar and Ne in carbon nanotubes. Mol Simul 29:677–684
36. Hinds BJ, Chopra N, Rantell R, Andrews R, Gavalas V, Bachas LG (2004) Aligned multiwalled carbon nanotube membranes. Science 303:62–65
37. McKeown NB, Budd PM, Msayib KJ, Ghanem BS, Kingston HJ, Tattershall CE, Makhseed S, Reynolds KJ, Fritsch D (2005) Polymers of intrinsic microporosity (PIMs): bridging the void between microporous and polymeric materials. Chem Eur J 11:2610–2620
38. Chen H, Sholl DS (2006) Predictions of selectivity and flux CH_4/H_2 separations using single walled carbon nano tubes as membranes. J Membr Sci 269:152–160
39. Gonzo EE, Parentis ML, Gottifredi JC (2006) Estimating models for predicting effective permeability of mixed matrix membranes. J Membr Sci 277:46–54
40. Cong H, Zhang J, Radosz M, Shen Y (2007) Carbon nanotube composite membranes of brominated poly(2,6-diphenyl-1,4-phenylene oxide) for gas separation. J Membr Sci 294: 178–185

41. Husain S, Koros WJ (2007) Mixed matrix hollow fiber membranes made with modified HSSZ-13 zeolite in polyetherimide polymer matrix for gas separation. J Membr Sci 288:195–207
42. Himeno S, Tomita T, Suzuki K, Nakayama K, Yajima K, Yoshida S (2007) Synthesis and permeation of a DDR-type zeolite membrane for separation of CO_2/CH_4 gaseous mixtures. Ind Eng Chem Res 46(21):6989–6997
43. van den Bergh J, Zhu W, Gascon J, Moulinj JA, Kapteijn F (2008) Separation and permeation characteristics of a DDR zeolite membrane. J Membr Sci 316:35–45
44. Yoo Y, Lai Z, Jeong HK (2009) Fabrication of MOF-5 membranes using microwave-induced rapid seeding and solvothermal secondary growth. Microporous Mesoporous Mater 123:100–106
45. Li S, Carreon MA, Zhang Y, Funke HH (2010) Scale-up of SAPO-34 membranes for CO_2/CH_4 separation. J Membr Sci 352:7–13
46. Aroon MA, Ismail AF, Montazer-Rahmati MM, Matsuura T (2010) Effect of chitosan as a functionalization agent on the performance and separation properties of polyimide/multi-walled carbon nanotubes mixed matrix flat sheet membranes. J Membr Sci 364:309–317
47. Mustafa A, Kusworo TD, Busairi A, Ismail AF (2010) The effect of functionalization carbon nanotubes (CNTs) on the performance of PESCNTs mixed matrix membrane. Int J Sci Eng 1(1):15–20
48. Bétard A, Bux H, Henke S, Zacher D, Caro J, Fischer RA (2012) Fabrication of a CO_2-selective membrane by stepwise liquid-phase deposition of an alkylether functionalized pillared-layered metal-organic framework $[Cu_2L_2P]_n$ on a macroporous support. Microporous Mesoporous Mater 150:76–82
49. Li T, Pan Y, Peinemann KV, Lai Z (2013) Carbon dioxide selective mixed matrix composite membrane containing ZIF-7 nano-fillers. J Membr Sci 425–426:235–242
50. Bernardo P, Clarizia G (2013) 30 years of membrane technology for gas separation. Chem Eng Trans 32:1999–2004
51. Baldus W, Tillman D (1986) Conditions which need to be fulfilled by membrane systems in order to compete with existing methods for gas separation. In: Barry TR (ed) Membranes in gas separation and enrichment. Special Publication No. 62. Royal Society of Chemistry, London, pp 26–42
52. Bernardo P, Drioli E, Golemme G (2009) Membrane gas separation: a review/state of the art. Ind Eng Chem Res 48:4638–4663

Chapter 2
Fundamentals of Gas Permeation Through Membranes

2.1 Gas Permeation Through Membranes

Gas permeation is a technique for fractionating gas mixtures by using nonporous polymer membranes having a selective permeability to gas according to a dissolution–diffusion mechanism. The membrane gas separation process is driven by a pressure difference across the membrane. The membrane may be either in the form of a flat sheet or a hollow fiber. In general, hollow fibers are preferred as they achieve a higher effective membrane area within a given module volume.

2.1.1 *Technical Terms Used in Gas Permeation Membrane Science*

To understand the fundamentals of membrane gas separation, one should be familiar with some laws, processes or words that are commonly used.

Graham's law (Thomas Graham in 1848): Graham's law states that the rate of diffusion of a gas is inversely proportional to the square root of its molecular weight. This formula can be written as:

$$Rate_A / Rate_B = \left(M_B / M_A\right)^{1/2} \tag{2.1}$$

where $Rate_A$ is the rate of diffusion of the first gas (volume or number of moles per unit time), $Rate_B$ is the rate of diffusion for the second gas, M_A is the molar mass of gas A, and M_B is the molar mass of gas B.

Fick's first law: Fick's first law relates the diffusive flux to the concentration under the assumption of steady state. It postulates that the flux goes from regions of high

A.F. Ismail et al., *Gas Separation Membranes*,
DOI 10.1007/978-3-319-01095-3_2

concentration to regions of low concentration, with a magnitude that is proportional to the concentration gradient (spatial derivative).

Fick's second law: Fick's second law predicts how diffusion causes the concentration to change with time.

Henry's law: Henry's law is one of the gas laws formulated by William Henry in 1803. It states that at a constant temperature, the amount of a given gas that dissolves in a given type and volume of liquid is directly proportional to the partial pressure of that gas in equilibrium with that liquid.

Diffusion: Diffusion is the process by which molecules spread from areas of high concentration, to areas of low concentration. When the molecules are even throughout a space, that space is said to be in "EQUILIBRIUM". There are three main types of diffusion through a membrane: Knudsen diffusion, molecular sieving/molecular diffusion, and solution diffusion (Fig. 2.1).

Knudsen diffusion: Knudsen diffusion may take place in a microporous inorganic membrane or through pinholes in dense polymeric membranes. It is a means of diffusion that occurs in a long pore with a narrow diameter (2–50 nm) because molecules frequently collide with the pore wall. This mode of transport is important when the mean free path of the gas molecules is greater than the pore size. In such situations the collisions of the molecules with the pore wall are more frequent than the collision between molecules. Separation selectivities with these mechanisms are proportional to the ratio of the inverse square root of the molecular weights. This mechanism is often prominent in macroporous and mesoporous membranes [1].

Molecular diffusion: In molecular diffusion, the mean free path of the gas molecules is smaller than the pore size and diffusion occurs primarily through molecule-molecule collisions. In molecular diffusion, the driving force is the composition gradients. If pressure gradient is applied in such pore regimes, bulk (laminar) flow occurs, as given by Poiseuille's equation. Such transport is often referred to as *Poiseuille flow* or viscous flow [1].

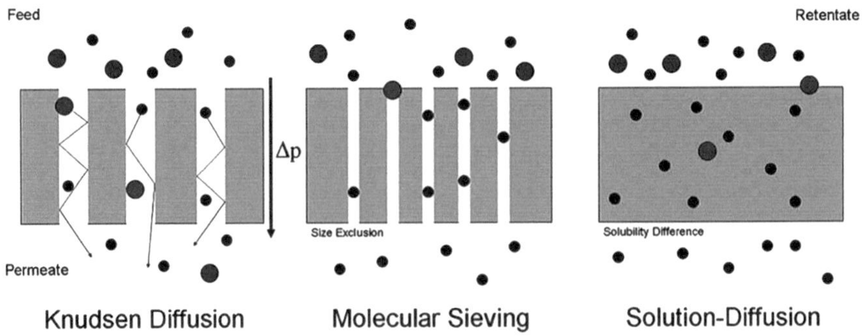

Fig. 2.1 Schematic representation of three of the different possible mechanisms for membrane gas diffusion—Knudsen diffusion, molecular sieving, and solution diffusion/surface diffusion

Surface diffusion/solution diffusion: This diffusion occurs when the permeating species shows a strong affinity for the membrane surface and adsorb along the pore walls. In this mechanism, separation occurs due to the differences in the amount of adsorption of the permeating species. Surface diffusion often occurs in parallel with other transport mechanisms such as Knudsen diffusion.

Configurational or micropore diffusion: This type of diffusion may be considered surface diffusion in the limit where the pore size becomes comparable to the molecular size. In this mechanism, diffusion is perceived as an "activated" process and separation is a strong function of molecular shape and size, pore size, and interactions between the pore wall and gas molecules. This process mainly works in microporous zeolite/zeolitic membranes and carbon molecular sieves.

Concentration gradient: Concentration gradient is the difference between concentrations in a space.

Pores: Pores are miniature openings or passageways in the membrane. Table 2.1 shows the classification of pore sizes in membranes.

Microporous membrane: A thin, porous film or hollow fiber having pores ranging from 0.01 to 10 μm

Capillary condensation: Capillary condensation is one form of surface flow where one of the gases condensable. The pores are completely filled by the condensed gas at certain critical relative pressures, especially in mesopores and small macropores. Due to the formation of menisci at both ends of the pore, transport can take place through hydrodynamic flow driven by a capillary pressure difference between the two ends. This mechanism of gas transport can be thought of as the ultimate limit of the process of adsorption as pressure is increased. In theory, capillary condensation can be used to achieve very high selectivities because the formation of the liquid layer of the condensable gas will block and prevent the flow of the non-condensable gas.

Free volume: The free-volume in a polymer is the space not occupied by polymer molecules. The occupied volume is generally taken to include the van der Waals volume multiplied by a factor (typically 2.2) to take into account that even for a perfect crystal at absolute zero, there is a limit to the packing density achievable. On this basis, fractional free volume, fv, can be calculated as [2].

$$f_v = \left(V - 1.3V_w\right) / V \tag{2.2}$$

Table 2.1 IUPAC (International Union of Pure and Applied Chemistry) classification of pores as a function of their size

Microporous		Mesopores	Macropores
<2 nm			
Ultra-micropores	Super-micropores	2–50 nm	>50 nm
<0.7 nm	>0.7 nm		

where V is the specific volume of the polymer (i.e., reciprocal of density) and V_w is the specific van der Waals volume. This equation is widely used in the membrane literature. However, it should be noted that in other contexts the occupied volume is taken also to include the effects of molecular vibrations, and in such cases is temperature dependent.

Selectivity: This term means that the membrane lets in some compounds while keeping others out. For example, a cell membrane can keep out ions while letting in small hydrophobic compounds. The best measure of the ability of the membrane to separate two gases, A and B, is the ratio of their permeabilities, $\alpha_{A/B}$, also called the membrane selectivity.

Mean free path: The motion of a molecule in a gas is complicated. Besides colliding with the walls of the confinement vessel, the molecules collide with each other. The mean free path is the average distance travelled by a molecule between collisions with another molecule. The mean free path of a molecule is related to its size; the larger its size the shorter its mean free path.

Kinetic diameter: Kinetic diameter of a gaseous molecule can be calculated by following equations based on ideal gas laws. Table 2.2 shows the molecular weight and kinetic diameter (Å) of a few gases encountered in membrane gas separation.

$$l = 1 / \pi d^2 n \tag{2.3}$$

where l is the mean free path length of the molecule, d is the kinetic diameter of the molecule and n is number of molecules per unit volume.

Molecular sieve: Molecular sieve is a material with very small holes of precise and uniform size. These holes are small enough to block large molecules while allowing small molecules to pass. Many molecular sieves are used as desiccants. Some examples include activated charcoal and silica gel. According to IUPAC notation, microporous materials have pore diameters of less than 2 nm (20 Å) and macroporous materials have pore diameters of greater than 50 nm (500 Å); the mesoporous category thus lies in the middle with pore diameters between 2 and 50 nm (20–500 Å) [3].

Molecular sieve effect: With respect to porous solids, the surface associated with pores communicating with the outside space may be called the *internal surface*. Because the accessibility of pores may depend on the size of the fluid molecules, the extent of the internal surface may depend on the size of the molecules comprising

Table 2.2 Molecular weight and kinetic diameter (Å) of gases encountered in membrane gas separation

Molecule	Molecular weight	Kinetic diameter (Å)
CO_2	44	3.3
O_2	32	3.46
N_2	28	3.64
H_2O	18	2.65
CH_4	16	3.8
H_2	2	2.89

the fluid, and may be different for the various components of a fluid mixture. This effect is known as the molecular sieve effect.

Glass transition temperature: The glass transition temperature is a function of chain flexibility. The glass transition occurs when there is enough vibrational (thermal) energy in the system to create sufficient free-volume to permit sequences of 6–10 main-chain carbons to move together as a unit. At this point, the mechanical behavior of the polymer changes from rigid and brittle to tough and leathery—the behavior we define as "plastic behavior". Actually, the glass transition temperature is more important in plastics applications than the melting point, because it tells us a lot about how the polymer behaves under ambient conditions. The melting temperature is often referred to as the "first-order transition"—that is where the polymer changes state from solid to liquid. Technically, only crystalline polymers have a true melting point, the temperature at which the crystallites melt and the total mass of plastic becomes amorphous. Amorphous polymers do not have a true melting point; however, they do have a first-order transition where their mechanical behavior transitions from a rubbery nature to a viscous rubbery flow [4].

2.1.2 Membrane Separation Principles

Membrane separation methods can be divided into classes according to their separation characteristics: (i) separation by sieving action; (ii) separation due to a difference in affinity and diffusivity; (iii) separation due to a difference in charge of molecules; (iv) carrier-facilitated transport; and (v) the process of (time-)controlled release by diffusion.

Gas is made to pass through the membrane by applying a pressure difference on either side of the membrane. This pressure difference causes a difference in dissolved gas concentration between the two faces of the membrane and, hence, a diffusion gas flows through the membrane. Gas permeation through the membrane occurs with the following three steps:

1. Absorption of the permeating species into the polymer.
2. Diffusion through the polymer.
3. Desorption of the permeating species from the polymer surface and removal.

The gas permeation is affected by:

1. Solubility and diffusivity of the small molecule in the polymer.
2. Chain packing and side group complexity, polarity, crystallinity, orientation, fillers, humidity and plasticization.

Permeance: The state or quality of a material or membrane that causes it to allow liquids or gases to pass through it. The productivity of a gas separation membrane is expressed in terms of its permeance the amount of permeate that passes through a certain membrane area in a given time for a particular pressure difference.

Values of permeance are often quoted in units of GPU [1 GPU = 10^{-6} cm^3(STP) cm^{-2} s^{-1} $cmHg^{-1}$].

Permeability: Permeance multiplied by the thickness of the membrane gives permeability P (sometimes called permeability coefficient). It is a characteristic of the material. In principle, permeability is independent of membrane thickness for a homogeneous membrane, but in practice values can depend both on thickness of the membrane and on its history. Values of permeability are often quoted in units of Barrer [1 Barrer = 10^{-10} cm^3 (STP) cm cm^{-2} s^{-1} $cmHg^{-1}$ = 3.35×10^{-16} mol m m^{-2} s^{-1} Pa^{-1}]. Permeability measurements are frequently made on thick (20–100 μm) homogeneous membranes, but for practical application a very thin active layer (<1 μm) is desired in order to increase the permeance. For a suitable polymer, a so-called phase inversion process may be used to generate asymmetric membranes with a thin, dense surface layer and a highly porous sublayer. Alternatively, a thin separating layer may be coated onto a porous substrate, giving a composite membrane [2].

The gas permeability of the membrane can be measured also by means of two chamber cell [5], which is discussed in detail in Chap. 7.

Different mechanisms may be involved in the transport of gases across membranes depending on their physical properties such as porous, non-porous, glassy, and rubbery.

2.1.3 Gas Permeation Through Porous Membranes

For gas separation, the selectivity and permeability of the membrane material determines the efficiency of the gas separation process. Based on flux and selectivity, a membrane can be classified broadly in two classes: (1) porous and (2) nonporous.

A porous membrane is a rigid, highly voided structure with randomly distributed interconnected pores. The separation of materials by porous membrane is mainly a function of the permeant character and membrane properties, such as the molecular size of the membrane polymer, pore size, and pore size distribution. A porous membrane is very similar in its structure and function to a conventional filter. In general, only those molecules that differ considerably in size can be separated effectively by microporous membranes. Porous membranes for gas separation do exhibit very high levels of flux but inherit low selectivity values. Microporous membranes are characterized by the average pore diameter d, the membrane porosity, and tortuosity of the membrane. Porous membranes can be utilized for gas separation. The pore diameter must be smaller than the mean free path of gas molecules. Under normal condition (100 kPa, 300 K) diameter is about 50 nm. The gas flux through the pore is proportional to the molecule's velocity, i.e., inversely proportional to the square root of the molecule mass (Knudsen diffusion). Flux through a porous membrane is much higher than through a nonporous one, 3–5 orders of magnitude. Separation efficiency is moderate—hydrogen passes four times faster than oxygen.

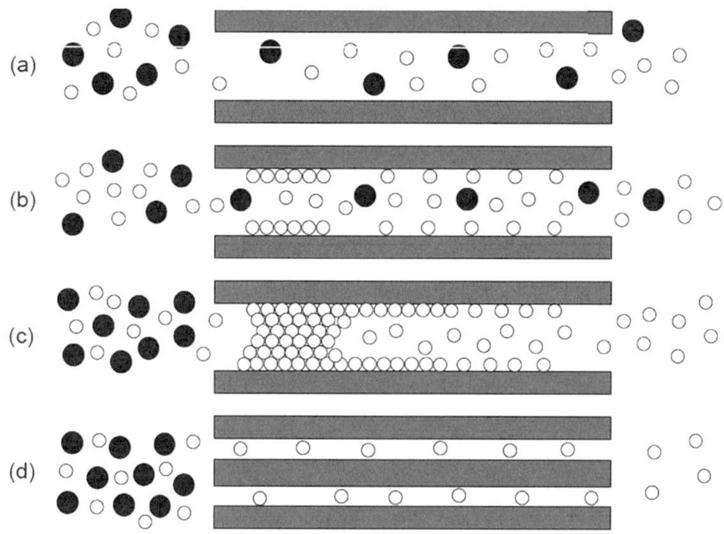

Fig. 2.2 Four types of diffusion mechanism

Four types of diffusion mechanism can be utilized to effect separation by porous membranes (see Fig. 2.2) [6].

The four diffusion mechanisms above are:

1. Knudsen (or free molecule) diffusion.
2. Surface diffusion.
3. Capillary condensation.
4. Molecular sieving.

In some cases, molecules can move through the membranes by more than one mechanism. Knudsen diffusion gives relatively low separation selectivity compared to surface diffusion and capillary condensation. Shape selective separation or molecular sieving can yield high selectivities. The separation factor of these mechanisms depends strongly on the pore size distribution, temperature, pressure and interaction between gases being separated, and the membrane surfaces.

Knudsen (or free molecule) diffusion: The Knudsen number (Kn) is defined as the ratio of the mean free path of the gas molecules (average distance between collisions) (λ) and a representative physical length scale (e.g., the pore radius), (r).

$$K_n = \lambda / r \tag{2.4}$$

The mean free path is given by

$$\lambda = (\eta / P)\left\{(\pi k_B T / 2M)^{1/2}\right\} \tag{2.5}$$

where η is the viscosity of the gas, k_B the Boltzmann constant, T the temperature, M the molecular weight, and P the pressure.

If the pore radius is used as the representative physical length scale, the mean free path lengths are substantially higher than the pore radius when the Knudsen number is larger than 10. The result is that mainly the lighter molecules permeate through the pores. Selectivity is limited and can be calculated with the square root of the ratio of the molar masses of the gases involved. The smaller the K_n, the larger the pores become (relative to the mean free path of the gas molecules). For Knudsen numbers < 1 the dominant transport mechanism is *viscous flow,* which is non-selective. When K_n is inbetween, the permeation of gas through porous membrane consists of Knudsen diffusion and Poiseuille flow [7].

For the Knudsen flow, the molar flow rate of gas, G_{Kn}, through a pore with a radius of r is given by

$$G_{Kn} = \left(8r\left(P_1 - P_2\right)\right) / \left(3L\left(2\pi MRT\right)^{1/2}\right) \tag{2.6}$$

where P_1 and P_2 are pressures on high pressure and low pressure side of the membrane, and L is the pore length.

For the viscous flow, the molar flow rate of gas, G_{vis}, is given by

$$G_{vis} = r_2\left(P_1 - P_2\right) / 16\,L\mu RT \tag{2.7}$$

where μ is the gas viscosity.

Knudsen separation can be achieved with membranes having pore sizes smaller than 50 nm. Table 2.3 presents the ideal separation factors of various pairs of gases based on Knudsen flow. However, the actual separation factor is found to be smaller. This is attributed to back diffusion, to non-separative diffusion, concentration polarization on the feed or on the permeate side, and the occurrence of viscous flow (in large pores).

Surface diffusion (Fig. 2.2b) can occur in parallel with Knudsen diffusion. Gas molecules are adsorbed on the pore walls of the membrane and migrate along the surface. Surface diffusion increases the permeability of the components adsorbing more strongly to the membrane pores. At the same time, the effective pore diameter is reduced. Consequently, transport of non adsorbing components is reduced and selectivity is increased. This positive contribution of surface diffusion only works for certain temperature ranges and pore diameters.

Table 2.3 Calculated separation factors based on Knudsen flow of selected binary gas mixtures

Gas pair	Separation factor
H_2/N_2	3.73
H_2/CO	3.73
H_2/H_2S	4.11
H_2/CO_2	4.67
H_2/SO_2	5.64
N_2/O_2	1.07
O_2/CO_2	1.17

Capillary condensation (Fig. 2.2c) occurs if a condensed phase (partially) fills the membrane pores. If the pores are completely filled with condensed phase, only the species soluble in the condensed phase can permeate through the membrane. Fluxes and selectivities are generally high for capillary condensation. The appearance of capillary condensation, however, strongly depends on gas composition, pore size, and uniformity of pore sizes.

Molecular sieving (Fig. 2.2d) occurs when pore sizes become sufficiently small (3.0–5.2 Å), leading to the separation of molecules that differ in kinetic diameter: the pore size becomes so small, that only the smaller gas molecules can permeate through the membrane.

There are several ways to prepare porous polymeric membranes, such as solution casting, sintering, stretching, track etching, and phase separation. The final morphology of the membrane obtained will vary greatly, depending on the properties of the materials and process conditions utilized.

Various mechanisms have been distinguished to describe the transport in membranes: transport through bulk material (dense membranes); Knudsen diffusion in narrow pores; viscous flow in wide pores or surface diffusion along pore walls.

2.1.4 Gas Permeation Through Nonporous Membranes

In dense polymeric membranes, solution diffusion is widely accepted to be the main mechanism of transport [1]. This mechanism is generally considered to be a three-step process. In the first step the gas molecules are adsorbed by the membrane surface on the upstream end. This is followed by the diffusion of the gas molecules through the polymer matrix. In the final step the gas molecules evaporate on the down-end stream.

Sir Thomas Graham [8] proposed the transport of gases in dense, nonporous polymers based on the solution-diffusion mechanism. Under the driving force of a pressure difference across a membrane, penetrant molecules dissolve in the upstream (or high pressure) face of a membrane, diffuse across the membrane, and desorb from the downstream (or low pressure) face of the membrane. Thus, according to the solution-diffusion model, the permeation of the module is controlled by two major parameters: diffusivity coefficient (D) and solubility coefficient (S) (see Eq. (2.8)). Diffusion is the rate-controlling step in penetrant permeation. The rate-controlling step in diffusion is the creation of gaps in the polymer matrix sufficiently large to accommodate penetrant molecules by thermally stimulated, random local segmental polymer dynamics [9]. The permeation of a gas through the dense polymer (membrane) can be described by using the sorption–diffusion theory [10].

The productivity of a membrane is defined by the permeability of the gas through the membrane. The permeability of a gas A is given by:

$$P_A = D_A S_A \tag{2.8}$$

where D_A and S_A represent the diffusion and solubility coefficients of component A, respectively. Permeability can also be expressed as the flux normalized by film thickness (l) and the transmembrane pressure (Δp_A), as shown by Eq. (2.9):

$$P_A = (Flux_A \cdot l) / (\Delta p_A) \tag{2.9}$$

A commonly accepted unit for gas permeability is the Barrer, where 1 Barrer = 10^{-10} (cm^3 (STP) cm)/(cm^2 s cmHg). When the thickness is difficult to define (as is often the case with asymmetric membranes), the pressure normalized flux, or permeance (P_A/l) is used instead. In this case, the gas permeation unit (GPU) is used, which is defined as 1 GPU = 10^{-6} (cm^3 (STP))/(cm^2 s cmHg). The ratio of the permeabilities can be used to signify the permselectivity of separation of the desired component within the mixture. The ideal selectivity of the membrane, thus, is the ratio of the permeabilities or permeances of the individual gases. For a mixture of gas A and B the ideal selectivity is described by:

$$\alpha_{A/B} = P_A / P_B \tag{2.10}$$

Nonporous or dense membranes have high selectivity properties but the rate of transport of gases through the medium is usually low. An important property of a nonporous dense membrane is that even permeates of similar sizes may be separated if their solubility in the membranes differ significantly. Non-porous membranes primarily consist of polymer membranes. The non-porous structure of the polymer is related to the non-continuous passages present in the polymer chain matrix. These passages are created and destroyed due to thermally induced motion of the chains. Therefore, the transport of a penetrant is based on its movement through these passages. The effects of penetrant activity (driving force) and operating conditions then play an important role in governing the gas transport rate and separation property of the membrane. In the area of membrane-based gas separation, the general approach of the solution-diffusion mechanism has been used for modelling of non-porous membranes.

2.1.5 Gas Permeation Through Asymmetric Membranes

The membrane used in practice has a particular structure that is called asymmetric, which combines high permeability and good mechanical strength. This structure has a thin dense and selective skin (0.1–1 μm thick) supported by a thick microporous substrate (50–200 μm). Such membranes come either in flat shape or in the form of hollow fibers with their skins outside.

The permeation of a simple gas through an asymmetric membrane with a small fraction of surface defects was found to be a combination of Knudsen flow and viscous flow in the porous part and solution-diffusion flow in the nonporous part. Theoretical analysis is given in the following equation [11].

$$\left(\frac{P}{L}\right)_i = P_k^0 \frac{1}{\sqrt{M_i T}} \frac{\varepsilon r}{L_p} + P_v^o \frac{1}{\mu_i T} \frac{\varepsilon r^2}{L_p} \bar{p} + P_s^o \exp\left(-\frac{E_i}{RT}\right) \frac{1}{L} \quad (2.11)$$

where P_k^0 corresponds to Knudsen flow, P_v^0 corresponds to viscous flow and P_s^0 corresponds to pore flow; R is a gas constant; T is absolute temperature; and M are the viscosity and molecular weight of the gas, respectively; P_s^0 is the pre-exponential factor and E is the activation energy of permeation in the polymer material; r is the mean pore size; ε is the surface porosity; L_p is the pore length and L is the skin-layer thickness; and is average gas pressure. Equation (2.11) is based on the following assumptions:

1. The porous part of the membrane is constructed by cylindrical pores with a mean pore size r and pore length of L_p.
2. The nonporous part has a uniform skin with a thickness of L.

The relative contribution of the each flow to the total gas permeability will depend on the gas permeability in the dense membrane, the membrane pore size, porosity of the membrane, and operating pressure and temperature. It should be noted that the accuracy of Eq. (2.11) is dependent on the slope of the equation. Error may be introduced if the slope is too steep.

Shilton et al. [12] discussed resistance modelling of gas permeation through asymmetric polysulfone hollow fiber membranes. The structural information was used to interpret the relationship between spinning conditions and fiber properties. Dope concentration determines the general morphology of the fiber, such as porosity (voidage fraction), thickness of the active layer and order of magnitude of surface porosity (fraction of surface area that is pores), and thus the permeability and level of selectivity are likely to be achieved on coating.

Henis and Tripodi [13] discussed the resistance modelling for gas permeation through asymmetric membranes. They reported an asymmetric membrane becomes useless for gas separation when defects are formed in the skin layer, but defect gas separation properties akin to those of the solid polymer could be achieved once coated with a silicone layer. The performance of these hypothetical coated membranes was predicted by their resistance model.

Fouda et al. [14] discussed the limitations in the Henis and Tripodi model when fitting actual gas permeation results to possible membrane structures. They then introduced the Wheatstone bridge model which better explained the permeation data.

2.2 Diffusion Theory of Small Molecules in Nonporous Polymer Membranes

Small molecules diffuse faster in solids than large molecules. For example, diffusivities for several components (in cm^2/s) in low density polyethylene at 25 °C are given in Table 2.4 [15, 16]. From Table 2.4 it is clear that the diffusion rate decreases with permeant size.

Table 2.4 Diffusivities for several components (in cm^2/s) in low density polyethylene at 25 °C

Gas	Diffusivity (cm^2/s)
Helium	6.8×10^{-6}
Hydrogen	4.74×10^{-6}
Nitrogen	0.320×10^{-6}
Propane	0.0322×10^{-6}

Both high permeability coefficients and high selectivities of membrane are desirable for increasing process capacity and purity of product; however, there is an inherent trade-off between permeability and selectivity, whereby polymers with high permeability typically have low selectivity, and vice versa [17]. This well-known trend was analyzed in detail by Robenson, who empirically deduced the concept of an upper bound for a variety of gas separations [18]. The upper bound represents the most favorable combinations of permeability and selectivity characteristics of polymer membranes reported in the literature, and are described by the following equation:

$$\alpha_{A/B} = (\beta_{A/B}) / (P_A^{\lambda A/B}) \tag{2.12}$$

where $\alpha_{A/B}$ is selectivity, P_A is permeability and $\beta_{A/B}$ and $\lambda_{A/B}$ are empirically determined parameters that depend on the gas pair of interest. A theoretical model that describes the upper bound behavior of polymer membranes, based on fundamental considerations, was developed by Freeman [19].

As mentioned earlier, the mechanism for small molecule transport through dense polymeric materials is described by the solution-diffusion model, whereby the penetrant molecules dissolve at the upstream (i.e., high pressure) interface, diffuse through the material, and then desorb at the downstream (i.e., low pressure interface). According to this mechanism, the permeability coefficient for gas A, which is its flux normalized by membrane thickness and driving force, is equal to the product of the solubility coefficient, S_A, and diffusion coefficient, D_A, as shown in Eq. (2.8). This simple expression for permeability is developed for the case where downstream pressure is negligible as compared to upstream pressure and Fick's law controls mass flux through the polymer:

$$\alpha_{A/B} = P_A / P_B = (D_A / D_b) \times (S_A / S_B) \tag{2.13}$$

The diffusion of small molecules is a thermally activated process that is often described by the Arrhenius equation:

$$\ln D_A = \ln D_{0A} \ (E_{DA} / RT) \tag{2.14}$$

where D_{0A} is the front factor, E_{DA} is the activation energy of diffusion, R is the ideal gas constant, and T is the absolute temperature. Small molecule diffusion coefficient data in polymers can be represented by Eq. (2.14) [17].

A functional relationship between activation energy and the front factor for diffusion in polymeric materials can be presented by the following equation [20, 21]

$$\ln D_{0A} = \left[a \left(E_{DA} / RT \right) \right] - b \tag{2.15}$$

where a and b are constants that are independent of gas type. The value of 'a' is 0.64 (independent of polymer type), and 'b' has the value of 9.2 for rubbery polymers and 11.5 for glassy polymers. These values correspond to diffusion coefficients in units of cm^2/s.

2.3 Diffusion Models for Rubbery Polymers

It is well known that the mechanism of diffusion is very different in rubbery and glassy polymers, due to the fact that glassy polymers are not in a true state of equilibrium. The difference in mechanism is reflected in the significant differences observed in the dependence of the diffusion coefficient, as well as the permeability and solubility coefficients, on the penetrant gas pressure or concentration in polymers and on the temperature. For example, the diffusion coefficients for light gases in rubbery polymers are often independent of concentration. By contrast, in glassy polymers the diffusion coefficients are highly nonlinear functions of concentration and reach a constant value at sufficiently high concentration.

A macromolecular material (such as rubber or a synthetic material having similar properties) returns rapidly to approximately the initial dimensions and shape after substantial deformation by a weak stress and release of the stress. In rubbery polymers, such as polyisoprene, the sorption selectivity term is dominant. Permeability increases with increasing permeant size, and a large molecule permeates preferentially. When used to separate an organic vapor from nitrogen, rubbery membranes permeate organic vapor preferentially.

Figure 2.3 shows permeability as a function of molar volume of permeant for a rubbery and a glassy polymer, illustrating the different balance between sorption and diffusion in these polymer types [22]. Rubbery polymers tend to have much higher permeability than glassy polymers (Fig. 2.3). The higher the permeability, the smaller the membrane area required to permeate a given volume flow of gas. Rubbery membranes provide better selective purge capability.

Figure 2.4 Illustrates a schematic representation of the various models proposed for polymer microstructure for the transport of small permeant molecules through the matrix.

- Figure 2.4a illustrates a bundle of parallel polymer chains and inclusion of gas molecules. In order to move into the polymer matrix the gas molecule pushes the polymer chain and jumps into a new position.
- Figure 2.4b shows that the polymer segments are in a normal and an activated state. In the activated state the polymer chain accommodates a diffusing mole-

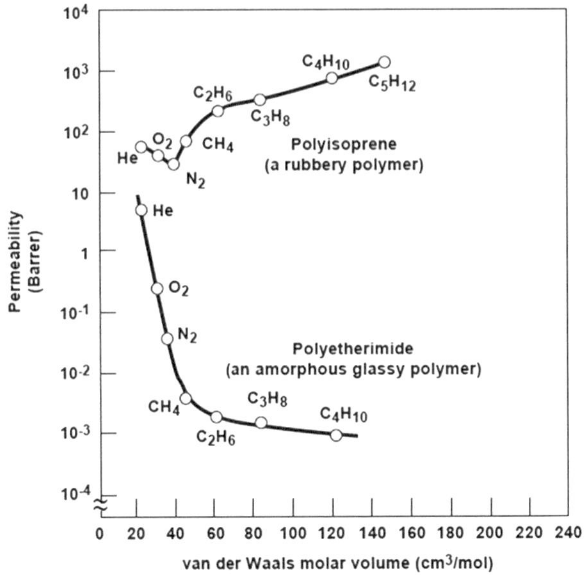

Fig. 2.3 Permeability as a function of molar volume of permeant for a rubbery and a glassy polymer, illustrating the different balance between sorption and diffusion in these polymers

cule, allows it to diffuse, and then returns to the normal configuration after the jump of the molecule. It also shows the normal and activated states of the polymer segments.

- Figure 2.4c shows the model proposed by Pace and Datyner [23]. The model accounts for the structure of the polymer contributing to gas diffusions and incorporates some of the features of models Fig. 2.4a, b.

In the model illustrated in Fig. 2.4c, it is assume that non-crystalline polymer regions possess an appropriate semi-crystalline order with chain bundles. These bundles are parallel along distances of several nanometers and can be considered as tubules. The tubules, consisting of parallel chains, facilitate the movement of the permeant, and the transport occurs by leaps between these tubules. These jumps occur when the thermal motions of local segments of the polymer chain open up a sufficiently large channel to a neighboring gap. The gas molecule/particles can then diffuse through this channel. Once the channel closes, the jump is successfully concluded. According to this model, the selectivity of a membrane material depends on the control of these leap channels. Large openings or high flexibilities cause large diffusion coefficients and low apparent energies of activation for diffusion, whereas more limited motions permit the passage of smaller species much more readily than the large particles [7].

Other explanations by various molecular mechanisms for gas diffusion in polymers are given in literature, but all are based on the concept of available free volume as diffusing channels. The 'hole' or lattice vacancy theory assumes that a certain

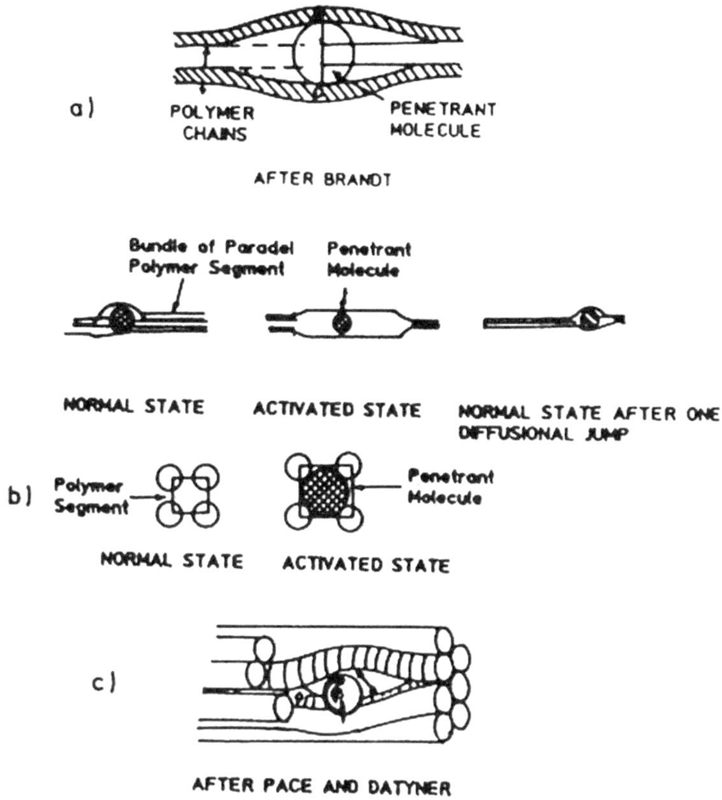

Fig. 2.4 Models for the transport of small permeant molecules in polymers [7]

amount of work must be done on the matrix to create or expand a hole to accommodate the diffusing penetrant [24]. The activated complex theory proposes that the permeant must be given sufficient energy to overcome the potential energy barrier of the membrane [25]. Pace and Datyner [23] suggested that in the fluctuation theory of volume, fluctuations are considered to provide spaces for movement of a molecule in a chemical potential gradient.

These three approaches lead to mathematical expressions which are similar in functional form and contain the empirical Arrhenius relations. Thus, Diffusivity (D) is given by

$$D = D_0 \exp\left(-E_D / RT\right) \tag{2.16}$$

and permeability

$$P = P_0 \exp\left(-E_p / RT\right) \tag{2.17}$$

where D_0 and P_0 are the pre-exponential factors, dependent on diffusion and pressure, and E_D and E_p are the apparent activation energy. These interactions between the permeant molecule and polymeric material indicate strong dependences of physical and chemical permeabilities on the structure and selectivity. A number of other empirical or semiemperical correlations of permeability coefficients, as well as diffusion and solubility coefficients, have been reported in the literature [7].

2.4 Diffusion Models for Glassy Polymers

In glassy polymers, the sorption of gases becomes a complex process, which has been described by a combination of Henry's law and Langmuir expressions. This has been referred to as "dual mode sorption theory" [26]. Diffusion in glassy polymers is usually an activated process, and Arrhenius relations may be used to express the permeability, diffusivity, and solubility coefficients.

In glassy polymers, such as polyetherimide, the rigid nature of the polymer chains means the permeability falls with increasing permeant size, and small molecules permeate preferentially. When used to separate an organic vapor from nitrogen, amorphous glassy membranes preferentially permeate nitrogen. The vapor/permeant gas selectivity of most glassy polymers is very dependent on organic vapor partial pressure. At low vapor concentrations (or partial pressures), the selectivity of the membrane approaches the selectivity predicted by the ratio of the pure gases. However, as the organic vapor concentration (partial pressure) increases, the amount of vapor sorbed in the polymer also increases. The vapor plasticizes the polymer, which becomes rubbery. In the plasticized material, the nitrogen permeability increases, but the organic vapor permeability increases even more. The membrane then switches from being a glassy, permanent-gas-selective membrane to being a rubbery, organic vapor-selective membrane. Examples of the few exceptions to this behavior are the Teflon® AF (DuPont) and Hyflon™ AD (Solvay Solexis) [27]. These polymers, because of the inert nature of their perfluoro chemistry, have exceptionally low sorption for most organic vapors and so retain their glassy nature, even in the presence of high concentrations of organic vapors. For glassy polymers, the gas transport properties depend on the amount and distribution of free volume and on chain mobility. The free volume in a polymer sample is the space not occupied by polymer molecules (see Sect. 2.1.1—Free volume).

The glass transition may be thought of as occurring when there is sufficient fractional free volume for large scale polymer motions to occur. In a rubbery state, free volume increases markedly with increasing temperature. If a polymer is quenched from the rubbery into the glassy state, excess free volume is trapped. Over time, excess free volume may be lost, a process referred to as physical aging. This can be particularly pronounced in very thin films, as used for gas separation [28, 29]. Long-term aging has a profound influence on transport properties and represent a major problem in the application of glassy polymer membranes. Membrane for high permeability needs high free volume. If the glassy polymer membrane has too much free volume (free volume elements are essentially interconnected), it will exhibit adsorption behavior like a molecular sieve [30]. In other words, it contains microspores (pores with dimensions <2 nm), as defined by IUPAC, in the context of adsorption studies [31]. It should be noted that in the membrane literature the word "microporous" is also often applied to materials with much larger pores (macroporous in the IUPAC sense). On the basis of nitrogen adsorption studies, certain polyphenylene oxides possess interconnected microcavities and are named as "intrinsic microporosity" [32].

2.5 General Membrane Transport Equations

Figure 2.5 shows infinite dilution diffusion coefficients vs. van der Waals volume of penetrant. As free volume is much lower in glassy polymers than in rubbery polymers, the diffusion tends to be slower [33].

Figure 2.5 clearly illustrates that free volume is much lower in glassy polymers than in rubbery polymers and, therefore, diffusion tends to be slower. For rubbery-rubbery interfaces, repetition models can be used to describe the chain motions near the interface. The diffusion is found to depend on chain concentration, and error functions are used for concentration profiles across the interfaces, but for glassy polymer interfacial diffusion. The mutual diffusion coefficient of the glassy-rubbery interdiffusion is strongly dependent on the chain composition but this composition is sometimes inconsistent. Lin et al. [34] studied the microstructure and chain diffusion behavior of a rubbery/glassy polymer interface (PS/PPO) using the depth-resolved technique of secondary ion mass spectroscopy (SIMS).

Fundamentals of transport phenomena in polymeric systems are discussed by George and Thomas [35]. The transport of small molecules through a polymer membrane occurs due to random molecular motion of individual molecules. The driving force behind the transport process—which involves sorption, diffusion, and permeation—is the concentration difference between the two phases separated by the membranes. The transport process slowly tries to equalize the concentration difference or the chemical potential of the penetrant in the phases separated by the membrane. This process can be described in terms of Fick's first law of diffusion, according to which the flux J, in the direction of flow is proportional to the concentration gradient ($\partial c/\partial x$) as

$$J = -D\left(\frac{\partial c}{\partial x}\right) \tag{2.18}$$

Here D is the diffusion coefficient. Equation (2.18) is applicable to the diffusion in the steady state, i.e., when the concentration does not vary with time. On the other hand, Fick's second law describes the nonsteady state for the transport process, which is given by the rate of change of the penetrant concentration ($\partial c/\partial x$) at a plane within the membrane, i.e., Eq. (2.19)

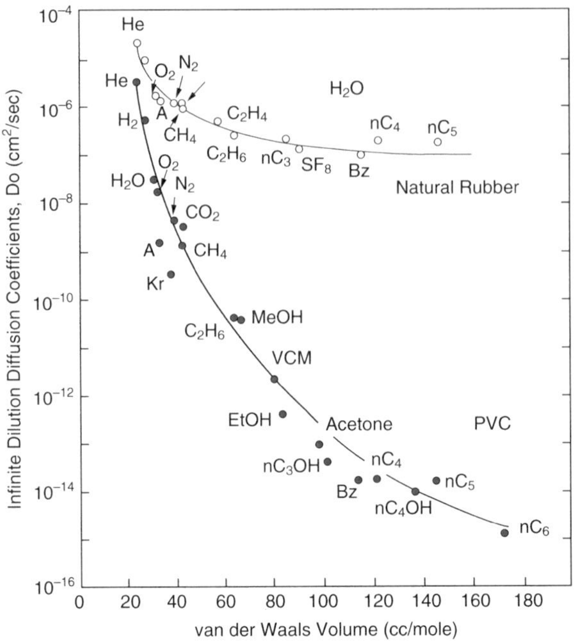

Fig. 2.5 Permeation in glassy polymers vs. rubbery polymers

$$\frac{\partial c}{\partial t} = D\left(\frac{\partial c^2}{\partial x^2}\right) \tag{2.19}$$

This is an ideal case in which the membrane is isotropic and the diffusion coefficient is independent of distance, time and concentration. Depending on the boundary conditions, many solutions are available for Eq. (2.19).

Strong polymer–penetrant interaction occurs with many organic penetrant molecules and hence D is dependent on concentration. Therefore, Eq. (2.19) becomes Eq. (2.20)

$$\frac{\partial c}{\partial t} = D\frac{\partial D(c)(\partial c / \partial x)}{\partial x} \tag{2.20}$$

Analytically this cannot be solved easily and hence another form, Eq. (2.21), is commonly used.

$$\frac{\partial c}{\partial t} = D(c)\frac{\partial c^2}{\partial x^2} + \frac{\partial D(c)}{\partial c}\left(\frac{\partial c}{\partial x}\right)^2 \tag{2.21}$$

Generally experiments are conducted over relatively small intervals of c and the term $(\partial D(\mathbf{c})/\partial \mathbf{c})$ is negligible compared to $D(c)$. Then we obtain a mean or integral diffusion coefficient as given by Eq. (2.22)

$$\bar{D} = \int_{c_1}^{c_2} D(c)\mathrm{d}c / c_1 - c_2 \tag{2.22}$$

where c_1 and c_2 are the concentrations of penetrant at the low and high concentration faces of the film, respectively.

In the steady state, diffusion flow is constant and the diffusion coefficient is independent of concentration. Then Eq. (2.18) may be integrated to give Eq. (2.23)

$$J = \frac{D(c_1 - c_2)}{h} \tag{2.23}$$

where h is the membrane thickness. The penetrant distribution between the ambient penetrant and the polymer phase is described by the Nernst distribution law Eq. (2.24)

$$c = KC \tag{2.24}$$

where c is the sorbed concentration; C the ambient penetrant concentration in contact with the polymer surface and K dependent on temperature and c. In the case of transport of gases and vapors, pressure p is used instead of ambient penetrant concentration. According to Henry's law Eq. (2.25)

$$c = Sp \tag{2.25}$$

where S is the solubility coefficient. The combination of Eqs. (2.23) and (2.25) will give the well-known permeation Eq. (2.26)

$$J = \frac{DS(p_1 - p_2)}{h} \tag{2.26}$$

where p_1 and p_2 are the ambient pressures on two sides of a film of thickness h. The product DS is called the permeability coefficient P, so that Eq. (2.27)

$$P = DS \tag{2.27}$$

In terms of permeability, the flux Eq. (2.26) can be written as Eq. (2.28)

$$J = \frac{P(p_1 - p_2)}{h} \tag{2.28}$$

The transport behavior for a given penetrant varies from one polymer to another. Transport properties depend on the free volume within the polymer and on the segmental mobility of the polymer chains. The segmental mobility of the polymer chains is affected by the extent of unsaturation, degree of cross-linking, degree of crystallinity, and nature of substituents.

2.6 Models for Gas Transport in Nanocomposite Membranes

Cong et al. [36] described the following four mechanisms for gas transportation through nanocomposite membranes.

2.6.1 *Maxwell's Model*

Adding impermeable inorganic nanoparticles to a polymer reduces the gas permeability. Maxwell's model, developed to analyze the steady-state dielectric properties of a diluted suspension of spheres, is often used to model permeability in membranes filled with roughly spherical impermeable particles.

$$P_c = P_p\left[\left(1-\phi_f\right)/\left(1+0.5\phi_f\right)\right] \tag{2.29}$$

where P_c and P_f are the permeability of the nanocomposite and the pure polymer matrix, respectively, and ϕ_f is the volume fraction of the nanofiller.

The numerator represents the loss of membrane solubility due to loss of polymer volume available for sorption. The denominator represents a decrease in diffusivity due to increase in the penetrant diffusion path length. Both factors act to decrease permeability with increasing particle volume fraction.

The weakness with Maxwell's model is that it neglects the interactions between the nanofillers and the polymer chains, and the nanofillers and the penetrants. In most nanocomposite membranes, these interactions are strong, and significantly change the diffusivity and solubility of penetrants. Maxwell's model partly explains the gas permeability loss in some nanocomposite membranes.

2.6.2 *Free-Volume Increase Mechanism*

The effect of polymer free volume on penetrant diffusion coefficients is often modeled by the statistical-mechanical description of diffusion in a liquid of hard spheres. This model provides the following expression of penetrant diffusion coefficients (D);

$$D = A\exp\left(-\gamma V^{*}/V_f\right) \tag{2.30}$$

where A is a pre-exponential factor weakly dependent on temperature, γ an overlap factor introduced to avoid double-counting free volume elements, V^* the minimum free volume element size that can accommodate a penetrant molecule (and is closely associated with penetrant size), and V_f is the average free volume in the media accessible to penetrants for transport. According to Eq. (2.30), an increase in polymer free volume is expected to enhance penetrant diffusion.

The free volume increase mechanism provides a qualitative understanding of the interaction between polymer-chain segments and nanofillers; the nanofillers may disrupt the polymer-chain packing and increase the free volume between the polymer chains, enhancing gas diffusion and, in turn, increasing gas permeability.

2.6.3 Solubility Increase Mechanism

The solubility increase mechanism is based on the interaction between the penetrants and the nanofillers. Functional groups such as hydroxyl, on the surface of the inorganic nanofiller phase, may interact with polar gases such as CO_2 and SO_2 and increase the penetrants' solubility in the nanocomposite membranes. This, in turn, increases the gas permeability. Eq. (2.31) shows the gas permeability (P) based on the Arrhenius equation

$$P = P_0 \exp\left(-E_p / RT\right); \quad E_p = E_d + \Delta H_s \tag{2.31}$$

where P_0 is a pre-exponential factor, E_p the apparent activation energy equal to the activation energy of diffusion (E_d) plus enthalpy of sorption (ΔH_s), R is the ideal gas constant, and T is absolute temperature. It was found that ΔH_s decreased for CO_2 permeation by incorporating TiO_2 nanoparticles due to the interaction between CO_2 and nanoparticles, and CO_2 increase.

2.6.4 Nanogap Hypothesis

Cong et al. [37] found that unmodified silica dispersed rather heterogeneously in the membranes (brominated poly(2,6-diphenyl-1,4-phenylene oxide, $BPPO_{dp}$)) and greatly improved the diffusivities and permeabilities of CO_2 and CH_4 without changing the CO_2/CH_4 selectivity compared with the pure $BPPO_{dp}$.

This finding could not be explained well by the chain-unpacking-caused free-volume increase, which occurs when surface modified silica particles are well dispersed in the polymer matrix, suggesting that in the $BPPO_{dp}$/silica composite membrane, the increased gas permeability does not result from the disrupted polymer-chain packing. Rather, the authors proposed that due to the poor compatibility of the unmodified silica surface and the polymer, the polymer chains could not tightly contact the silica nanoparticles, thus forming a narrow gap surrounding

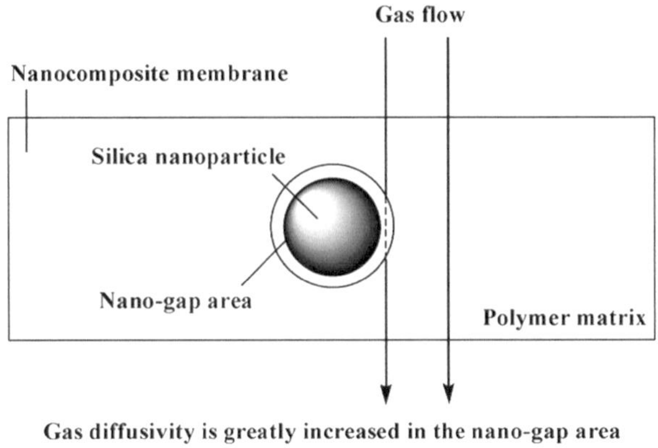

Fig. 2.6 Illustration of nanogap formation in the BPPOdp/silica nanocomposite membranes

the silica particles (Fig. 2.6). The gas diffusion path is shortened and, thus, the apparent gas diffusion and permeability increase. This also explains why the addition of the nanoparticles enhances gas permeability but does not affect the gas selectivity. Once the nanoparticle surface is compatible with the polymer, the nanogaps do not form any more due to tight contact between the polymer and the filler particles.

2.7 Facilitated Transport Membranes

Facilitated or carrier-mediated transport is a coupled transport process that combines a (chemical) coupling reaction with a diffusion process. The solute has first to react with the carrier to form a solute–carrier complex, which then diffuses through the membrane to finally release the solute at the permeate side. The overall process can be considered as a passive transport since the solute molecule is transported from a high to a low chemical potential. In the case of polymeric membranes, the carrier can be chemically or physically bound to the solid matrix (fixed carrier system), whereby the solute hops from one site to other. Mobile carrier molecules have been incorporated in liquid membranes, which consist of a solid polymer matrix (support) and a liquid phase containing the carrier molecules [38, 39].

Facilitated transport membranes rely on the chemical reaction occurring between the gas of interest and a component of the membrane (carrier). The reacted species is readily carried across the membrane, whereas the diffusion of non-reactive gases is inhibited (Fig. 2.7). The active carrier is generally basic in nature, given that carbon dioxide is acidic. The driving force for gas transportation remains the partial pressure difference across the membrane; however, the facilitator carrier increases both the permeability and selectivity of the membrane through the increased loading. The facilitator carrier can be either fix-sited within the polymer matrix, or

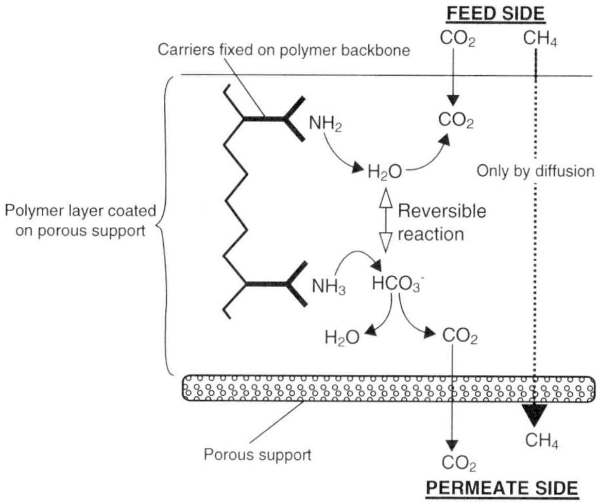

Fig. 2.7 Schematic of fixed carrier facilitated transport membranes for transport of carbon dioxide [40]

mobile. An illustrated schematic of a fix-sited carrier, polyvinylamine, in operation is shown in Fig. 2.7.

The immobilized liquid nature of the facilitator membrane poses practical problems, such as leakage and evaporation of water, as well as loss of the facilitator through degradation. Hence, their performance over long periods of time presents problems for large scale applications. Much of the work has been done on reducing these aging effects. In this regard, Sirkar et al. disclosed the use of dendrimers as carriers [41].

Ho and Dalrymple [42] reported new cross-linked PVA containing silver nitrate membranes. The membrane showed high olefin flux and olefin/paraffin selectivity. The mass transfer resistance due to complexation and decomplexation was discussed for the facilitated transport of butenes and propylene in the membrane. For the membrane, butene fluxes increased with increasing water content in the membrane until it was saturated with water. Butene fluxes increased linearly with silver nitrate content in the PVA membrane, which indicated that Ag^+ in the membrane was mobile, rather than fixed geometrically. According to Ho and Dalrymple, the flux N through a facilitated can be expressed as:

$$N = (p_1 - p_2) / \{(R_c + R_d) + (1/P)\} = \Delta p / \{(R_c + R_d) + (1/P)\} \quad (2.32)$$

where p_1 is the partial pressure of the olefin in the upstream, p_2 is the partial pressure of the olefin in the downstream low pressure side of the membrane, R_c is the mass transfer resistance due to the complexation, R_d is the mass transfer resistance due to decomplexation, l is the membrane thickness, and P is the *true* permeability excluding the mass transfer resistance due to complexation and decomplexation. Δp is the partial pressure difference across the membrane for the olefin.

A novel transport membrane module for gas separation was proposed by Teramoto et al. [43] in which a carrier solution was forced to permeate the membrane. Both a feed gas and a carrier were supplied to the lumen side (high-pressure side, feed side) of the capillary ultrafiltration membrane module, and flow upward.

References

1. Javid A (2005) Membranes for solubility-based gas separation applications. Chem Eng J 112:219–226
2. Budd PM, McKeown NB (2010) High permeable polymers for gas separation membranes. Polym Chem 1:63–68
3. Rouquerol J, Avnir D, Fairbridge CW, Everett DH, Haynes JM, Pernicone N, Ramsay JDF, Sing KSW, Unger KK (1994) Recommendations for the characterization of porous solids (Technical Report). Pure Appl Chem 66:1739–1758
4. Geoffroy R (2004) What is the glass transition temperature, Tg? Polymer Services Group. POLYSERV@cox.net, posted: 12 May 2004 (Edited 27 Sep 2004)
5. Peighambardoust SJ, Rowshanzamir S, Amjadi M (2010) Review of the proton exchange membranes for fuel cell preparation. Int J Hydrogen Energy 35:9349–9384
6. Kluiters SCA (2004) Status review on membrane systems for hydrogen separation. Intermediate report EU project MIGREYD NNE5-2001-670, ECN-C-04-102
7. Pandey P, Chauhan R (2001) Membrane for gas separation. Prog Polym Sci 26:853–893
8. Graham T (1866) On the absorption and dialytic separation of gases by Collid Septa. Philos Mag J Esc 32:401–420
9. Stern SA (1994) Polymers for gas separations: the next decade. J Membr Sci 94:1–65
10. Dai Y, Johnson JR, Karvan O, Sholl DS, Koros WJ (2012) Ultem®/ZIF-8 mixed matrix hollow fiber membranes for CO_2/N_2 separation. J Membr Sci 401–402:76–82
11. Wang D, Teo WK, Li K (2000) Permeation of H_2, N_2, CH_4, C_2H_6, and C_3H_8 through asymmetric polyetherimide hollow-fiber membranes. J Appl Polym Sci 86:698–702
12. Shilton SJ, Bell G, Ferguson J (1996) The deduction of fine structural details of gas separation hollow fiber membranes using resistance modelling of gas separation. Polymer 37:485–492
13. Henis JMS, Tripodi MK (1981) Composite hollow fiber membranes for gas separation: the resistance model approach. J Membr Sci 8:233–246
14. Fouda A, Chen Y, Bai J, Matsuura T (1991) Wheatstone bridge model for the laminated polydimethylsiloxane/polyethersulfone membrane for gas separation. J Membr Sci 64: 263–271
15. Sengbusch GV (1994) Future of membranes, technological and economical aspects. Vortrag: Synthetic Membranes in Science and Industry, Tüebingen
16. Gas separation with membranes—Preamble—Mecadi GmbH, technologyreport.mecadi.com/ Chapter 2. Gas separation with membranes
17. Rowe BW, Robeson LM, Freeman BD, Paul DR (2010) Influence of temperature on the upper bound: theoretical considerations and comparison with experimental results. J Membr Sci 360:58–69
18. Robeson LM (1991) Correlation of separation factor versus permeability for polymeric membranes. J Membr Sci 62:165–185
19. Freeman BD (1999) Basis of permeability/selectivity trade off relations in polymeric gas separation membranes. Macromolecules 32:375–380
20. Barrer RM (1942) Permeability in relation to viscosity and structure of rubber. Trans Faraday Soc 38:322–330
21. Van Amerongen GJ (1946) The permeability of different rubbers to gases and its relation to diffusivity and solubility. J Appl Phys 17:972–985

22. Baker RW (2001) Membranes for vapor/gas separations. MTR Inc., Menlo Park, CA
23. Pace RJ, Datyner A (1979) Statistical mechanical model for diffusion of simple penetrants in polymers: I. Theory. J Polym Sci Polym Phys Ed 17:437–451
24. Kumins CA, Kwei TK (1968) Free volume and other theories. In: Crank J, Park GS (eds). Diffusion in polymers. Academic Press, New York, pp 107–25
25. Brandt WW (1959) Model calculation of the temperature dependence of small molecule diffusion in high polymers. J Phys Chem 63:1080–1084
26. Kesting RE, Fritzsche AK (1993) Polymeric gas separation membranes. John Wiley & Sons Inc., New York
27. Pinnau I, He Z, Da Costa AR, Amo KD, Daniels R (2002) A gas separation using C3+-hydrocarbon- resistant membranes. U.S. Patents 6,361,582 and 6,361,583, 26, Mar 2002
28. Huang Y, Wang X, Paul DR (2006) Physical aging of thin glassy polymer films: free volume interpretation. J Membr Sci 277:219–229
29. Huang Y, Paul DR (2007) Effect of film thickness on the gas-permeation characteristics of glassy polymer membranes. Ind Eng Chem Res 46:2342–2347
30. Budd PM, McKeown NB, Fritch D (2005) Free volume and intrinsic microporosity in polymers. J Mater Chem 15:1977–1986
31. Everett DH (1972) Manual of symbols and terminology for physicochemical quantities and units, Appendix II. Definitions, terminology and symbols in colloid and surface chemistry. Pure Appl Chem 31:577–638
32. Ilinitch OM, Fenelonov VB, Lapkin AA, Okkel LG, Terskikh VV, Zamaraev KI (1999) Intrinsic microporosity and gas transport in polyphenylene oxide polymers. Microporous Mesoporous Mater 31:97–110
33. Wood-Adams PM (2006) Permeation in glassy polymers vs rubbery polymers. GCH 6101-Polymer-Diffusion
34. Lin HC, Tsai IF, Yang ACM, Hsu MS, Ling YC (2003) Chain diffusion and microstructure at a glassy-rubbery polymer interface by SIMS. Macromolecules 36:2464–2474
35. George SC, Thomas S (2001) Transport phenomena through polymeric system. Prog Polym Sci 26:985–1017
36. Cong H, Radosz M, Towler BF, Shen Y (2007) Polymer-inorganic nanocomposite membrane for gas separation. Sep Purif Technol 55:281–291
37. Cong H, Hu X, Radosz M, Shen Y (2007) Brominated poly(2,6-diphenyl-1,4-phenylene oxide) and its SiO_2 nanocomposite membranes for gas separation. Ind Eng Chem Res 46:2567–2575
38. Mulder M (1996) Basic principles of membrane technology. Kluwer Academic Publisher, Dordrecht, The Netherlands
39. Figoli A, Sager WFC, Mulder MHV (2001) Facilitated oxygen transport in liquid membranes: review and new concepts. J Membr Sci 181:97–110
40. Haegg MB, Kim TJ, Li B (2005) Membrane for separating CO_2 and process for the production thereof. WO05089907, 29 Sept 2005
41. Chen H, Kovvali, SA, Sirkar KK (2003) Improved membrane separation of carbon dioxide. WO 2003008070 A1, 30 Jan 2003
42. Ho WS, Dalrymple DC (1994) Facilitated transport of olefins in Ag^+-containing polymer membranes. J Membr Sci 91:13–25
43. Teramoto M, Kitada S, Ohnishi N, Matsuyama H, Matsumiya N (2004) Separation and concentration of CO_2 by capillary-type facilitated transport membrane module with permeation of carrier solution. J Membr Sci 234:83–94

Chapter 3
Gas Separation Membrane Materials and Structures

A membrane is a layer of material which serves as a selective barrier between two phases and is impermeable to specific particles, molecules, or substances when exposed to the action of a driving force. Some components are allowed passage by the membrane into a permeate stream, whereas others are retained by it and accumulate in the retentate stream. Membranes can be of various thicknesses, with homogeneous or heterogeneous structures. Membrane can also be classified according to their pore diameter. There are three different types of pore sizes based on the IUPAC (International Union of Pure and Applied Chemistry) classification: microporous (d_p<2 nm), mesoporous (2 nm<d_p<50 nm), and macroporous (d_p>50 nm) [1, 2]. Membranes can be neutral or charged, and the transport through a membrane can be active or passive. The latter can be facilitated by pressure, concentration, chemical or electrical gradients. Membranes can be generally classified into synthetic membranes and biological membranes.

The chemistry and the structure of membranes play an important role in their separation characteristics. Ideally, membranes should be defect-free in large scale, exhibiting high thermal, chemical, and mechanical stabilities, selectivity, and permeability. Another way of classifying membranes is by the type of material used: inorganic and polymeric. Inorganic membranes include metal, metallic oxide, glass, silicate and zeolite/zeolitic type materials, and so on. Compared with polymeric membranes, inorganic membranes have many advantages: such as high thermal, chemical, and mechanical stabilities, less plasticization, and better control of pore size and pore size distribution, which allows better control of selectivity and permeability. For example, molecular sieve membranes are inorganic membranes developed in recent years with unique properties such as uniform micropores, ion exchange, tunable-Si/Al ratios, high temperature stability, solvent resistance, a wide range of hydrophilicity–hydrophobicity, catalytic activities, and so on. Hence, molecular sieves have excellent membrane material to achieve separation and catalytic reaction, simultaneously. For this reason, they have been widely used as membrane separators, membrane reactors, and as sensors.

A.F. Ismail et al., *Gas Separation Membranes*,
DOI 10.1007/978-3-319-01095-3_3

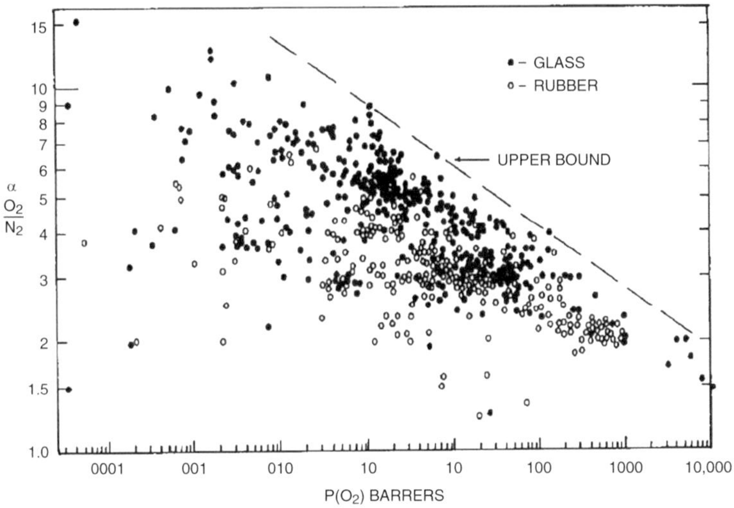

Fig. 3.1 Literature data for O_2/N_2 separation factor versus O_2 permeability [3]

For polymeric membrane materials, there is a trade-off between selectivity and permeability. Robeson [3] has suggested that this trade-off is represented as an upper boundary to the selectivity vs. permeability plot of many polymeric materials. This upper boundary can clearly be seen in Fig. 3.1 for a range of polymeric membrane materials involved in O_2/N_2 separations. Overcoming this upper boundary is the focus of many recently awarded patents in polymeric membranes, because achieving both high oxygen permeability and selectivity is desirable (Fig. 3.1).

As stated earlier, high permeability and selectivity are not the only properties that are desired. For membrane materials to be viable, they need to be thermally and chemically robust, resistant to plasticization and aging to ensure continual performance over long time periods, and at the same time they need to be cost-effective to be manufactured for industrial applications.

3.1 Membrane Materials for Gas Separation

The selection of membrane materials is one of the most important tasks in membrane separation technology. One way to choose membrane material for gas separation is based on its chemical properties. Chemical interactions between membrane materials and gaseous penetrants determine the separation efficiency between the components of gas mixtures [3, 4].

Despite the many advantageous characteristics of inorganic membranes, organic polymeric membranes are first surveyed here, since most of the industrial membranes are currently manufactured from polymeric materials.

3.1.1 Polymeric Membranes

Polymeric membranes take the form of polymeric interphases, which can selectively transfer certain chemical species over others. There are several mechanisms that could be deployed in their functioning. Knudsen diffusion and solution diffusion are prominent mechanisms (see Chap. 2). Polymeric membranes are of particular importance in gas separation applications. Polymeric membrane materials are generally characterized by their transport properties such as permeability and selectivity. Permeability is a measure of the productivity of the membrane and selectivity is a measure of separation efficiency.

Usually nonporous polymeric membranes are used for gas separation. The vapors and gases are separated due to their different solubility and diffusivity in the polymers. The permeability or permeation coefficient of such nonporous membranes can generally be expressed as the solubility, S, of the gas in the membrane polymer multiplied by the diffusivity, D, of the gas in the polymer, i.e., $P=D \cdot S$. In such cases, permeation is said to occur by a "solution-diffusion" model. Polymers in the glassy state are generally more effective for separation, and predominantly differentiate gases based on their different diffusivities. Small molecules of penetrants move among polymer chains according to the formation of local gaps by thermal motion of polymer segments. In a polymeric membrane, pores and channels have a wide range of sizes and topologies. Free volume, the fraction of the volume not occupied by the electronic clouds of the macromolecule, plays a significant role in the transport properties of low molecular weight species and gases. In other words, this is the volume that is not occupied by polymer chains due to conformational constraints. Within this free volume, transient gaps are formed which can accommodate gas molecules. According to the driving force, the gas molecules have to be transported by successive movement between transient gaps close to the feed side to those close to the permeate side. Thus, movement necessary for the transport of the gas molecules between the microvoids is possible due to thermal motion of segments of the polymer chains. Moreover, the distribution of the effective micropore size, if the free-volume elements are interconnected, is likely to have a significant influence on the properties of the membrane. Some experimental techniques (probe methods such as Positron Annihilation Lifetime Spectroscopy, Inverse Gas Chromatography, Xe NMR, and small angle neutron diffraction) can be used to determine the average radius (1–10 Å) and the distribution of the free volume elements. Free volume pockets or voids are crucial for a variety of dynamic processes in membranes. The diffusivity of a penetrant gas depends mainly on its molecular size.

Porous membranes can also be utilized for the gas separation. The pore diameter must be smaller than the mean free path of gas molecules. Under normal conditions (100 kPa, 300 K), this is about 50 nm. In this case, the gas flux through the pore is proportional to the molecule's velocity, i.e., inversely proportional to the square root of the molecule's mass. This is known as Knudsen diffusion. Gas flux through a porous membrane is much higher than through a nonporous one by 3–5 orders of magnitude. The separation efficiency is moderate; hydrogen diffuses four times

Table 3.1 Comparison of glassy polymers in CO_2/CH_4 separation

Polymer	Selectivity A = {P(CO_2)/P(CH_4)}	Permeability P(CO_2) {Barrer}
Cellulose derivatives	3	4,550
Polycarbonate	11–33	75–15
Polyimides	15–25	110–65

faster than oxygen. Porous polymeric or ceramic membranes for ultrafiltration serve the purpose. Note that when the pores are larger than the limit viscous flow occurs, and hence no separation will take place.

Polymeric membrane materials can be divided into rubbery and glassy polymers. Rubber is an example of an elastomer type polymer, which has the ability to return to its original shape after being stretched or deformed. The rubbery polymer is coiled when in the resting state. The elastic properties arise from its ability to stretch the chains apart, but when the tension is released the chains snap back to the original position. Long range motions of chains in rubbery polymers are not possible in glassy polymers. In other words, the chains of glassy polymers are rigid. On increasing the temperature glassy polymers will soften and become rubbery. Glassy polymers are not elastomer type polymers.

Glassy polymers show very attractive separation characteristics, e.g., high selectivity combined with medium/low permeability. On the other hand, rubbery polymers show comparatively low selectivity and high permeability for common gas pairs such as O_2/N_2, H_2/CH_4, and CO_2/CH_4 [5]. The correlation for the CO_2/CH_4 separation is shown in Table 3.1 [5].

The high selectivity of a glassy polymer is due to its lower free volume, a narrower distribution of the free volume, as well as lower flexibility of the polymer chains compared to those of rubbery polymers. Rubbery polymers present high permeabilities and their selectivities are mainly influenced by differences in the condensability of the gas species. When applied to separate an organic vapor from nitrogen, rubbery membranes are preferentially permeable to the organic molecules. An amorphous polymer will behave as a rubbery state when above its glass transition temperature (T_g). It will present a relatively large amount of free volume, owing to transient voids between the highly mobile polymer chains. At below its T_g, the polymer will behave as a rigid glass, which will result in a reduction of the fractional free volume. Thus, there will be an insufficient space for the large-scale cooperative movements of the polymer's backbone.

Many polymers have been investigated as gas separation membrane materials, but up to now only a handful have found commercial success. These include rubbery polymers, such as poly(dimethylsiloxane), and glassy polymers, such as polysulfone, poly(phenylene oxide), cellulose acetate, and polyimides. For gas separation membranes, materials are required to offer high permeability as well as good selectivity for a desired separation.

For a given polymer, how to make a high separation performance membrane by studying the membrane formation protocol is an important topic in research.

Higher permeability decreases the amount of membrane area to treat a given amount of gas, thereby decreasing the capital cost of membrane units. Higher selectivity results in higher purity product gas. A polymer that exhibits good selectivity generally has low permeability and vice versa. For glassy polymers, the gas transport properties depend on the amount and distribution of free volume and on chain mobility. The most highly permeable polymers have rigid, twisted macromolecular backbones that give rise to microvoids [6]. Examples include substituted polyacetylene, perfluoropolymers, addition-type polynorbornene, polymers of intrinsic microporosity (PIMs), and some polyimides. High permeability membranes may also be produced by thermal rearrangement of precursor polymers.

Although thousands of polymers exhibit permselective properties for gas mixtures, only a few glassy polymers are useful in making asymmetric gas separation membranes, particularly polysulfone (PSf), polyether sulfone (PES), polyimide (PI) and polyetherimide (PEI). For example, PSf is one of the most promising polymers for separating oxygen and nitrogen with respect to its permeability and selectivity, even though other polymers such as polyacetylenes, polyaniline, poly(acrylene ethers), polyacrylates, polycarbonates, polyetherimides, poly(ethylene oxide), poly(phenylene oxide), poly(pyrrolone) [7], and cellulose acetate [8] can also be used for gas separation membranes.

Examples of structures of polymers used in fabrication of gas separation membranes are displayed in Fig. 3.2.

Though a large number of polymeric materials have been developed for gas separations, the number of polymers used in commercial systems is still limited [9]. Rubbery membranes reject lighter gases such as nitrogen, methane, and hydrogen, and permeate heavier hydrocarbon components. The main rubbery and glassy polymers employed for gas separation are listed in Table 3.2 [10].

Further to discussions in Chap. 2, the chemical structures of high free volume polymers are elaborated upon below in Fig. 3.3 [6]. Membrane permeability and selectivity data for those polymers are plotted in Fig. 3.4 for the O_2/N_2 and CO_2/CH_4 gas pairs.

Table 3.3 shows the polymers used for gas separation membranes. It should be noted that new polymers have been further developed on the basis of these polymers.

3.1.1.1 Silicone Rubber

Silicone rubber is an elastomer (rubber-like material) composed of polysiloxane that contains silicon together with carbon, hydrogen, and oxygen. Silicone rubbers are often one- or two-part polymers, and may contain fillers to improve properties or reduce cost. Silicone rubber is generally non-reactive, stable, and resistant to extreme environments and temperatures from −55 °C to +300 °C while still maintaining its useful properties. Compared to organic rubbers, however, silicone rubber has a very low tensile strength. For this reason, care is needed in designing products to withstand even low imposed loads. The material is also very sensitive to fatigue from cyclic loading.

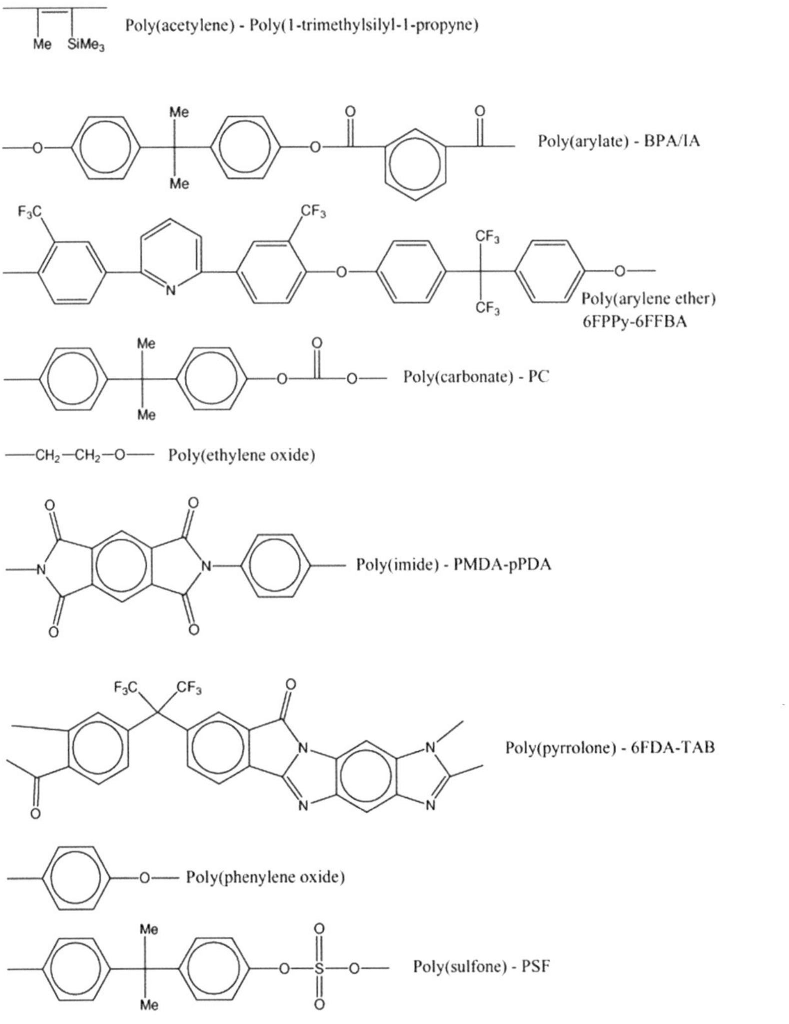

Fig. 3.2 Example of polymeric structures [7]

Table 3.2 Most important glassy and rubbery polymers used in industrial membrane gas separation

Rubbery polymers	Glassy polymers
Poly(dimethylsiloxane)	Cellulose acetate
Ethylene oxide/pyroline oxide-amide copolymers	Polycarbonates
	Polyperfluorodioxoles
	Polyimides
	Poly(phenylene oxide)
	Polysulfone

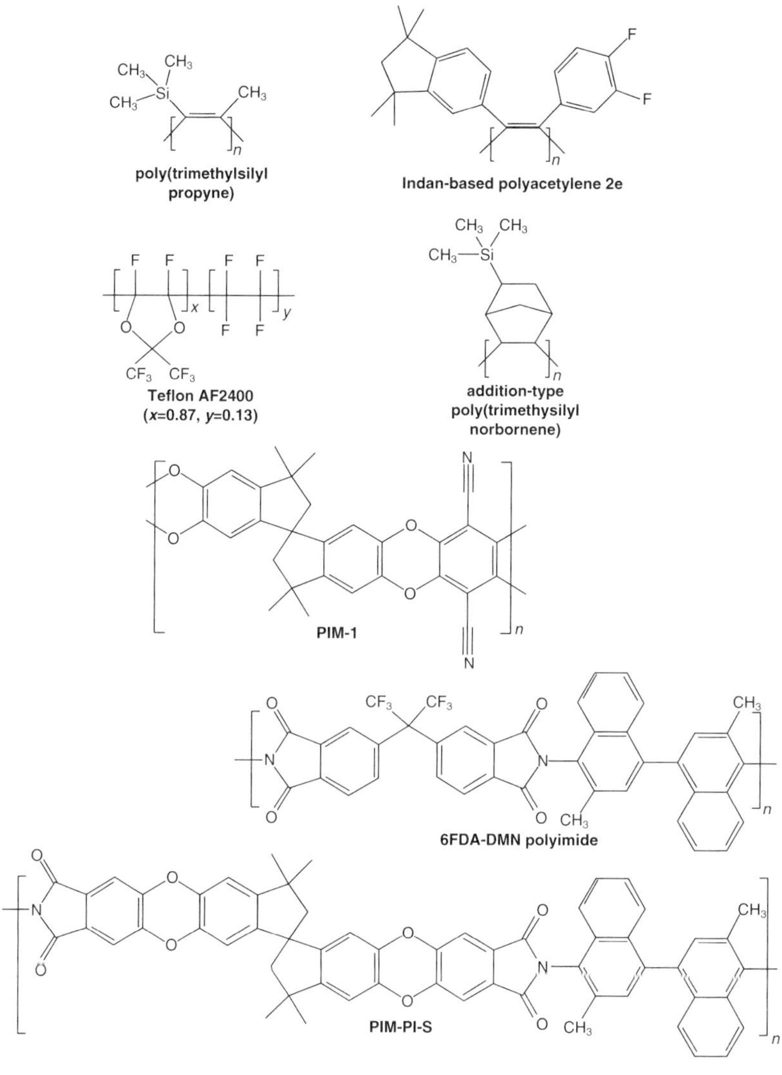

Fig. 3.3 Chemical structures of important high free volume polymers

Polysiloxanes (Fig. 3.5) differ from other polymers in that their backbones consist of Si–O–Si units unlike many other polymers that contain carbon backbones. The C–C backbone unit has a bond length of 1.54 Å and a bond angle of 112°, whereas the siloxane backbone unit Si–O has a bond length of 1.63 Å and a bond angle of 130°. Polysiloxane is very flexible due to large bond angles and bond lengths when compared to those found in more basic polymers such as polyethylene.

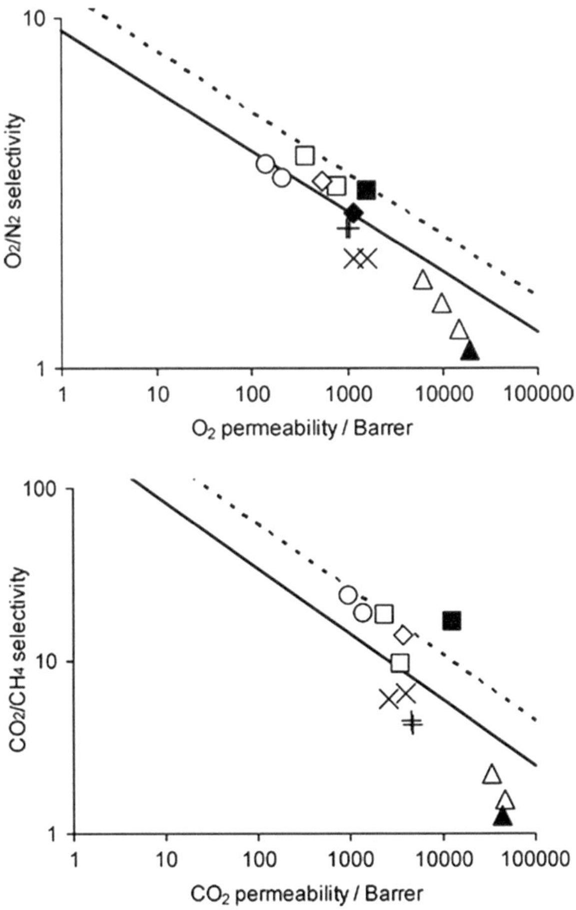

Fig. 3.4 Double logarithmic plots of selectivity versus permeability for (*top*) O_2/N_2 and (*bottom*) CO_2/CH_4. (*Solid line*) Robeson's 1991 upper bound and (*dashed line*) 2008 upper bound, and data for (*open triangle*) poly(trimethylsilyl propyne) (PTMSP), (*filled triangle*) indan-based polyacetylene 2e, (*times*) Teflon AF2400, (*plus*) addition-type poly(trimethylsilyl norbornene), (*open square*) PIM-1, (*filled square*) PIM-1 after methanol treatment, (*open circle*) 6FDA-DMN polyimide, (*open diamond*) PIM-PI-8, and (*filled diamond*) PIM-PI-8 after methanol treatment [6]

Polysiloxanes also tend to be chemically inert, due to the strength of the silicon–oxygen bond. The application of silicone rubber in synthetic membranes for gas separation is of special interest due to its intrinsic high permeation rate. The high permeability of silicon rubber has been attributed to the flexibility of the siloxane linkages in the polymer. This is accompanied by a low permselectivity. Silicone rubber coated composite membranes are commonly utilized in industrial membranes for gas separation [11], removal of organic vapors from air, pervaporation of various organic species, olefin separation, etc. [12]. Silicone rubber has also been used as a gutter layer as well as sealing layer to make thin film composite (TFC) membranes of potentially high-performance materials, if formation of defect-free ultrathin membranes for such types of polymers is difficult. The substitution of polysiloxane with different groups has been reported to improve the intrinsic selectivity, which was usually accompanied by a general decrease in penetrant solubility and

Table 3.3 Polymers used for gas-separation membranes

Polymer
Silicon rubber
Cellulose acetate
Polyether sulfone
Polyimides and polyetherimides
Polypropylene
Polyetherketone
Poly(norbornene)s
Poly(2,6-dimethyl-1,4-diphenyl oxide) (PPO)
Perfluoropolymers
Polycarbonates
Polysulfone
Polyperfluorodioxoles
Polyacetylenes
Polyaniline
Polyalkynes
Polybenzimidazole
Polysaccharides
Polyvinylidene fluoride PVDF

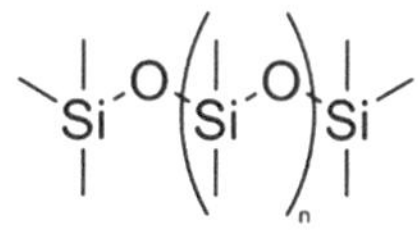

Fig. 3.5 Silicone rubber (polysiloxane)

permeability. Effects of various alkyl group substitutions on gas permeability of the resulting polysiloxanes were studied by Stern et al. [13]. Effects of the incorporation of amide in multiblock copolymers with poly(dimethylsiloxane) (PDMS) are well documented by Furuzono et al. [14].

The use of silicone rubber (SR) based thin film composite (TFC) membranes is increasing significantly in various industries due to the rubber's thermal and chemical stability. Achalpurkar et al. [12] studied the gas permeation properties of dense and thin film composite membranes (TFC) based on amine substituted silicone rubber (ASR) and unsubstituted silicone rubber (SR). The ASR dense membrane exhibited higher CO_2 (15 %) as well as CH_4 (12 %) permeability as compared to the SR dense membrane, while the permeability for other gases (He, H_2, N_2, and O_2) was decreased up to 15 %. Accordingly, CO_2-based selectivities in the ASR dense membrane increased by 22–34 %. The permeance of TFC membranes based on different UF supports decreased in the order of decreasing porosity and increasing solution concentration (ASR). It was also noted that the gas permeance of TFC membranes based on SR and ASR remained unaffected with increasing pressure (up to 4.2 kg/cm^2), except for a nominal increase in CO_2 permeance.

3.1.1.2 Cellulose Acetate

Cellulose acetate membrane, initially developed for reverse osmosis [15], is the most widely used and tested for natural gas sweetening (Table 3.4), as in UOP's membrane systems [16].

Gantzel and Merten [8] applied a cellulose acetate membrane, used for desalination by RO, for the permeation of different gases. They reported the permeances in units of GPU = 10^{-10} cm^3 (STP)/cm^2-s-cm of Hg for He, 10.6; Ne, 1.9; O_2, 0.71; Ar, 0.37; CH_4, 0.34, and N_2, 0.31; all at 22 °C. Cynara-NATCO produces hollow fiber modules with cellulose triacetate for natural gas sweetening in an offshore platform in the Thailand gulf (830,000 Nm^3 h^{-1}, which is the biggest membrane system for CO_2 removal) [10].

Cellulose acetate based membranes for CO_2 separation from CH_4 have been commercialized since the mid-1980s, and arose directly from the use of cellulose acetate reverse osmosis membranes [17]. A critical factor in the performance of cellulose acetate membranes is the degree of acetylation, which is the substitution of the hydroxyl groups on the glucoside repeating unit with acetyl groups. The size difference between the hydroxyl and acetyl group reduces the efficiency of chain packing, as the degree of acetylation increases, and improves chain flexibility and mobility because of reduced internal molecular hydrogen bonding between chains. This improves the intrinsic gas permeability, as shown in Fig. 3.6, with little change in the selectivity.

Table 3.4 Intrinsic permeation properties of cellulose acetate [10]

Permeability (Barrer)	Separation factor (*i*/CH_4) (–)				
CO_2	H_2O	CO_2	H_2S	N_2	C_2H_6
8.9	500	20–25	50	1	0.42
8.9	500	21	19	1	0.42

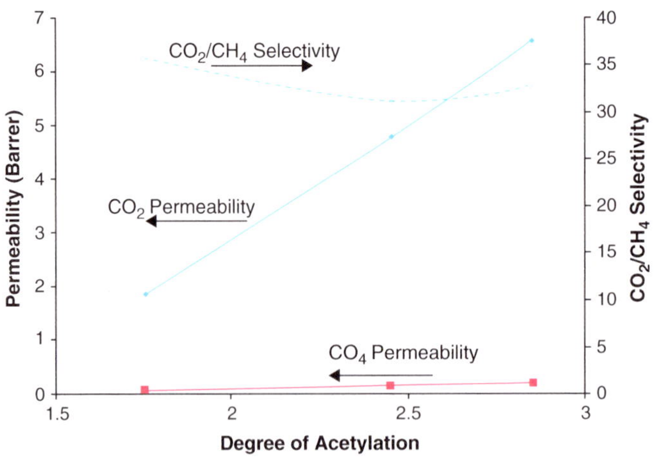

Fig. 3.6 Cellulose acetate membrane permeability (Barrer) and CO_2/CH_4 selectivity as a result of different degrees of acetylation, at 35 °C and 1 atm [18]

Table 3.5 Cellulose acetate membrane performance dried from different solvent exchange, at 60 °C and 300 psi [19]

Drying solvent	Surface tension (mN/m)	CO_2 permeance (GPU)	CO_2/CH_4 selectivity
Heptane	20.1	18	24
Octane	21.6	144	27
Nonane	22.9	174	33
Decane	23.9	156	36
Undecane	24.7	216	28
Cycloheptane	27.1	23	19
Toluene	28.4	57	22
Xylene	30.1	113	30

The other critical factor is fabrication of the membrane, and importantly the approach taken to achieve a dry film. During drying of water wet membranes, evaporation of water leads to capillary forces collapsing the microporous structure, creating a denser active layer. In the case of solvent exchange, water—a high surface tension solvent—is gradually replaced by a low surface tension solvent that has less impact on the membrane structure when evaporated. The effect of different solvents in the exchange on the membrane performance is shown in Table 3.5 [19].

The evaporation temperature for the preparation of wet reverse osmosis membranes is also critical for gas membrane performance. Minhas et al. [20] reported around 80 °C provided the best morphology for selectivity, but generally with low permeability, depending on the combination of solvents used for solvent exchange.

In natural gas processing via CA membranes, plasticization occurs in the membrane, either by CO_2 or heavy hydrocarbon components in feed gas. Plasticization increases the flexibility of the polymer chains and alters the separation performance as well as reduces the mechanical strength and speeds up aging effects. These effects are the cause of catastrophic membrane failure. Houde et al. [21] noticed that CO_2 plasticization pressure at around 10 atm for a dense membrane of 76 μm thickness, while Donohue et al. [22] observed the plasticization at <5 atm for asymmetric cellulose acetate membranes.

To achieve better separation and performances, some research work on cellulose acetate polymer derivatives has been reported. Some examples in the literature incorporate zeolites, transition metal complexes, and silicon species into the membrane or form composites with poly(methyl methacrylate) or poly(ethylene glycol) [20]. Another approach has been to cross-link cellulose acetate, which produces a more resistant structure that reduces the influence of plasticization [23–26]. Rahman et al. [27] also reported that the structure and separation properties of the asymmetric CA membranes depend on the variation in casting shear rate. Cellulose acetate membranes can also be used for the separation of ammonia from ammonia–nitrogen and ammonia–hydrogen systems [28].

Tanioka et al. [29] synthesized a freeze-dried CA membrane with high permeability and high separation factor. It was reported that the gas permeabilities for such a membrane were inversely proportional to the square root of molecular weight, suggesting that the mechanism of gas flow through this membrane was Knudsen flow. Separation factors for H_2-He, Ar-Kr, N_2-Kr, He-Ne, H_2-Ne, He-Ar, Ar-H_2, He-Kr, Ar-He, and He-H_2 were measured. Separation efficiency largely depended upon the combinations of mixed gases.

Kim et al. [30] fabricated composite membranes with cellulose acetate and 2–6 wt% AMH-3 flakes and studied the CO_2/CH_4 gas separation. It was reported that performance of the CA membrane was significantly increased by incorporating only 2–6 wt% of silicate SAMH-3 flakes. There was a large increase in CO_2 permeability with maintenance of selectivity.

The first facilitated transport membranes fabricated from cellulose acetate (patent awarded to General Electric in 1967) [31] were related to CA films swollen by the inclusion of an aqueous carbonate solution. Carbon dioxide readily dissolves and reacts with water to form the bicarbonate anion.

$$H_2O + CO_{2(aq)} - H_2CO_{3(aq)} - H^+_{(aq)} + HCO^-_{3(aq)}$$

Carbonate acts as a carrier by increasing the amount of carbon dioxide absorbed:

$$CO_3^{2-} + CO_{2(aq)} + H_2O - 2HCO^-_{3(aq)} \left(+\ \text{Heat}\right)$$

Hence, this reaction occurs on the feed side of the membrane and the bicarbonate anion transports through to the permeate side, where the reverse reaction occurs and carbon dioxide is released.

3.1.1.3 Polycarbonate (PC)

Polycarbonate membranes possess good mechanical properties and are able to operate under extreme conditions of temperature and pressure. However, polycarbonate membranes sometimes possess lower gas flux or permeability, or gas selectivity than desired for certain applications. A number of polycarbonates (PCs) have been synthesized and their carbon dioxide/nitrogen gas transport properties studied. Polycarbonates are generally synthesized by reaction between a diol and phosgene under a variety of conditions. Like many polymers used for gas separation membrane materials, the ready availability of structural variants of bisphenol A has led to a large number of different PCs. Most PCs tend to have a carbon dioxide permeability of under 40 Barrer, and selectivity ranges from 15 to over 25. One notable exception to this is the polycarbonate TMHFPC, which has a carbon dioxide permeability of 111 Barrer and a carbon dioxide selectivity of 15.0 [7]. The structure of this polymer is given in Fig. 3.7.

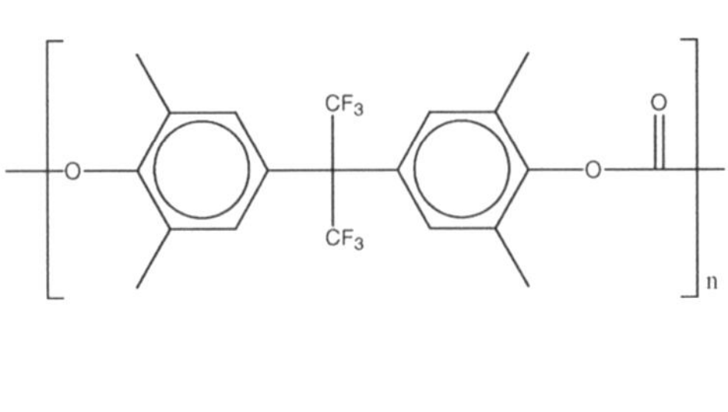

Fig. 3.7 TMHFPC

Ward et al. [32] reported that uniform and defect-free ultrathin polycarbonate membrane can produce 30 % oxygen-enriched air, which enables on-board generation of nitrogen-rich air to inert fuel tank vapor spaces and vent lines. Acharya et al. [33] reported that the permeability of hydrogen was greater than that of carbon dioxide through the polycarbonate (Bisphenol-A polycarbonate) membrane. They suggested that it could be due to the more diffusion energy required for carbon dioxide (linear shape of molecules) than hydrogen (spherical shape). In another study Vijay et al. [34] reported that the permeability of both gases, i.e., H_2 and CO_2, increased rapidly after etching PC membranes by α particles. The etched samples, which received a higher dose of particles showed better selectivity than the lower dose etched samples. Fu et al. [35] demonstrated that the permeability of gases in polycarbonate membranes was related to the forming of pores on the membrane surface.

Hacarlioglu et al. [36] reported that the gas separation properties of the polycarbonate-polypyrrole MMMs systems were highly dependent on the synthesis method (electrochemical or chemical) and also membrane casting conditions (casting solvent type). The MMM of electrochemically synthesized polypyrrole-polycarbonate (ECPPY-PC) showed considerably higher permeabilities compared with both pure PC and the chemically synthesized polypyrrole-polycarbonate (CPPY-PC) films, with a loss in selectivities, whereas the CPPY-PC membranes had good separation properties. A mixed matrix membrane structure, which can be cast as permselective gas separation membranes with attractive gas separation performances, could be obtained with polycarbonate as matrix and zeolite 4A filler [37].

López-González et al. [38] measured the transport of pure oxygen, nitrogen, and carbon monoxide through the membranes prepared from poly[bisphenol A carbonate-co-4,4′-(3,3,5-trimethylcyclohexylidene) diphenol carbonate]. The permeabilities of oxygen, nitrogen and carbon monoxide at 35 °C and $p_0 = 1$ atm were 8.75, 1.87, and 2.91 Barrer, respectively. The values of the diffusion coefficient in 10^{-8} cm^2/s, measured in the same conditions, were 8.26, 2.81, and 3.08, respectively. By replacing the two methyl groups of bisphenol A by other molecular groups, higher permeability was achieved without a substantial diminution of permselectivity. The permeability and the diffusion coefficients followed Arrhenius behavior.

3.1.1.4 Poly(norbornene)s

Norbornenes are important monomers in ring-opening metathesis polymerizations (ROMP). Polynorbornenes are polymers with high glass transition temperatures. Norbornene derivatives may be polymerized either by ring-opening metathesis polymerization (ROMP) or by addition polymerization at the double bond. Most ROMP polynorbornenes show modest permeabilities; however, addition-type poly(trimethylsilyl norbornene), prepared with nickel as catalyst and methyl aluminoxane as co-catalyst, exhibited high free volume and high permeability [6].

Yampol'skii et al. [39] studied the two fluorine-containing glassy polynorbornenes (permeabilities, diffusion, and solubility coefficients, etc.). It was observed that these studied polymers had very high free volumes, and thus high permeabilities.

Addition-type poly(norbornene)s with siloxane substituents were synthesized using the n^3(allyl)(n^5-cyclopentadienyl)palladium, (Cp)Pd(ally) system [40, 41]. These synthesized polymers had very high T_g of up to 255 °C, and their films displayed high flexibility and optical clarity. The gas permeabilities of the polymer films varied with the content and structure of the siloxane substituents, and the films for the polymers with branched side groups showed high oxygen permeability ($P(O_2)=66–360$ Barrer). The films obtained for the polymers in the three arm-siloxane group (–Si(OSiMe$_3$)$_3$) displayed high oxygen permeability in the range of 39–239 Barrer. Asymmetric poly(norbornene) membranes were first prepared by a dry/wet-phase inversion technique. The membranes consisted of the thin skin layer supported by the porous substructure, and the surface skin layer had a unique nanoporous structure related to the nature of the side groups. The findings of this study suggest that the structure of the side group linked to the chain is an important factor for the regulation of the material properties and presents a new possibility for the application of poly(norbornene) materials.

Dorkenoo et al. [42] studied the permeation of gases in polynorbornenes and concluded that for noncondensable gases such as H_2 and He, the selectivity over N_2 decreased when the length of the pendant group increased, but remained relatively stable for the more condensable gases (O_2 and CO_2). The permeability correlates well to the inverse of the fractional free volume of the polymers.

3.1.1.5 Poly(2,6-Dimethyl-1,4-Diphenyl Oxide) (PPO)

Poly(2,6-dimethyl-1,4-diphenyl oxide) (PPO) exhibits excellent film formation and gas separation properties, and is resistant against a number of chemical agents, strong acids and bases. It belongs to the thermally stable glassy polymers having high glass transition temperatures. PPO is an example of a fairly permeable polymer having alternating aromatic cycles and C–O linkage in the main chain. Among the many aromatic polymers that possess high glass transition temperatures (T_g), PPO shows the highest permeability to gases. It is not evident why this polymer is much more permeable than polysulfone or bisphenol A polycarbonate, having similar structures of repeating units. A possible reason could be the absence of polar groups attached to the main chain. Whatever the reason for its properties, PPO drew attention as a permeable and rather permselective material.

The main application of modified and unmodified PPO membranes is directed to gas separation because of PPO's high permeability to gases. A number of electrophilic substitution reactions have been conducted on PPO to improve its gas permeation characteristics, such as bromination, carboxylation, methyl esterified carboxylation, sulfonylation, acylation, and silylation, introduction of trialkyl-silyl, hydroxyethylene, and ethyleneoxytrialkyl-silyl groups to be polymer backbone [43]. Permeability of PPOs as high as 100–200 Barrer have been reported for gases such as H_2 and CO_2 [44, 45]. Membranes from PPO are also known to be moderately permselective and it ranks among the hydrophobic polymers. Table 3.6 shows the permeability data of PPO for CO_2, CH_4, O_2, and N_2 gases [46].

Table 3.6 Permeability data of PPO for CO_2, CH_4, O_2, and N_2

Gas permeability (Barrer)[a]				Perm selectivity	
CO_2	CH_4	O_2	N_2	CO_2/CH_4	O_2/N_2
90.0	5.4	16.7	3.7	16.7	4.5

[a]Permeabilities at permeate pressure (absolute) ≊ 100.0 kPa

Khulbe et al. [47] studied a PPO dense membrane prepared at different temperatures (22, 4 and −10 °C) via ESR, AFM and gas permeation. It was reported that the morphology of the surfaces of the membrane, the shape of its spin, and the selectivity of gases depend on the temperature of the evaporation of solvent used. The permeation rate of CO_2 increased with the decrease in the temperature used in the preparation of the membrane. However, methane permeation rate increased in the membrane prepared at −10 °C. It was suggested that Langmuir sites could be favorable for the CH_4 permeation.

Hamad et al. [48] reported that a brominated high molecular weight PPO membrane showed higher permeability of CO_2, CH_4, O_2, and N_2 gases when compared with non-brominated PPO. However, the permeability ratios for CO_2/CH_4 and O_2/N_2 changed only a little. Solubility of gases increased on increasing the bromination level. The trends in the gas permeability data and IR spectra obtained in this study both conform with and support the mechanism proposed in the literature that a higher degree of bromination is needed to enhance the permeability of gases by stiffening the PPO backbone, which increases the rate of diffusional jumps. It was also reported by Hamad and Matsuura [46] that the main effect of simultaneous sulfonation and bromination of PPO was: (1) to increase the gas permeability, and decrease the gas permeability ratio, in comparison to sulfonated PPO (SPPO), while on the other hand, (2) to decrease the gas permeability, and increase the gas permeability ratio, in comparison to brominated PPO (PPOBr). The observed trend in sulfonated brominated PPO membranes, when the degree of bromination was increased while the same degree of sulfonation was maintained, was the occurrence of a minimum in both gas permeability and diffusivity at 37.4 % degree of bromination. All these effects due to sulfonation of PPO, or bromination of PPO, or simultaneous sulfonation and bromination of PPO, were believed to be the direct result of the manipulation in polymer backbone stiffness, the packing density, and the free volume fraction.

Yu et al. [49] reported that sulfonated poly(2,6-diphenyl-1,4-phenylene oxide) ($BSPPO_{dp}$) as a new material for CO_2/N_2 separation. Compared with PPO membrane, the CO_2 permeability (58 Barrer) and CO_2/N_2 permselectivity (36) of the $BSPPO_{dp}$ membrane were higher by 1.2 and 2.5 times, respectively, than the PPO membrane. The addition of silica nano particles in the $BSSPO_{dp}$ membrane resulted in an increase in CO_2 permeability while maintaining CO_2/N_2 selectivity. The CO_2 permeability increased as a function of the silica content in the membrane. The separation mechanism for CO_2/N_2 in the membranes was attributed to the gas solubility effect rather than the gas diffusivity.

3.1.1.6 Polyimides (PI)

An extensive review of the gas separation properties of polyimides was published in 1996 [50]. Aromatic polyimides, usually obtained from bifunctional carboxylic acid dianhydrides and primary diamines, are widely used in different branches of industry such as electronics, insulators, high-temperature adhesives, and photoresistance. The successful applications of such high-performance polyimides are mainly due to excellent physicochemical properties such as high temperature stability, exceptional mechanical strength, low thermal expansion coefficiency, dimensional stability, superior insulation properties, and radiation and chemical resistance. In the field of membrane science and technology, polyimide membranes have been extensively studied because of their excellent separation performances, particularly for gas separation [51].

Many polyimides exhibit good selectivity, but at the expense of permeability. Polyimides are a class of polymers that display high permeability for CO_2 and good selectivity against CH_4. They are supposed to be an alternative to cellulose acetate because they are also easy to prepare as asymmetric membranes while demonstrating good thermal and chemical stability. PIs are rigid, have high melting point and high T_g, and are thermally stable polymers obtained by polycondensation reactions of dianhydrides and diamines. A wide range of polyimides exist with good gas separation properties based on varying both the diamine and dianhydride. Some polyimides, particularly those incorporating the group 6FDA (2,2-bis(3,4-dicarboxyphenyl) hexafluoropropane dianhydride), possess both high selectivities and high permeabilities due to following reasons:

1. The CF_3 group considerably increases the stiffness of the chain, allowing the membrane to more effectively separate molecules on the basis of steric bulk.
2. Effective chain packing is reduced by the large CF_3 groups, which leads to an increase in the permeability.

6FDA-based polyimide membranes for gas separation have been studied mainly with dense membrane containing various diamine moieties [52]. Kim et al. [52] synthesized 6FDA-based polyimides with polar hydroxyl or carboxyl group in diamine, such as 6FDA-BAPAF, 6FDA-DAP, and 6FDA-DABA by the thermal imidization method. The corresponding composite membranes were then prepared by the dip-coating technique using a poly(ether sulfone (PES)) membrane as the supporting layer. Chemical structures of the 6FDA-based polyimides synthesized by Kim et al. are shown in Fig. 3.8. 6FDA-TrMPD polyimide was used for the preparation of a reference membrane material to compare solubility.

Table 3.7 shows the gas permeation characteristics of the composite membranes with different 6FDA-based polyimides.

The 6FDA-DAP polyimide membrane exhibited higher permeance and selectivity than other polyimides. CH_4 permeance for the 6FDA-BAPAF membrane was obtained at a much higher level than expected; hence, CO_2/CH_4 selectivity for the 6FDA-BAPAF polyimide membrane was low. CO_2/N_2 selectivities for the composite membranes prepared using different polyimides above were equal to those for

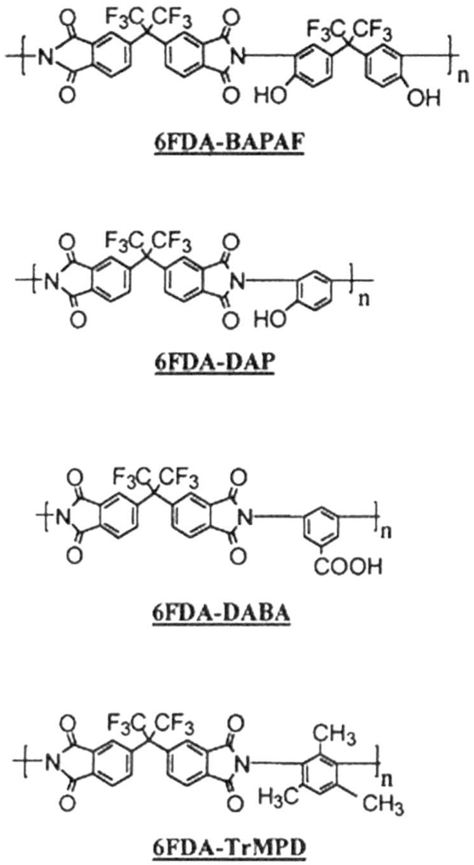

Fig. 3.8 Chemical structure of 6FDA-based polyimides

Table 3.7 Gas permeation characteristics of the composite membranes with different 6FDA-based polyimides

	Permeance (GPU)					α[a]	
Polyimides	CO_2	N_2	CH_4	H_2	O_2	CO_2/N_2	CO_2/CH_4
6FDA-BAPAF	24.60	1.23	1.10	46.93	4.86	20.11	22.78
6FDA-DAP	38.57	1.32	0.49	73.80	6.52	29.26	78.82
6FDA-DABA	26.30	0.93	0.56	52.85	4.85	28.28	46.96

[a]α: ideal selectivity. Coating solvent: 2-methoxyethanol. Coating polymer concentration: 1 %. GPU = 10^{-6} $cm^3(STP)cm^{-2}\,s^{-1}\,cmHg^{-1}$

other dense or asymmetric 6FDA-based polyimide membranes in the literature, as shown in Table 3.8 [52].

Li et al. [53] developed a delamination-free, dual layer, asymmetric composite hollow fiber membrane. 6FDA-durene-1,3-phenylenediamine (mPDA) (50:50) copolyimide was used to form the outer asymmetric separating layer, while PES was employed to yield the inner interpenetrated porous supporting layer.

Table 3.8 Gas selectivity of 6FDA-based polyimide membranes in the literature [52]

Polyimide	Membrane type	α (selectivity) CO_2/N_2	α (selectivity) CO_2/CH_4
6FDA-3BDAF	Dense	27	48
6FDA-IPDA	Dense	23	43
6FDA-DAFO	Dense	22	60
6FDA-APPS	Asymmetric	29	39

Pure gas permeance test results revealed that the O_2/N_2 selectivity of the dual layer asymmetric hollow fiber was about 4.6, which is very close to the intrinsic value of the outer-layer material (4.7) with an O_2 permeance of around 28 GPU at room temperature.

The insertion of bulky substituents into the peripheral polymer backbones should disrupt efficient packing of the polymeric chains, leading to an increase in the permeabilities and a loss of selectivities. A number of different polymers have been investigated for this effect. Some bulky novel polyimides structures are given in Fig. 3.9.

A range of polyimide membranes are given in Table 3.9. Among these, Matrmid 5218 is the most often reported in the literature. The chemical structure of Matrimid is given in Fig. 3.10.

Figure 3.11 displays the carbon dioxide permeability and carbon dioxide/nitrogen selectivities of a variety of polyimides [7] and Table 3.10 provides permeability and selectivity data of pure and mixed gas for Matrimid 5218 [54].

Like cellulose acetate, polyimides are known to be susceptible to plasticization by CO_2. For asymmetric Matrimid, the plasticization pressure is <4 bar, while dense 6FDA-DAD at 35 °C is plasticized by 14 bar CO_2 [55]. However, plasticization pressure is dependent on the thickness of the membrane. Polyimides are more sensitive to plasticization than cellulose acetate. This could be due to the more ordered structure in the polyimide film compared to cellulose acetate, which makes it more susceptible to, structural changes. White [17] observed that hexane plasticized polyimide membranes and that naphthalene fouled the polyimide membrane surface. Various approaches have been made to improve polyimide performance in terms of gas separation properties, as well as resistance to plasticization by CO_2 and other components in natural gas. Approaches, such as hyper-branching and cross-linking, preparation of mixed matrix membranes, and blending with other polymers such as polysulfone and polyethersulfone have been studied. Bos et al. [56] observed an increase in the plasticization pressure of Matrimid when blended with Therrmid FA-700 from around 10 bar to around 20 bar.

Faiz and Li [57] concluded that of all the investigated polyimide and copolyimide membranes, 6FDA-based membranes showed the best performances for olefin/paraffin separation due to loose polymer chain packing. However, the membrane performances seemed to deteriorate with time since the glassy state of the polymer was not in an equilibrium state initially, and shifted to a denser packing state with aging time.

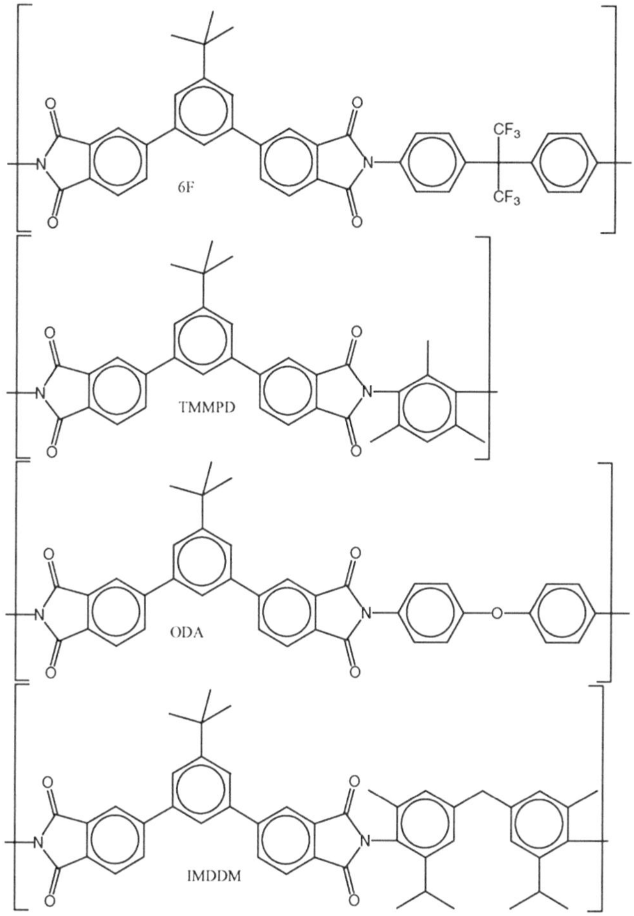

Fig. 3.9 Structure of some bulky novel polyimide [7]

Tin et al. [54] modified the Matrimid® 5218 membrane by cross-linking. A room temperature cross-linking modification of Matrimid® 5218 has been done by immersing the films in a 10 % (w/v) of *p*-xylenediamine in methanol solution for a certain period of time. The chemical structure changes during the cross-linking process, i.e., after cross-linking, imide functional groups of Matrimid were converted to the amide functional groups. The gas permeabilities and ideal selectivities of He, O_2, N_2, CH_4, and CO_2 through cross-linked Matrimid films are summarized in Table 3.11. The table depicts the behaviors of pure gas permeabilities with immersion time. The gas permeabilities are found to attain the maximum values for 1-day cross-linking, followed by subsequent decrease with immersion time. It seems that the degree of cross-linking governs the gas transport properties. The selectivity of CO_2/CH_4 for cross-linked film is higher than the film without cross-linking. It was

Table 3.9 CO_2 permeability and CO_2/CH_4 selectivity for a range of polyimide membranes at 35 °C [19]

Polymer	CO_2 permeability (Barrer)	CH_4 permeability (Barrer)	CO_2/CH_4	Temp. (°C)	Press.
Matrimid 5218	6.5	0.19	34	35	10 bar
Matrimid 5218	5.39	0.15	36	35	2 bar
6FDA-TAPOB	7.4	0.098	75	25	1 bar
ODPA-TAPOB	0.63	0.0064	98	25	1 bar
PMDA-TAPOB	3.3	0.66	50	35	1 bar
6FDA-DATPA	23	0.68	34	35	10 bar
ODPA-IPDA	0301	0.0064	47	35	10 bar
6FDA-6FpDA	63.9	1.5	39.9	35	10 bar
6FDA-6FmDA	5.1	0.08	63.8	35	10 bar
DAD-6FDA	381	15.24	25	25	300 psi
DAM-6FDA	691	48.7	14.2	25	300 psi
DDBT-BPDA	8.20	0.24	34.2	50	10 bar

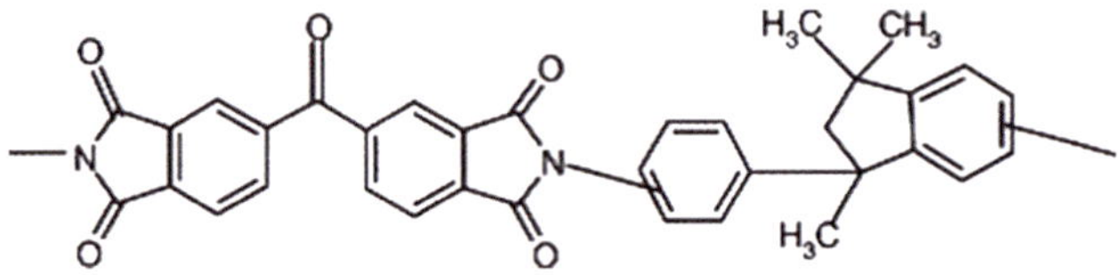

Fig. 3.10 Chemical structure of Matrimid

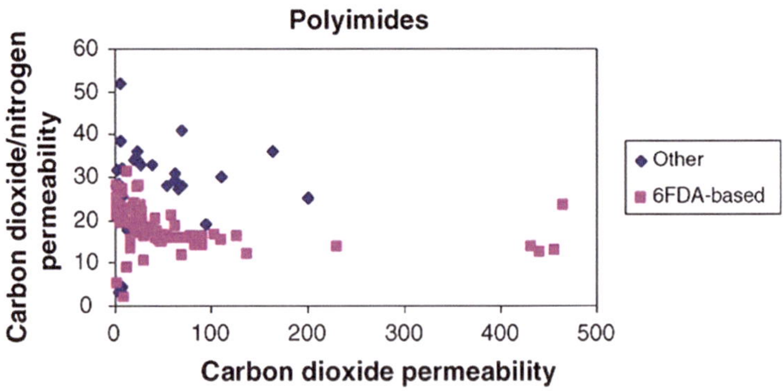

Fig. 3.11 Carbon dioxide permeability vs. selectivity for polyimides

also observed that Matrimid membranes undergo the plasticization phenomenon at 15 atm; however, plasticization is effectively suppressed by the cross-linking modification.

Asymmetric hollow fibers fabricated from commercially available polyimide material Matrimid® gave high CO_2/CH_4 separation factors (ranging up to 67), which were among the highest reported for purely polymeric hollow fibers without post-treatments and exceeded the commonly reported bulk values of Matrimid® [58].

Table 3.10 CO_2 and CH_4 permeability (Barrer) and CO_2/CH_4 selectivity in Matrimid 5218 under mixed gas conditions (10 % CO_2 in CH_4) at 7.5 bar and 35 °C in comparison with pure gas [19]

	CO_2 permeability (Barrer)	CH_4 permeability (Barrer)	CO_2/CH_4
Pure gas	11.5	0.23	49
Mixed gas	9.0	0.22	41

Table 3.11 Gas separation properties of original and cross-linked Matrimid dense films[a]

Immersion	Permeability (Barrer)					Selectivity			
time (days)	He	O_2	N_2	CH_4	CO_2	He/N_2	O_2/N_2	CO_2/CH_4	CO_2/N_2
0	22.2	1.7	0.25	0.19	6.5	87	6.6	34	25.6
1	26.2	1.9	0.29	0.20	7.4	91	6.5	36	25.6
3	25.0	1.6	0.24	0.18	6.0	105	6.9	34	25.2
7	22.1	1.5	0.21	0.15	5.1	107	7	33	24.6
14	21.7	1.4	0.19	0.14	4.7	112	7	34	24.1
21	19.4	1.1	0.15	0.10	3.4	128	7.4	32	22.2
32	17.5	0.9	0.13	0.07	1.9	140	6.9	38	15

[a] 1 Barrer = 1×10^{-10} $cm^3(STP) cm\ cm^{-2}\ s^{-1}\ cmHg$

Peng and Chung [59] spun a defect free as-spun Torlon® hollow fiber membrane with an ultrathin dense layer of around 540 Å from only a one polymer/one solvent binary system at reasonable take-up speeds of 10–50 m/min. The best O_2/N_2 perm-selectivity achieved was much higher than the intrinsic value of Torlon® 4000TF poly(amide imide) dense film, which was about 8.

Polyimide membranes were modified by immersing the films in the diamine/methanol solution for a stipulated period of time (cross-linking). A series of linear aliphatic cross-linking diamines reagents (ethylenediamine, propane-1,3-diamine, and butane-1,4-diamine) were used. This study demonstrated that diamine cross-linked membranes possess high separation performance and provide impressive separation efficiency for H_2/CO_2 separation. Both pure gas and mixed gas data were better than other polymeric membranes and above the Robeson's upper boundary curve (Fig. 3.12) [60].

It was suggested that this modification can alter the physiochemical structure of polyimide membranes with superior performances for H_2 and CO_2 separation. Liu et al. [61] developed an extremely simple room temperature chemical cross-linking technology for the modification of polyimide films for gas separation of He/N_2 and O_2/N_2. Using 6FDA-durene as an example, chemical modification was performed by immersing the dense 6FDA-durene films in a *p*-xylenediamine methanol solution for a certain period of time followed by washing with fresh methanol and drying at ambient temperature. Gas permeation properties of modified polyimides for He, O_2, N_2, and CO_2 were measured at 35 °C and 10 atm. The gas permeabilities of the cross-linked 6FDA-durene dense films are summarized in Table 3.12.

Table 3.12 indicates that the gas permeability decreased significantly in the order of $CO_2 > N_2 > O_2$ with an increase in the degree of cross-linking, which was mainly

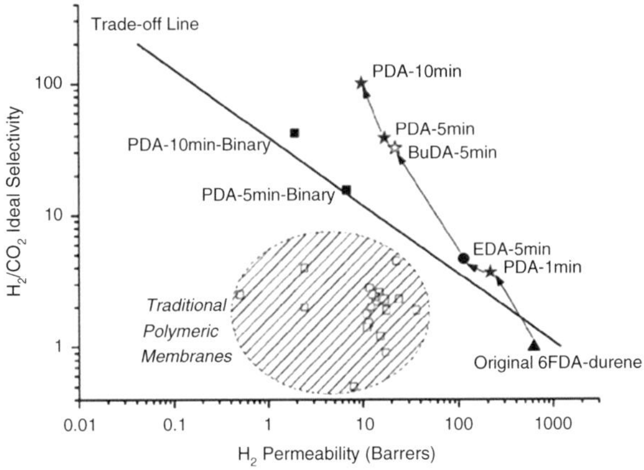

Fig 3.12 Both pure gas and mixed gas separation properties of H_2/CO_2 separation membrane derived from 6-FDA-durene with respect to the *upper bound curve* [60]

Table 3.12 Gas permeabilities of the cross-linked 6FDA-durene dense films

Immersion time (min)	*P* (Barrer)			
	He	O_2	N_2	CO_2
0	362	125	33.5	456
5	204	45.2	11.1	136
10	168	28.9	6.53	91.8
15	157	26.5	6.05	70.0
30	109	13.7	2.87	30.3
60	34.4	2.34	0.40	2.14

attributed to the significant decreases in diffusion coefficients, but CO_2/N_2 decreased from 12 to 5.4, which suggested that this cross-linking approach was most useful for the application of He/N_2 and O_2/N_2 separation.

Wang et al. [62] synthesized a series of novel fluorinated copolyimides and studied the permeation of pure and mixed gases through these membranes. The basic formula of copolyimides was 2,2′-bis(3,4′dicarboxyphenyl)hexafluoropropane dianhydride (6FDA)-2,6-diamine toluene (2,6-DAT)/1,3-phenylenediamine (mPDA). The permeability decreased with increasing mPDA content; however, the permselectivity of gas pairs such as H_2/N_2, O_2/N_2, and CO_2/CH_4 was enhanced with the incorporation of mPDA moiety. The permeabilities of H_2, O_2, N_2, CO_2, and CH_4 were found to decrease with increasing order of kinetic diameters of the penetrant gases. 6FDA-2,6-DAT/mPDA (3:1) copolyimide and 6FDA-2,6-DAT polyimide had high separation properties for H_2/N_2, O_2/N_2, and CO_2/CH_4. Their H_2, O_2, and CO_2 permeabilities were 64.99, 5.22, and 23.87 Barrer for 6FDA-2,6-DAT/mPDA (3:1) copolyimide, respectively, while for 6FDA-2,6-DAT polyimide they were 81.96, 8.83, and 39.59 Barrer, respectively. All copolyimides exhibited similar performances, lying on or above the existing upper boundary trade-off lines between permselectivity and permeability.

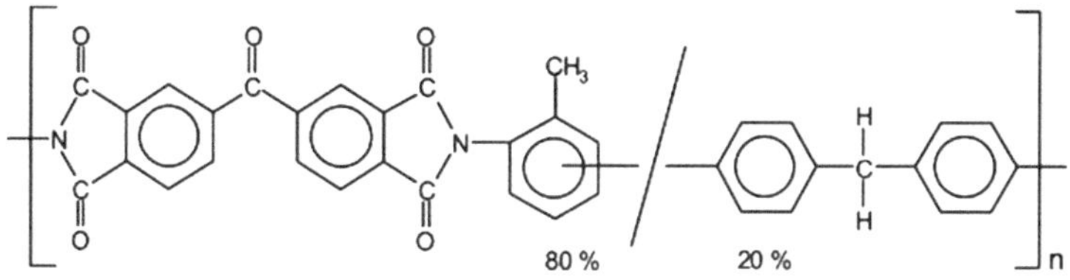

Fig. 3.13 Chemical structure of BTDA-TDI/MDI (P84) co-polyimide

Co-polyimide BTDA-TDI/MDI is a commercial polymer produced by Lenzing with the trade name P84. The chemical structure of this polymer is shown in Fig. 3.13. This material is known as non-plasticizible and its glass transition temperature is 315 °C.

Barsema et al. [63] prepared dense flat sheets as well as asymmetric hollow fiber membranes based on BTDA-TDI/MDI (P84) and used both for the separation of a CO_2–N_2 (80/20) mixture and permeation rates of pure gases. The permeation rates of He, CO_2, O_2, and N_2 were measured by the variable pressure method at different feed pressures and temperatures. It was revealed that P84 co-polyimide is a one of the most selective glassy polymers. It is a promising material for the preparation of gas separation membranes with high selectivities such as 285–300 for He/N_2, 45–50 for CO_2/N_2, and 8.3–10 for O_2/N_2. The permeability of CO_2 was relatively low (1 Barrer at 25 °C). The permeation of CO_2 through the asymmetric hollow fiber membranes increased with pressure indicating that the plasticization behavior of asymmetric membranes differs from the respective dense ones. However, no evidence of plasticization was observed when a CO_2/N_2 (80/20) mixture was fed to the hollow fiber membranes at a pressure up to 30 bar. In all cases, CO_2 permeance decreased with pressure while that of N_2 remained constant.

3.1.1.7 Polyetherimide

Polyetherimide (PEI) is an amorphous, amber-to-transparent thermoplastic with characteristics similar to the related plastic PEEK. Relative to PEEK, PEI is cheaper and lower in impact strength, but has a higher use temperature. The repeating unit of PEI is shown in Fig. 3.14.

The molecular formula of the repeating unit of PEI is $C_{37}H_{24}O_6N_2$ and the molecular weight is 592 g/mol. The glass transition temperature of PEI is 216 °C. Its amorphous density at 25 °C is 1.27 g/cm^3. It is prone to stress cracking in chlorinated solvents.

Ultem is a family of PEI products manufactured by SABIC as a result of acquiring the General Electric Plastics Division in 2007. Ultem 1000 (standard, unfilled polyetherimide) has a high dielectric strength, natural flame resistance, and extremely low smoke generation. Ultem has high mechanical properties and performs in continuous use to 340 °F (170 °C). Ultem 1000 has a thermal conductivity of 0.122 W/m K.

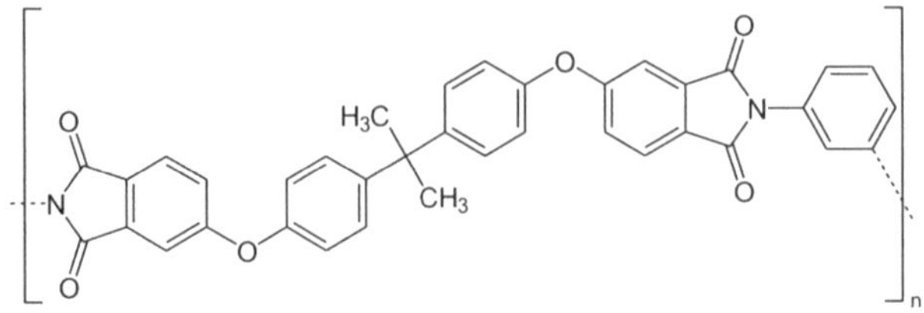

Fig. 3.14 Repeating unit formula of PEI

Kneifel and Peinemann [64] developed asymmetric PEI (Ultem® 1000) hollow fiber membranes both with a microporous skin and with a dense skin on the bore side of the fibers. The effect of the composition of the polymer solution, the spinning conditions and the post-treatment on the porosity of the skin, and on the shape and morphology of the fiber wall was investigated. Silicone composite membranes with an oxygen permeance up to $5.3 \times 10^{-9}\ m^3(N)\,m^{-2}\,s^{-1}\,Pa^{-1}$ and dense membranes with a helium/nitrogen selectivity of about 170, at a helium permeance up to $5.6 \times 10^{-10}\ m^3(N)\,m^{-2}\,s^{-1}\,Pa^{-1}$ were obtained.

Gas permselection properties for N_2, CH_4, Ar, CO_2, O_2, H_2, and He through the PEI membranes were investigated in some detail as a function of pressure and temperature. Wang et al. [65, 66] studied the permeation properties of pure H_2, N_2, CH_4, C_2H_6 and C_3H_8 through asymmetric PEI hollow fiber membranes. The PEI asymmetric hollow-fiber membrane was spun from an *N*-methyl-2-pyrrolidone/ethanol system via a dry-wet phase-inversion method, with water as the external coagulant and 50 wt% ethanol in water as the internal coagulant. The prepared asymmetric membrane exhibited sufficiently high selectivity (H_2/N_2 selectivity > 50) at 25 °C. The apparent separating layer thickness determined by the gas permeation was in the range of 370–500 Å. It was also reported by Wang et al. that on adding volatile organic compounds as additives into the dope solutions for the preparation of PEI hollow fiber, the hollow fiber membrane showed high selectivity for He/N_2 separation [67].

It has been reported that homogeneous and ultrathin (60 nm) uniform films of polyetherimide polymer (Ultem) can be fabricated on a smooth support, like a clean glass by spin coating. The thin Ultem film can be laminated and removed from the support without any damage using an in situ cast support film of PPO over an Ultem film. The films of Ultem and PPO have shown good adhesion without any interpenetrating regions. It has also been reported that the oxygen and nitrogen permeability was independent of the Ultem film thickness down to 60 nm. The helium permeability for the 150 and 60 nm films were about 20 % higher than 500 and 815 nm thick films [68].

Bruma et al. [69] synthesized a series of polyetherimides by polycondensation reactions of 2,2-bis[4,4-(3,4-dicarboxyphenoxy)phenyl]propane dianhydride with various aromatic diamines at high temperature. Polymer solutions in chloroform were processed into thin films, which were tested as gas separation membranes. Transport parameters for light gases were measured. The dependence of glass

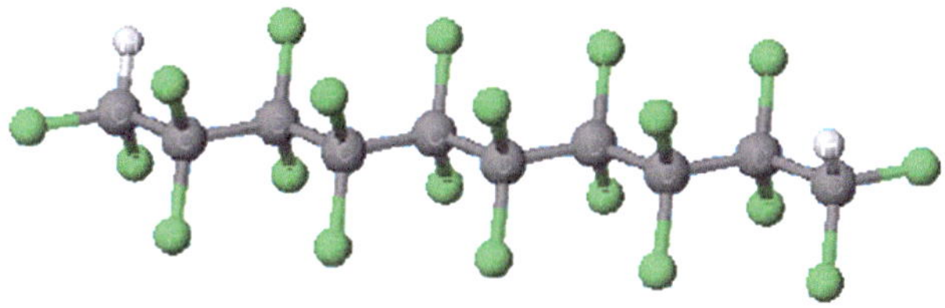

Teflon, $-(CF_2CF_2)-$

Fig. 3.15 Chemical structure of Teflon

transition and decomposition temperature on conformational rigidity parameters was calculated. All these properties, associated with easy processability, make these polymers potential candidates for practical applications as gas separation membranes.

3.1.1.8 Perfluoropolymers

Perfluoropolymers are a class of polymers that display unique resistance to hostile chemical and thermal environments, and in particular resistance to plasticization from both CO_2 and hydrocarbons. The reason for their unique properties is the high energy of the C–F bonds that exist in the substituent groups of the polymer backbone. The most common perfluoropolymer is poly(tetrafluoroethylene) (PTFE). PTFE and similar polymers, such as poly(hexafluoropropylene-tetrafluoroethylene), have poor gas separation properties because of a high degree of crystallinity when cast as a film. The discovery in the 1980s of amorphous perfluoro polymers, in particular Cytop©, Teflon AF©, and Hyflon AD© [70, 71] that utilize bulky fluorine substituent groups to inhibit chain packing has enabled efficient gas separation membranes to be fabricated. Teflon is one brand name for a number of fluorinated polymers (Fig. 3.15) and its chemical name is polytetrafluoroethylene (PTFE). This is a polymer with repeating chains of $-(CF_2-CF_2)-$ in it. Teflon was first discovered accidentally by Roy J. Plunkett (1910–1994) in 1938 at the DuPont research laboratories in N.J. and introduced as a commercial product in 1946. Teflon resists many chemicals. This includes ozone, chlorine, acetic acid, ammonia, sulfuric acid, and hydrochloric acid. The only chemicals known to affect these coatings are molten alkali metals and highly reactive fluorinating agents.

There are a number of commercially available perfluoropolymers (Teflon AF, Hyflon AD, Cytop) that form amorphous and glassy films. Highest permeabilities were obtained for Teflon AF2400 and Teflon AF1600, which are copolymers of 2,2-bistrifluoromethyl-4,5-difluoro-1,3-dioxole and tetrafluoroethylene with dioxole mole fractions of 0.87 and 0.65, respectively [72, 73].

The permeability and selectivity performance of some common perfluoropolymers are listed in Table 3.13 [19].

Table 3.13 Permeability and selectivity of perfluorinated glassy polymers for natural gas separation [20]

Membrane	CO_2 permeability (Barrer)	CH_4 permeability (Barrer)	CO_2/CH_4	Temp. (°C)	Press. (bar)	Refs.
Telfon AF2400	2,200	390	5.7	35	27	[71]
Telfon AF1600	520	80	6.5	–	–	[71]
Hyflon AD80	150	12	13	–	–	[71]
Hyflon AD60	130	10	13	–	–	[71]
Cytop	35	2.0	18	–	–	[71]
Poly-(perfluoro (2-methylene-4-methyl-1,3-dioxolane)	67	2	33.5	25	7.8	[74]
Cyclic perfluorodimethylene-bis(perfluorovinyl ether)	8.2	0.21	39	–	–	[75]
PDD-CTFE	70.4 (GPU)	0.88 (GPU)	80.0	60	–	[76]
Pdd-TFE-MA	18.6 (GPU)	0.69 (GPU)	27.0	60	–	[76]

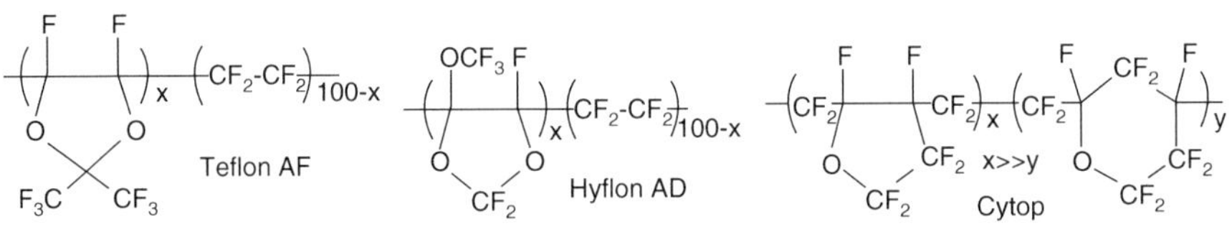

Fig. 3.16 Repeat units of glassy and amorphous perfluoropolymers used in the preparation of gas separation membranes. Teflon AF2400: $x = 87$, $T_g = 240$ °C. Teflon AF1600: $x = 65$, $T_g = 160$ °C. Hyflon AD80X: $x = 80$, $T_g = 134$ °C. Hyflon AD60X: $x = 60$, $T_g = 130$ °C. Cytop, $T_g = 108$ °C [10]

Amorphous Teflon AF2400 possesses many advantages as a membrane material, including good film forming properties, extremely high chemical stability, low susceptibility to swelling, and—unlike other high permeability polymers—no detectable aging. Beside this, AF2400 is insoluble in common organic solvents but shows good solubility in perfluorinated solvents, which are used in preparation of composite membranes [77]. Teflon AF2400 and Teflon AF1600 (DuPont) are the most permeable among perfluoropolymers (Fig. 3.16 and Table 3.14), a family with excellent thermal and chemical resistance, melt stability, good mechanical properties, and usable at broad temperature range. Hyflon AD60X is a compromise of a moderately high selectivity and still interesting permeability, in comparison with the more permeable but less selective Teflon AF. It is particularly suitable for use in dense GS membranes.

In perfluoropolymers the solubility selectivity substantially changes; therefore, they represent materials suited for challenging separations (e.g., olefin/paraffin or natural gas treatment).

Table 3.14 Permeation properties of glassy and amorphous perfluoropolymer membranes

	Permeability (Barrer)						Selectivity
Polymer	O_2	N_2	CH_4	CO_2	C_2H_6	C_3H_8	CO_2/CH_4
Teflon AF2400[a]	1600	780	600	3,900	370	200	6.5
Teflon AF1600	270	110	80	520			6.5
Hyflon AD80	67	34	12	150			13
Hyflon AD60	57	20	10	130			13
Cytop	16	5.0	2.0	35			18

[a]Pure gases, feed pressure, 3.5 bar; thickness, 20 μm

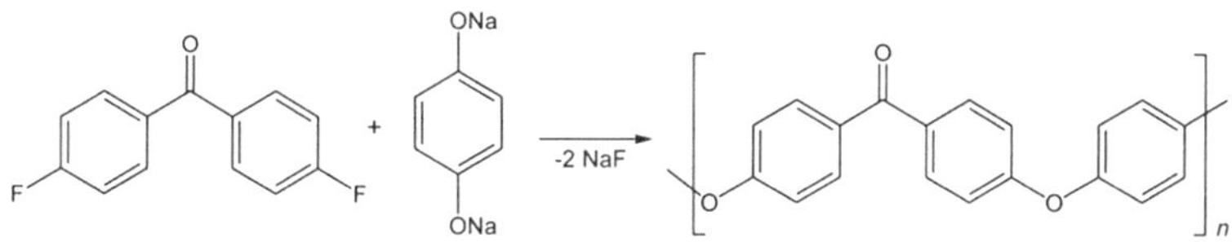

Fig. 3.17 Polyether ether ketone formation

3.1.1.9 Poly(Ether Ether Ketone) (PEEK)

Polyether ether ketone (PEEK) is a colorless organic thermoplastic polymer in the polyaryletherketone (PAEK) family. PEEK polymers are obtained by step-growth polymerization by the dialkylation of bisphenolate salts (Fig. 3.17). Typical is the reaction of 4,4′-difluorobenzophenone with the disodium salt of hydroquinone, which is generated in situ by deprotonation with sodium carbonate. The reaction is conducted at around 300 °C in polar aprotic solvents—such as diphenyl sulfone.

Poly(ether ether ketone)-WC is a phenolphthalein-based poly(ether ether ketone), having a lactone group sticking out of the backbone. PEEK polymer is amorphous and soluble in chlorohydrocarbons, amides, and ethers; therefore, it is well suited for the preparation of polymeric membranes by phase separation techniques. CO_2/N_2 and O_2/N_2 selectivity of asymmetric PEEK-WC membranes prepared by the dry phase inversion technique are 33 and 6, respectively, which are comparable to typical commercial membranes, such as composite polyimide membranes. On the other hand, CO_2 and O_2 permeance are 2.3×10^{-11} m^3 m^{-2} s^{-1} Pa^{-1} and 4.3×10^{-12} m^3 m^{-2} s^{-1} Pa^{-1}, respectively, which are slightly lower than the typical commercial membranes [78].

3.1.1.10 Polyurethane (PU)

Polyurethane (PUR and PU) is a polymer composed of a chain of organic units joined by carbamate (urethane) links. While most polyurethanes are thermosetting polymers that do not melt when heated, thermoplastic polyurethanes are also available. The properties of polyurethane are greatly influenced by the types of isocyanates and polyols. Figure 3.18 shows the PU synthesis.

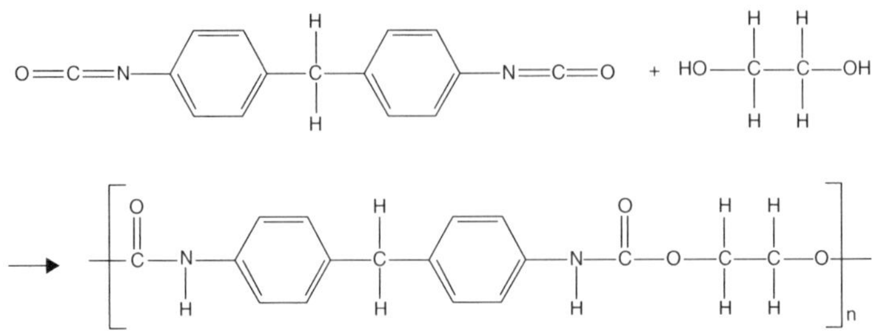

Fig. 3.18 Polyurethane synthesis, wherein the urethane groups –NH–(C=O)–O– link is the molecular units

Polyurethane (PU) membranes own high gas permeability but low selectivity. Some efforts were made to increase gas selectivity by modifying the polymer structure of these membranes but the effort was unsuccessful. It was reported that increasing the amount of oxygen carrier salt (cosalen) into a polycarbonate membrane increase the selectivity of oxygen to nitrogen, especially at low temperature [79].

Chen et al. [80] also reported the effects of oxygen carrier salt, namely cosalen. The dual mode analysis showed that the gas separations in PU membranes were dominated by gas diffusion rather than gas sorption. The selectivity of O_2/N_2 was 8.6 and the oxygen permeability was 1.1 Barrer for the PU membrane with 5 wt% cosalen at 5 °C. The key issue for improving the gas separation performance of a polyurethane membrane is to increase the diffusivity ratio but not the solubility ratio.

Sadeghi et al. [81] fabricated polyurethane-silica membranes by a solution blending and casting method. It was reported that CO_2/N_2, CO_2/CH_4, and O_2/N_2 permselectivities increased from 24.96, 9.56, and 2.17 for pure polyurethane to 41.26, 13.43, and 2.58 for polyurethane–silica (20 wt%). The gas permeation properties of prepared nano-composite membranes showed the decrease in gas permeability of membranes with silica content, but an increase in CO_2/N_2 and CO_2/CH_4 selectivities. The permeation of gases was also modeled by the modified Higuchi model. New constants for the Higuchi model were obtained for studied gases. The experimental data and modified model showed good agreement.

Talakesh et al. [82] studied the effect of the structure of polyether-based polyurethane (PU) membranes on their gas separation properties. In this regard, a series of polyurethanes were synthesized based on hexamethylene diisocyanate (HDI) and 1,4-butanediol as hard segments and different soft segments such as poly(tetramethylene glycol) (PTMG, 2,000 g/mol), poly(ethylene glycol) (PEG, 2,000 g/mol), and PTMG/PEG mixture. The results of gas permeation experiments showed that by increasing the ether group content in the polymer structure, permeability of the pure gases decreased, while CO_2/N_2 ideal selectivity increased. The obtained results also indicated that the permeability of CO_2 decreased from 132.52 Barrer in PU0 (PU containing 100 wt% of PTMG in soft segment) to 20 Barrer in PU100 (the PU containing 100 wt% of PEG in soft segment), respectively. CO_2/N_2 selectivity increased from 28 to 90. Trade-off evaluation also showed

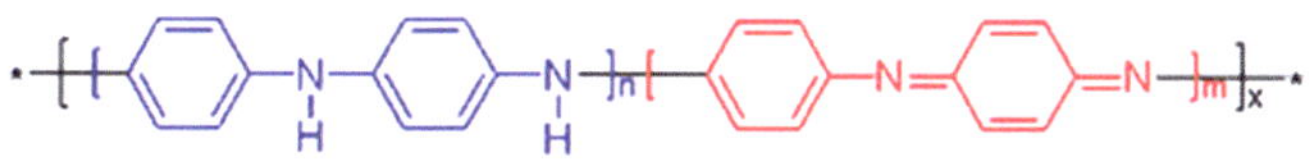

Fig. 3.19 Main polyaniline structures $n+m=1$, x=degree of polymerization

that the commercialization potential of the studied membranes for CO_2/N_2 and even CO_2/CH_4 separation increased with the PEG content in the polymer. In particular, the PU membrane which contained 75/25 wt% ratio of PEG/PTMG had the highest potential for commercialization.

3.1.1.11 Polyaniline (PANi)

Polyaniline (PANi) ($\{[C_6H_4NH]_{2n}[C_6H_4N]_{2m}\}_x$) is a conducting polymer of the semi-flexible rod polymer family. Its structural formula is shown in Fig. 3.19.

Polymerized from the inexpensive aniline monomer, polyaniline can be found in one of three idealized oxidation states.

1. Leucoemeraldine—white/clear and colorless $(C_6H_4NH)_n$.
2. Emeraldine—green for the emeraldine salt, blue for the emeraldine base ($\{[C_6H_4NH]_2[C_6H_4N]_2\}_n$).
3. (Per)nigraniline—blue/violet $(C_6H_4N)_n$.

PANi, which belongs to an important member of the family of electrically conducting polymers, has been studied extensively as a membrane due to its distinct electrochemical properties and environmental stability. The Martin group [83, 84] and the Anderson group [85] showed that electronically conductive polymers (example PANi) are promising membrane materials for industrial gas separation. Adding dopants to PANi leads to a decrease in gas permeability, while removal of these dopants would produce extremely high permeability [86]. Polypropylene-supported polyaniline membranes—photografted with 2-hydroxyethyl methacrylate and glycidyl methacrylate to produce hydrophilicity and reactivity and then reacted with diamines to provide basicity—have been prepared and used for the separation of carbon dioxide and methane. After solvation with water, these membranes exhibit a permeability of around 3,400 Barrer and a separation factor up to 490 [87]. Kuwabata and Martin [88] recommended that polyaniline must be regarded as a promising material for O_2/N_2 separation because its combination of $\alpha_{O_2/N_2}=15$ and $P_{O_2}=0.16$ Barrer places it above Robeson's upper bound [3].

Gas permeation experiments of O_2 and N_2 were performed with conducting polyaniline (PANi) composite membranes prepared by using a porous nylon membrane as a support. Lee et al. [89] reported that PANi composite membranes can be easily obtained by a novel solvent welding process. Doping, dedoping, and redoping kinetics of PANi composite membranes were studied by calculating the [Cl]/[N] content using elemental analysis. After doping and dedoping processes, the permeability of a dedoped PANi membrane decreased while selectivity slightly increased, probably because the changes in morphology of PANi. *d*-Spacing of the PANi film decreased from 4.89 to 3.67 Å. As redoping continued, the *d* spacing decreased, resulting in a dramatic increase in selectivity of the PANi membrane. The highest O_2/N_2 selectivity and permeability obtained from PANi redoped was 28 and 0.13 Barrer, respectively.

Illing et al. [90] reported transport rates (permeability) and ideal separation factors for several gas pairs through dense polyaniline membranes. PANi membranes were modified by doping into hydrochloric acid (4 M), dedoped by similar treatment with ammonia liquor (1 M) and slightly redoped with HCl solution. The ideal separation factors for all gas pairs tested were found to be independent of the polyaniline membrane thickness whereas the permeability of the single gases showed significant variations. The highest selectivities $\alpha_{(A/B)}$ found were 7.6 for the gas pair H_2/CO_2 (in the case of the dedoped membrane) and 10 for the gas pair H_2/CO_2, 6 for O_2/N_2 and 200 for H_2/N_2 (in the case of the redoped membrane). Hasbullah et al. [91] further developed an emeraldine base (EB) PANi integrally skinned asymmetric hollow fiber membrane for gas separation application. The macromolecular orientation was formed by the synergistic effect due to spin-line stresses, which improved the performance of the PANi hollow fiber for gas separation. The gas flux was significantly decreased while the selectivity was increased with increase in air gap from 2.5 to 50 cm. Membranes with longer air gaps (50 cm) showed promising ideal gas separation properties for H_2/N_2 (105.6), O_2/N_2 (10.2), CO_2/N_2 (13.3), and H_2/CO_2 (7.9) and the H_2 and O_2 permeance of about 5.0 and 0.49×10^{-6} cm^3 (STP) cm^{-2} s^{-1} cmHg, respectively.

3.1.1.12 Polysulfone (PSf) and Polyethersulfone (PES)

Polysulfone (PSf) and polyether sulfone (PES) are widely used for preparation of gas separation membranes. Polysulfone describes a family of thermoplastic polymers. These polymers are known for their toughness and stability at high temperatures. They contain the subunit aryl-SO_2-aryl, the defining feature of which is the sulfone group. Polysulfones were introduced in 1965 by Union Carbide. Figure 3.19 shows the repeating unit of polysulfone.

Polyethersulfone (PES) is a similar polymer having the structure of a repeating unit as shown in Fig. 3.20. The greatest characteristic of PES is that it has by far better high-temperature properties than conventional engineering plastics. Specifically, PES remains in satisfactory condition in long-term continuous use without causing any dimensional change or physical deterioration at temperatures as high as 200 °C. Hence, both PSf and PES are high performance engineering polymers. They have good stability, permeability, selectivity, high critical pressure of plasticization, and low cost (Fig. 3.21).

One of the most widely investigated glassy polymer membrane materials for CO_2/CH_4 separation is PSf. Its Polysulfone pure- and mixed-gas permeation properties have been extensively explored for gas separation due to PSf's low price, chemical stability, and mechanical strength. Compared to CA, PSf has lower CO_2

Fig. 3.20 Polysulfone repeating unit

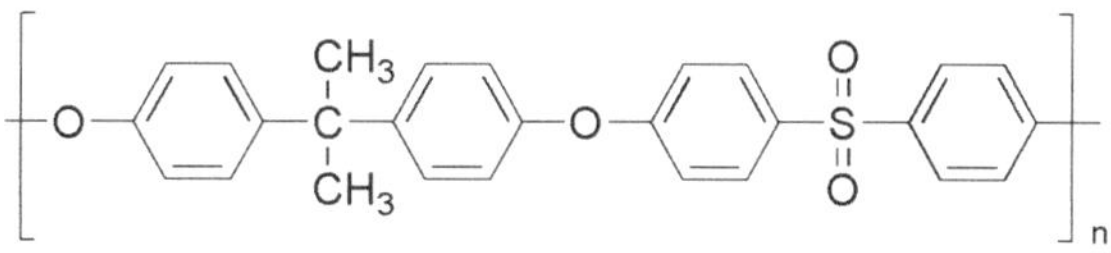

Fig. 3.21 Polyether sulfone

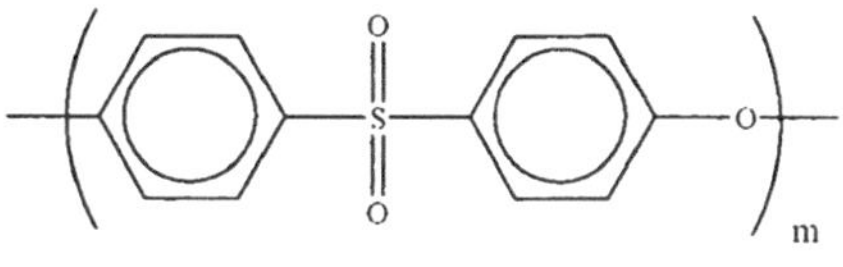

permeability and CO_2/CH_4 selectivity, but higher plasticization pressure [92]. There are three types of PSf membrane that can be used for CO_2/CH_4 separation: dense, asymmetric, and composite. Dense and asymmetric membranes consist of PSf only while composite membranes consist of PSf and other polymers in different layer. Manufacturing processes of these types of membranes are different and versatile. For dense and asymmetric membranes there are three major process: dry, wet, and dry/wet. The most extensively studied polysulfone is PSf formed using bisphenol A. Most other polysulfones are structurally related to this polymer. Two routes for the synthesis of PSf are displayed in Fig. 3.22.

The chemical structure and physical properties of the membrane material influence the permeability and permselectivity. For example, substitution of bulky groups in the side chains appears to have a greater influence on diffusivity than substitution of these groups in the polymer back bone.

To functionalize PSf bromination is an effective route for increasing the reactivity of the polymer, leading to the potential for more structural variation. A number of polysulfones have been modified by reaction with butyl-lithium followed by addition of a pendent group [93, 94].

By substitution of bisphenol A with a different diol, a large number of PSf derivatives have been synthesized. These display a wide range permeabilities and selectivities of carbon dioxide or other gases. The structures of these polymers are displayed in Figs. 3.23 and 3.24.

McHattie et al. [95] reported that replacing phenylene hydrogens of polysulfone with a methyl group had a significant effect on gas transport as well as other properties. It was revealed that the effect of the substituent on chain mobility and chain packing was related to the gas transport properties. Permeability measurements were made for He, H_2, O_2, N_2, CH_4, and CO_2 at 35 °C over a range of pressures up to 20 atm. Sorption experiments were also done for N_2, CH_4, and CO_2 under the same conditions. The permeability coefficients of these polymers for all of the gases rank in order: TMPSF (tetramethyl bisphenol A polysulfone) ≫ PSF (unsubstituted bisphenol A polysulfone) ≫ DMPSF (dimethyl bisphenol A polysulfone) ≫ DMPSF (dimethyl bisphenol Z polysulfone).

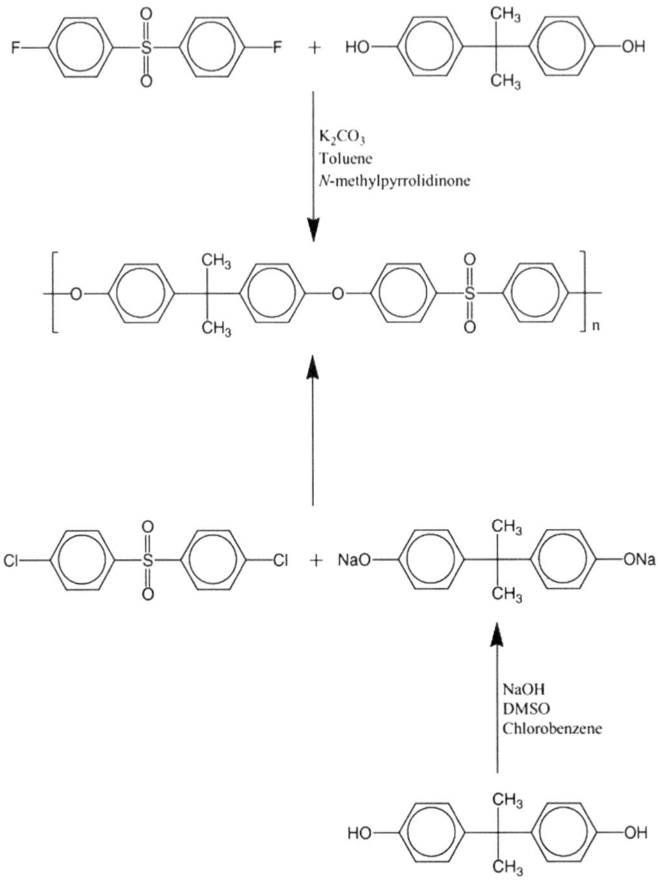

Fig. 3.22 Synthesis of PSf

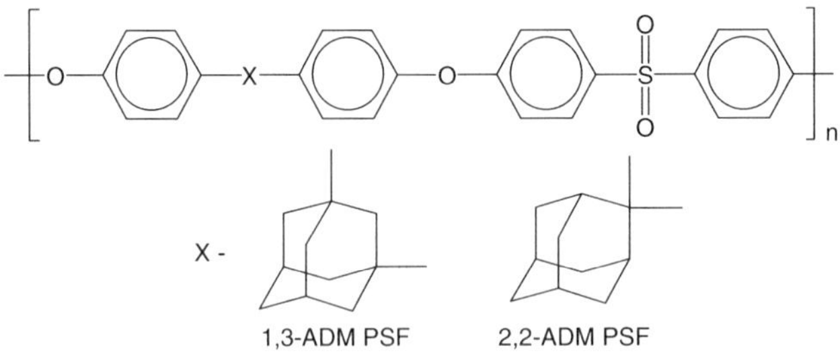

Fig. 3.23 Adamantane-based polysulfone membrane

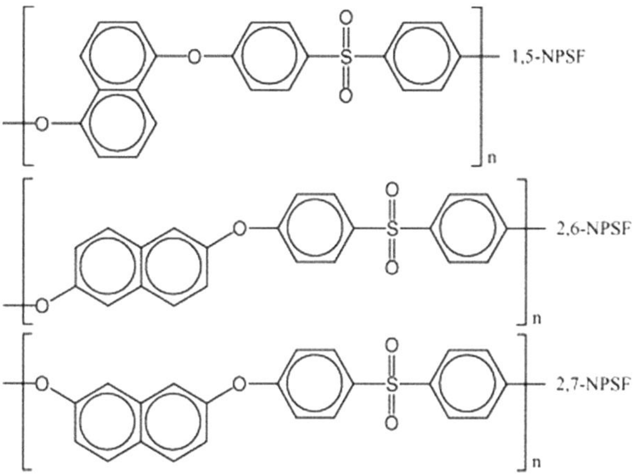

Fig. 3.24 Naphthalene-based polysulfone membrane

Marchese et al. [96] showed that composite membranes with appropriate H_2 separation performance can be obtained by flooding for a short time (1 min) the surface of an asymmetric polysulfone membrane with a solution of 6 % Sylgard 182 in cyclohexane. They achieved the ideal separation factors of 43.24 and 34.04 for H_2/N_2 and H_2/CH_4, respectively.

Wang et al. [97] introduced polysulfone (PSf) hollow fiber membranes with high gas separation performance from *N*-methyl-2-pyrrolidone(NMP)/H_2O and NMP/ethanol solvent systems. Water was used as the external coagulant. The internal coagulants used included water, ethanol, 2-propanol, a mixture of water and ethanol and of water and 2-propanol. The separation performance of the membranes prepared from the NMP/water solvent system was better than that of the membranes spun from a NMP/EtOH system. The O_2 permeance of the membranes prepared was in the range of 20–30 GPU with the O_2/N_2 selectivity of 5–6.5 at 25 °C. The air gap had significant influence on the hollow fiber separation performance. The selectivity decreased with a decrease of coagulation bath temperature.

Ahn et al. [98] used PSf/silica nanoparticle MMMs to study gas permeabilities as well as diffusion and solubility coefficients of hydrogen, helium, oxygen, nitrogen, methane, and carbon dioxide as a function of silica volume fraction via a time-lag method. The effect of silica nanoparticles in PSf membranes on gas permeability was compared with a prediction using the Maxwell model. The O_2 permeability was approximately four times higher and CH_4 permeability was over five times greater than the pure PSf membrane. The performance, comprising permeability versus selectivity of PSf/silica MMMs for O_2/N_2 and CO_2/CH_4, followed a similar slope to that of the trade-off upper bound with increasing silica content.

Weng et al. [99] prepared nanocomposite membranes using MWCNTs with poly(A-co-4-nitrophthalic anhydride-co-1,3-phenylene diamine), and (PBNPI) as the polymer matrix. They extended this approach and demonstrated that at high MWCNTs concentrations, the permeabilities of H_2 and CH_4 improved significantly

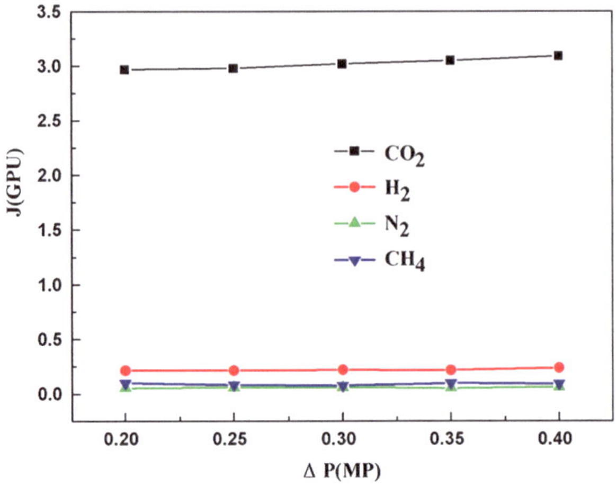

Fig. 3.25 Pressure difference dependence of permeability of various gases for the "water-swollen" dual layer cellulose/PSf hollow fiber

from 4.71 to 14.31 and 0.7 to 1.78 Barrer, respectively. They also cited that the selectivity of H_2/CH_4 reached 8.04.

Dual-layer cellulose/PSf hollow fiber membranes were used for the dehydration of isopropanol and CO_2 separation [100]. The water swollen dual-layer cellulose/PSf hollow fiber membrane showed a much higher gas permeation rate and comparable selectivities of CO_2/H_2, CO_2/N_2, and CO_2/CH_4 when compared with a dry membrane. The CO_2 permeance of a dual layer cellulose/PDf hollow fiber membrane was about 3 GPU, which was five times higher than that of a single-layer cellulose hollow fiber membrane. Thus, the performance, especially the permeation rate of a newly developed dual layer cellulose/PSf hollow fiber membrane, was greatly improved by its reduced separation layer (layer of cellulose). Figure 3.25 shows the permeance and permselectivity of CO_2, H_2, N_2, and CH_4 in the "water-swollen" dual layer cellulose/PSf hollow fiber membrane. Table 3.15 also compares the permeance and selectivity between a single-layer cellulose hollow fiber membrane and a dual layer cellulose/PSf hollow fiber membrane. The CO_2 permeance of the dual-layer cellulose/PSf hollow fiber membrane was five times higher than that of the single-layer cellulose hollow fiber membrane. The increase in permeance was mainly because of the large decrease of cellulose layer thickness, which was 10–15 μm compared with the single layer symmetric dense cellulose membrane (almost 160 μm).

Arahman et al. [101] studied the effect of the addition of hydrophilic polymeric surfactant Pluronic F127, polyvinylpyrrolidone (PVP) and Tetronic 1307 on the performance of the PES hollow fiber membrane. The addition of 5 wt% polymeric surfactant on the polymer solution resulted in a membrane with improved length and macrovoid structure. All membranes had a skin layer on the surface and finger like macrovoid structure inside the hollow fiber. The Sponge formation both near

Table 3.15 Comparison of permeance and selectivity between single-layer cellulose hollow fiber membrane and cellulose/PSf dual layer hollow fiber membrane [100]

	Permeance (GPU)				Separation factor		
Membrane	CO_2	N_2	CH_4	H_2	CO_2/N_2	CO_2/CH_4	CO_2/H_2
Dual-layer	3.05	0.058	0.08	0.178	53	38.7	17.1
Single-layer	0.6	0.013	0.02	0.038	45.4	30	15.8

the inner and outer surfaces of the hollow fiber membrane was another impact caused by the addition of polymeric additives.

Studies on gas permeation properties of isotropic polyethersulfone (PES) dense films have shown that PES exhibits better selectivity for the commercially important gas pairs (CO_2/CH_4, He/CH_4, H_2/N_2, O_2/N_2) compared to bisphenol-A polysulfone and cellulose acetate [102]. Studies have revealed that this polymer has only a moderate permeability. In order to use this polymer to prepare commercially attractive gas separation membranes, fabrication of ultrathin-skinned asymmetric hollow fiber membranes was studied. However, these studies are limited.

Researchers from the University of Twente have studied preparation of PES asymmetric hollow fiber membranes from NMP alone and a NMP/glycerol solvent mixture under different spinning conditions [103–106]. The PES hollow fibers prepared at various spinning conditions exhibited very low gas permeation. The observed CO_2 permeance was less than 12 GPU at 24 °C.

In the late 1980s, PES hollow fiber membranes with good permeance and selectivity were prepared from a spinning solution containing 1:1 molar mixtures of propionic acid and NMP and high polymer concentration (more than 35 wt%) [107]. The oxygen permeance was reported to be 13.1 GPU with a O_2/N_2 selectivity of 5.1 at 50 °C. The skin layer structure of the membrane was examined by Fritzsche et al. [108].

Systematic studies on the preparation and characterization of PES hollow fiber membranes spun from moderate polymer concentrations (25–30 wt%) and solvent systems containing various alcohols as non-solvent additives (NSA) have been done by Wang et al. [109, 110]. These studies show that NSA plays a dominant role in determining membrane structure and gas separation properties. The PES hollow fiber membranes with the best combination of gas permeability and selectivity were prepared using ethanol as an additive. The studies also demonstrated that good NSAs should possess good affinity and diffusivity with the coagulant. Wang et al. [109] fabricated ultrathin silicone-coated PES asymmetric hollow fiber membranes with high permeances and ideal selectivities for gas pairs of He/N_2, CO_2/N_2, and O_2/N_2, from NMP/H_2O solvent systems with a mass ratio of 8.4:1. The observed permeance and selectivity were higher than those of the PES hollow fibers spun from NMP/alcohols and NMP/propionic acid solvent systems reported in the literature. The macrovoids on the membrane wall could be reduced significantly by choosing a suitable internal coagulant with a moderate non-solvent strength, such as a mixture of alcohol and water.

Kim et al. [111] used PESf hollow fiber membrane to recover sulfur hexafluoride (SF_6) from a N_2/SF_6 binary mixture gas. The highest SF_6 purity in recovered gas was 50.4 vol% when the pressure difference, temperature, and stage cut was highest in experimental conditions, but the recovery ratio marked the lowest value.

Jiang et al. [112] fabricated almost defect-free Matrimid/PES dual-layer hollow fibers with an ultrathin outer layer of about 10×10^{-6} m (10 μm). These dual-layer membranes showed impressive CO_2/CH_4 selectivity of around 40 in tests using a gas mixture.

Defect-free high performances for O_2/N_2 separation PES membranes were reported by Ismail et al. [113]. Membranes were prepared by coating the porous PES membrane of a hyperthin-skin layer with silicon rubber. The combined effects of fabrication parameters in a dry/wet phase inversion process and a casting dope rheology, enabled improvement of membrane performance in O_2 and N_2 separation. The thinnest skin layer was 538 ± 95.6 Å.

3.1.1.13 Polybenzimidazole (PBI)

PBI is a heterocyclic polymer and is well known for its many excellent properties such as high thermal stability (over 550 °C), excellent mechanical properties and chemical stability, making it an outstanding candidate over common polymers. Because it has a rigid structure (T_g=420 °C) and stability at high temperatures, PBI could be best suited for H_2–CO_2 separation applications at high temperature. Its extremely rigid structure, as evident from its high T_g, should show resistance towards CO_2 plasticization and—unlike other polymers—may not lose its separation performance even at elevated temperatures. It has been studied for gas permeability at 200–270 °C and selectivity for H_2/CO_2 of about 20 was noted [114]. PBI-based composite membranes can function at significantly higher temperatures (>350 °C) than commercially available polymeric membranes (<150 °C). The membranes can maintain commercially attractive selectivity between H_2 and CO_2 even at 400 °C [115].

Composite membranes of PBI with proton exchanged AMH-3 (silicate) and swollen AMH-3 were characterized by electron microscopy and X-ray scattering, and tested for hydrogen/carbon dioxide ideal selectivity [116]. Proton-exchanged AMH-3 was prepared under mild conditions by the ion exchange of Sr and Na cations in the original AMH-3 using aqueous solution of DL-histidine. Swollen AMH-3 was fabricated by sequential interaction of dodecylamine following ion exchange in the presence of DL-histidine. Both silicate materials were introduced into a continuous phase of PBI as a selective phase. Mixed matrix nanocomposite membranes, prepared under certain casting conditions with only 3 wt% of swollen AMH-3, resulted in substantial increase of hydrogen/carbon dioxide ideal selectivity at 35 °C, i.e., by a factor of more than 2 compared to pure PBI membranes (40 vs. 15). Similar ideal selectivity was noted using higher loading (14 wt%) proton exchanged AMH-3 particles, suggesting that transport of hydrogen was faster than carbon dioxide in AMH-3 derived silicates. However, the ideal selectivity of MMMs approaches that of pure polymer as the operating temperature increases to 100 °C

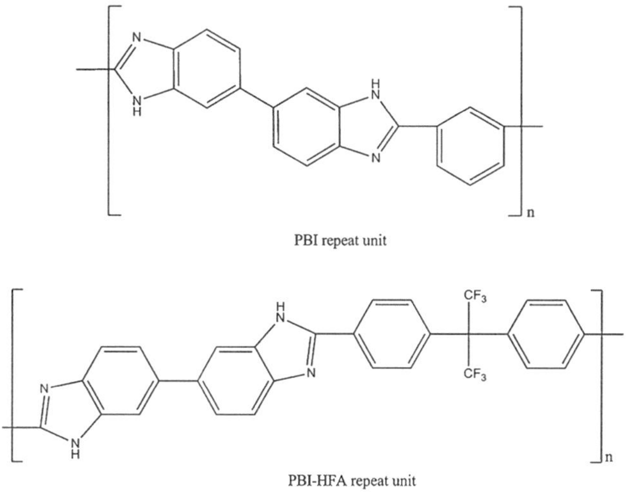

Fig. 3.26 Structure of repeat unit of PBI and PBI-HFA

and 200 °C. Dual layer hollow fiber prepared from PBI and Matrimid® after silicon rubber coating showed a H_2/CO_2 selectivity of about 11 [117].

Kumbharkar et al. [118] demonstrated the development of a polybenzimidazole (PBI)-based asymmetric hollow fiber membrane for H_2/CO_2 separation at high temperatures. High molecular weight PBI was synthesized in-house by a solution polycondensation method using polyphosphoric acid (PPA) as catalyst. Two different PBI, viz., PBI (based on isophthalic acid) and PBI-HFA [based on 4,4′-(hexafluoroisopropylidene)bis(benzoic acid)] were prepared. The chemical structures of the repeat unit of these PBIs are shown in Fig. 3.26. Defect free asymmetric hollow fiber membranes were successfully produced, which eliminated the step of silicon rubber coating, and the membranes were tested in the high temperature range of 100–400 °C. With an increase in temperature these membranes showed a relatively larger increase in H_2 permeance than CO_2 permeance, thereby enhancing the H_2/CO_2 selectivity. This was due to the high rigidity of PBI and smaller kinetic diameter of H_2 than CO_2, which led to relatively higher diffusion of the former than the later with an increase in temperature. The H_2 permeability at 400 °C was increased to 2.6 GPU by around eightfold over its permeability at 100 °C. The CO_2 permeability was increased by only around twofold at 400 °C as compared to its permeability at 100 °C. This significant improvement in permeance of H_2 led to H_2/CO_2 of 27.3, about 3.5 times higher than at 100 °C.

Young et al. [119] patented their invention as cross-linked polybenzimidazole membranes for gas separation. A cross-linked, supported polybenzimidazole membrane for gas separation was prepared by reacting polybenzimidazole (PBI) with sulfone-containing cross-linking agent 3,4-dichloro-tetrahydro-thiophene-1-1 dioxide. The cross-linking reaction product exhibited enhanced gas permeability to H_2, CO_2, N_2, and methane as compared to the unmodified analog, without significant loss of selectivity, at temperatures from about 20 °C to about 400 °C.

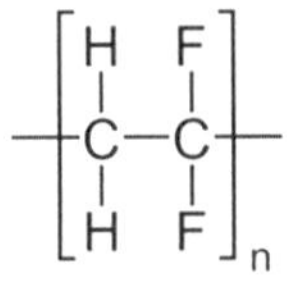

Fig. 3.27 Chemical structure of PVDF

3.1.1.14 Polyvinylidene Fluoride (PVDF)

Polyvinylidene fluoride or polyvinylidene difluoride (PVDF) (Fig. 3.27) is a highly non-reactive and pure thermoplastic fluoropolymer produced by the polymerization of vinylidene difluoride.

PVDF has a glass transition temperature (T_g) of about −35 °C and is typically 50–60 % crystalline. PVDF exists in several forms: alpha (TGTG′), beta (TTTT), and gamma (TTTGTTTG′) phases, depending on the chain conformations. PVDF is widely used as a basic polymer for the formation of hollow fibers. PVDF is a semicrystalline polymer containing a crystalline phase and an amorphous and/or rubbery phase. The crystalline phase provides thermal stability and the amorphous phase flexibility towards membranes. PVDF is stable while it is attacked by most of the corrosive chemicals and organic compounds including acids, alkaline, strong oxidants and halogens. In addition, the hydrophobicity of this polymer provides a potential application in membrane-based gas absorption and oil/water separation.

Porous PVDF hollow-fiber membranes with high porosity were fabricated using the immersion precipitation method [120]. Shen and Lua [121] fabricated three types of inorganic fillers, i.e., SiO_2, MCM-41, and zeolite 4A were incorporated into a PVDF matrix to prepare MMMs. The single gas (He, CO_2, O_2, and N_2) permeabilities of the resulting membranes were measured. The gas permeabilities of the three MMMs exhibited similar behaviors, especially at lower inorganic filler loadings, although the inorganic fillers had different pore structures and particle sizes. The highest permeabilities for CO_2 and O_2 were obtained by the PVDF/zeolite-4A 32 % composite membrane—3.26 and 0.41 Barrer, respectively—and the highest permeabilities for He and N_2 were obtained by the PVDF/MCM-41 32 % composite membrane—10.2 and 0.14 Barrer, respectively. These permeabilities are much higher than those of a pure PVDF membrane. The highest selectivities of 120.7, 33.1, and 4.6 for He/N_2, CO_2/N_2, and O_2/N_2, were obtained by three different membranes—PVDF/SiO_2 4 %, PVDF/SiO_2 32 %, and PVDF/SiO_2 24 %, respectively. These selectivities are also higher than those achieved by the pure PVDF membrane. However, the selectivities of the composite membranes showed differences and were dependent on the inorganic filler content and the specific gas pairs.

3.1.1.15 Poly(1-Trimethylsilyl-1-Propyne) (PTMSP)

Poly(1-trimethylsilyl-1-propyne) (PTMSP) is a glassy polymer. The chemical structure of 1-trimethylsilyl-1-propyne is illustrated in Fig. 3.28. PTMSP has one of the highest permeabilities recorded for glassy polymers for several permeants, including H_2, O_2, and CO_2. Poly[(1-trimethylsilyl)-1-propyne] (PTMSP) showed

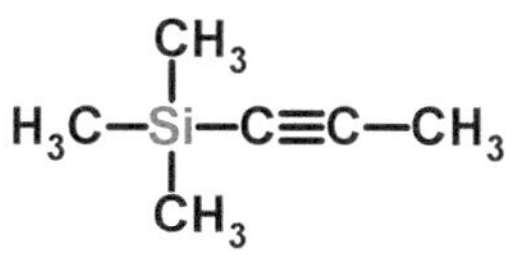

Fig. 3.28 Chemical structure of 1-trimethylsilyl-1-propyne monomer

Table 3.16 Fractional free volume (FFV) and permeation properties of "very high free volume" polymers

Polymer	FFV (%)	O_2 permeability (Barrer)	O_2/N_2 selectivity (–)	References
PTMSP	32–34	6,100	1.8	[126]
PMP	28	2,700	2.0	[127]
PIM-1	22–44	370	4.0	[122]
PIM-7		190	4.5	[122]

the highest permeability to gases—3.4×10^{-7} for nitrogen and 6.1×10^{-7} for oxygen—expressed by (cm^3 (STP) cm s^{-1} cmHg cm^{-2}), at room temperature [122].

PTMSP was first reported by Masuda et al. [123]. It is a glassy polymer that resembles rubber in its properties. Beyond PTMSP, other highly permeable polyactylenes have been reported, including poly(4-methylpentyne) PMP and poly{1-phenyl-2-(*p*-trimethylsilylphenyl)acetylene}. However, PTMSP remained the champion until 2008, when certain indan-containing poly(diphenylacetylene) derivatives were shown to exhibit even higher oxygen permeabilities [6]. The common feature of highly permeable acetylene-based polymers (polyalkynes) is the presence of bulky side groups that inhibit conformational change and force the backbone into a twisted shape. When these rigid, randomly coiled macromolecules are packed in solid state, the free volume distribution includes both small disconnected elements, as in conventional glassy polymers, and larger continuous microvoids. These results were discussed by molecular modeling as well as by positron annihilation lifetime spectroscopy (PALS).

These polyalkynes are known to be "very high free volume" polymers. Table 3.16 shows fractional free volume (FFV) and permeation properties of PTMSP and PIMs (polymers of intrinsic microporosity); the details of PIMs are discussed separately later. The polymer's high permeability stems from its large free volume, which is the space within the material that is not occupied by the polymer atoms. Due to its outstanding gas permeability and also vapor/gas selectivity PTMSP membranes became the focus by Nagai et al. [124] for gas separation. It was reported by Ichiraku et al. [125] that the permeability of PTMSP to light gases is higher than that of any other nonporous synthetic polymers at ambient temperature. Merkel et al. [126] reported an n-C_4H_{10}/CH_4 mixed gas selectivity of 35 in PTMSP membranes, which is the highest value reported for this gas-pair. However, practical utility of PTMSP is limited by a fast *physical aging* (gradual relaxation of non-equilibrium excess free volume in glassy polymers) and also by its solubility in many organic compounds, which results in membrane potential dissolution in process streams.

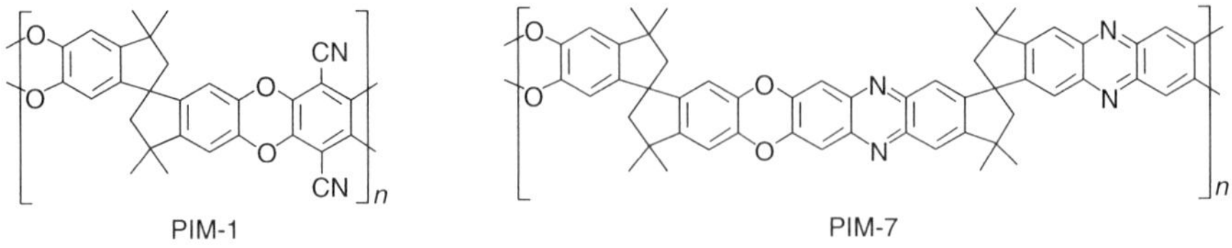

Fig. 3.29 Structure of PIM-1 and PIM-7

PTMSP has been shown much interest in the last 20 years because it has the highest known permeability of any polymer to gases and vapors [127]. This has been attributed to fast diffusion through the microvoids and the large excess free volume within the polymer matrix. However, the high permeability is coupled with low ideal selectivity (the ratio of the single gas permeabilities of two permeants), and numerous attempts have been made to overcome this trade-off. By using additives (organic or inorganic) the performances of the membrane can be changed. Permeability and positron annihilation measurements were carried out on PTMSP membranes, in pristine structure, as well as in the modified polymer after chlorination. It was found that permeability decreases in the chlorinated films; this can be due to the microscopic free volume decrease as probed by positronium [122].

The structure of PIM-1 and PIM-7 is given in Fig. 3.29.

Qui et al. [128] added a small organic filler trimethylsilylglucose (TMSG) to PTMSP and showed reduced permeabilities with increased selectivities, owing to the filling of the larger free volume elements in the polymer by TMSG and blocking the transport of gases through the microvoids. Merkel et al. [126] found that the mixed-gas *n*-butane/methane selectivity decreased with increasing filler concentration while the permeability of the two components increased.

Woo et al. [127] used PTMSP/MFI (silicalite-1) composite membranes for the separation of equimolar mixtures of *i*-butane and *n*-butane. The addition of 50 % MFI particles into PTMSP matrix showed increased permeability and simultaneously improved selectivity in the temperature range 25–200 °C. The best improvement was seen at 150 °C for the composites, giving almost threefold increase in permeability and 56 % higher *n*-butane/*i*-butane selectivity over the pure polymer. The composite membranes were also tested for separations of *n*-hexane/2,2-dimethylbutane and *p*-xylene/o-butane isomer separations.

Peter and Peinemann [129] developed a new multilayer composite membrane for gas separation, which consists of PTMSP as the gutter layer deposited on a poly(acrylonitrile) porous support and partially cross-linked Matrmid® 5218. The effect of PTMSP gutter layer on gas transport properties was compared with that of the PDMS sealing layer. It was observed that the gutter layer enhances both gas permeance and selectivities, whereas the sealing layer increases selectivities but with a small gas permeance decrease.

Vopiča et al. [130] developed a novel measuring procedure for mixed gas sorption tests on the n-C_4/CH_4 and CO_2/CH_4 mixtures in films of PTMSP. It was observed that the presence of CH_4 does not alter significantly the sorption of CO_2 and of n-C_4

in PTMSP, while the mixed gas solubility of CH_4 is lower than the pure gas value at the same CH_4 fugacity. The real CO_2/CH_4 solubility-selectivity of PTMS is similar to the ideal value at low CO_2 fugacity, but it becomes significantly higher, up to 4.5 times, at 25 bar of CO_2 fugacity. A quantitative rule can be drawn from this study, using data of several binary gas mixtures in glassy polymers. The ratio between actual mixed gas and pure ideal solubility selectivity of CO_2 over CH_4 is a single, monotonously increasing function of the ratio between the concentration of the two components, $c(CO_2)/c(CH_4)$, and becomes higher than unity as $c(CO_2) > c(CH_4)$. In other words, the competition effects depress the less abundant penetrant in the polymer, which is usually CH_4.

3.1.1.16 Polysaccharide

Polysaccharides are long carbohydrate molecules of monosaccharide units joined together by glycosidic bonds. They range in structure from linear to highly branched. Polysaccharides are often quite heterogeneous, containing slight modifications of the repeating unit. Depending on the structure, these macromolecules can have distinct properties from their monosaccharide building blocks. They may be amorphous or even insoluble in water. When all the monosaccharides in a polysaccharide are the same type, the polysaccharide is called a *homopolysaccharide* or *homoglycan*, but when more than one type of monosaccharide is present they are called *heteropolysaccharides* or *heteroglycans*.

Polysaccharides are an important class of biological polymers. Their function in living organisms is usually either structure- or storage-related. Starch (a polymer of glucose) is used as a storage polysaccharide in plants, being found in the form of both amylose and the branched amylopectin. In animals, the structurally similar glucose polymer is the more densely branched glycogen, sometimes called "animal starch." Glycogen's properties allow it to be metabolized more quickly, which suits the active lives of moving animals. Cellulose and chitin are examples of structural polysaccharides. Cellulose is used in the cell walls of plants and other organisms, and is said to be the most abundant organic molecule on earth. It has many uses such as a significant role in the paper and textile industries, and is used as a feedstock for the production of rayon (via the viscose process), cellulose acetate, celluloid, and nitrocellulose. Chitin has a similar structure, but has nitrogen-containing side branches, increasing its strength. It is found in arthropod exoskeletons and in the cell walls of some fungi. It also has multiple uses, including surgical threads.

Cellulose

In cellulose membranes the strong intermolecular and intramolecular H-bonding can lead to the dense packing of polymer chains. The dry cellulose membrane appears to be rigid. Within the cellulose matrix, water acts as a plasticizer, and decreases the *Tg* of the cellulose network. It is reported in the literature that the gas permeability of the dry cellulose membranes is low, but the "water-swollen" cellulose membrane shows a high permeation rate to CO_2 and excellent separation factors of CO_2 over N_2, CH_4, and H_2 [100].

Chitosan

Chitosan, poly[ß(1→4)-2-amino-2-deoxy-D-glucopyranose] is a linear polysaccharide obtained by deacetylation of chitin, ploy[ß(1→4)-2-acetamido-2-deoxy-D-glucopyranose]. Chitosan has proved to be a good biomedical material based on the properties of biocompatibility and biodegradability, and has also seen increased use as a functional polymer material in industries, especially in membrane technologies [131]. Non-porous chitosan membranes are applied in gas separation and pervaporation. The gas permeation properties of chitosan membranes have been reported in the literature; however, these references are few in number.

Chitosan membranes are dense and rigid in their fully dry state, and they show very low permeability to gases. However, higher permeation rates can be achieved after they are swollen by water. Xiao et al. [131] prepared cross-linked chitosan membranes via interfacial cross-linking in trimesoyl chloride (TMC)/hexane. The membrane with a higher degree of cross-linking showed a higher degree of swelling in water, and the degree of swelling decreased after gas separation and pervaporation. The TMC moieties changed the thermal properties of the chitosan membranes. Pure gas permeation was performed with CO_2 and N_2 at room temperature. The amino groups and transient gaps in the chitosan matrix, which influenced the permeation of CO_2 and N_2, were affected by TMC moieties from the cross-linking reaction. Chitosan-TMC membrane, which formed with a cross-linking time of 40 min (dry thickness was 145 μm) showed the best performance for the separation of CO_2/N_2, with a CO_2 permeability of around 163 Barrer and an ideal separation factor of around 42.

3.1.1.17 Polyvinyl Alcohol (PVA)

PVA is unique among polymers in that it is not built up in polymerization reactions of vinyl alcohol (Fig. 3.30). Instead, PVA is made by dissolving polyvinyl acetate (PVAc), in an alcohol such as methanol and treating it with an alkaline catalyst such as sodium hydroxide. The resulting hydrolysis, or "alcoholysis," reaction removes the acetate groups from the PVAc molecules without disrupting their long-chain structure. The chemical structure of the resulting vinyl alcohol repeating units is shown in Fig. 3.28.

When the reaction is allowed to proceed to completion, the product is highly soluble in water and insoluble in practically all organic solvents. Incomplete removal of the acetate groups yields resins less soluble in water and more soluble in certain organic liquids.

The applications of PVA-based membranes for catalysts or membranes for gas separation have been discussed by Papanceaa et al. [132]. Polyvinyl alcohol has

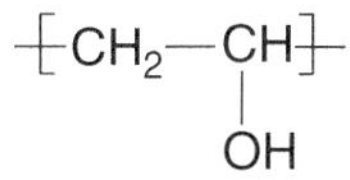

Fig. 3.30 Repeat unit of PVA

excellent film-forming, emulsifying and adhesive properties. It is also resistant to oil, grease and solvents. Zou and Ho [133] synthesized cross-linked PVA (containing amines) and reported that it was a good CO_2-selective membrane. The PVA membrane showed good selectivities for CO_2/N_2 and CO_2/CO separation. The membrane also showed good CO_2 permeabilities and CO_2/H_2 selectivities up to 170 °C. At 120 °C, the CO_2 permeability and CO_2/H_2 selectivity reached 8,200 Barrer (1 Barrer = 10^{-10} cm^3(STP) cm cm^{-2} s^{-1} cmHg) and 450, respectively.

Matsuyama et al. [134] studied PEI/PVA blend membranes for the facilitated transport of CO_2. The CO_2 permeance decreased with the increase in the CO_2 partial pressure, whereas the N_2 permeance was nearly constant. This suggested that only CO_2 was transported by the facilitated transport mechanism and also that PEI functioned efficiently as the carrier of CO_2.

Water-swollen hydrogel (WSH) membranes for gas separation were prepared by dip-coating asymmetric porous PEI supports with PVA-GA (glutaraldehyde) solution, followed by the cross-linking of the coated layer by a solution method, by Park and Lee [135]. It was observed that the behavior of gas permeation through a WSH membrane was parallel to the swelling behavior of the PVA/GA film in water. The permeance of carbon dioxide through the WSH membranes was 10^{-5} (cm^3 cm^{-2} s^{-1} cmHg) and a CO_2/N_2 separation factor was about 80 at room temperature.

3.1.2 *Copolymers and Polymer Blends*

A heteropolymer, also called a copolymer, is a polymer formed when two (or more) types of monomer are linked in the same polymer chain, as opposed to a homopolymer where only one monomer is used. If exactly three monomers are used, it is called a terpolymer. Copolymerization refers to methods used to chemically synthesize a copolymer.

Since a copolymer consists of at least two types of constituent units (called also structural units), copolymers can be classified based on how these units are arranged along the chain. These include (also, see Fig. 3.29):

- Alternating copolymers with regular alternating A and B units (see 2 in Fig. 3.29).
- Periodic copolymers with A and B units arranged in a repeating sequence (e.g., (A-B-A-B-B-A-A-A-A-B-B-B)$_n$).
- Statistical copolymers, which are copolymers where the sequence of monomer residues follows a statistical rule. If the probability of finding a given type monomer residue at a particular point in the chain is equal to the mole fraction of that monomer residue in the chain, then the polymer may be referred to as a truly random copolymer (see 3 in Fig. 3.29).

Fig. 3.31 Different types of copolymers

- Block copolymers comprise two or more homopolymer subunits linked by covalent bonds (see 4 in Fig. 3.29). The union of the homopolymer subunits may require an intermediate non-repeating subunit, known as a junction block. Block copolymers with two or three distinct blocks are called diblock copolymers and triblock copolymers, respectively [136].
- Graft copolymers are a special type of branched copolymer in which the side chains are structurally distinct from the main chain. Number 5 is Fig. 3.31 depicts a special case where the main chain and side chains are composed of distinct homopolymers. However, the individual chains of a graft copolymer may be homopolymers or copolymers. Note that different copolymer sequencing is sufficient to define a structural difference, and thus, an A-B diblock copolymer with A-B alternating copolymer side chains is properly called a graft copolymer.

Copolymers may also be described in terms of the existence or arrangement of branches in the polymer structure. Linear copolymers consist of a single main chain whereas branched copolymers consist of a single main chain with one or more polymeric side chains. Copolymers offer the potential to fine tune permeabilities. A copolymer will tend to have permeabilities which are intermediate compared with the homopolymers which make it up.

Polymer blends are an inexpensive route to the modification of polymer properties. Examples of the properties that may be altered upon blending are impact resistance, fatigue behavior, heat distortion, and improved processability [137]. In some blend systems, the effective property modification is dependent upon the miscibility or compatibility (i.e., the ability to form a homogeneous mixture) of the two homopolymers. Compatibility of polymers in a blend has been defined in a number of ways. The simplest definition of polymer compatibility in a blend is optical clarity upon preparation. Another definition of blend compatibility involves the glass transition temperature of the homogeneous polymers and the blend. Compatible blends must exhibit a single glass transition temperature (T_g) between the T_g's of the homopolymers, while incompatible blends will have two T_g's that correspond to those of the homopolymers [138]. A third definition of compatibility involves the use of infrared spectroscopy. Coleman and Painter [139] have proposed that if two polymers are compatible, then the IR spectra obtained from the blend should include band shifts and broadening when compared to the scaled addition of the infrared spectra of the homopolymers.

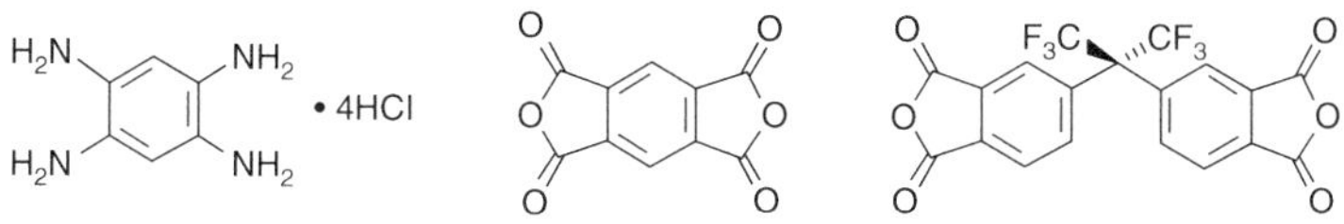

TABH PMDA 6FDA

Fig. 3.32 Structure of monomers

Polymer	Structure
6FDA–TAB	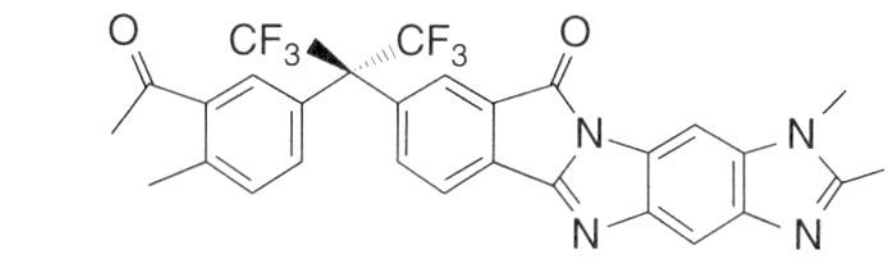
6FDA/PMDA–TAB (50/50) 6FDA/PMDA–TAB (25/75) 6FDA/PMDA–TAB (10/90)	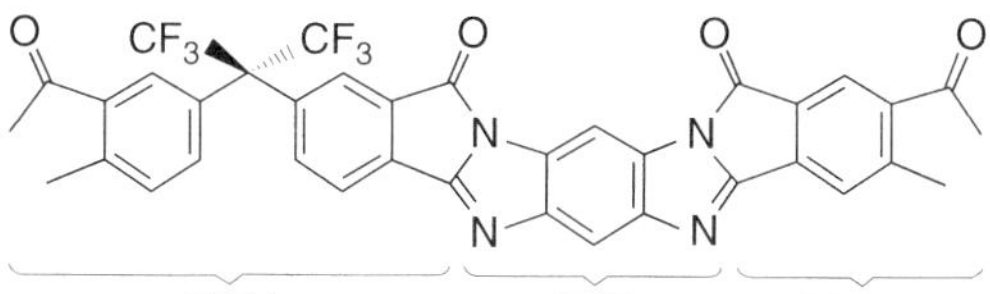

Fig. 3.33 Structures and compositions of 6FDA-TAB and the 6FDA/PMDA-TAB copolymers

The CO_2/N_2 gas separation properties of a large series of poly(ethylene oxide) (PEO) segmented copolymers with polyurethanes, polyamides and polyimides was studied by Yoshino et al. [140]. It was observed that the CO_2/N_2 separation properties depend on the hard-segment polymer. The contents of the hard and soft segments in the soft and hard domains, W_{HS} and W_{SH} respectively, were estimated from glass-transition temperatures with the FOX equation. The phase separation of the PEO domain depended on the kind of hard segment polymer, that is, W_{SH} was in the order PU > PA ≫ PI for PEO block length (n) of 45–52.

Zimmermann and Koros synthesized polypyrrolone copolymers comprising various compositions of 6FDA, PMDA and TAB for O_2/N_2 gas separation. The structure of monomers used to synthesize the polymers is shown in Fig. 3.32 [141].

Structures and compositions of 6FDA-TAB and the 6FDA/PMDA-TAB copolymers used in the gas transport study are shown in Fig. 3.33.

It was reported by Zeememan and Koros [148], on varying the fractions of 6FDA (bulky group) and PMDA (flat, packable group), that a *molecular jack* was created which altered the average interchain spacing and gas transport properties of these materials. All of the materials showed O_2/N_2 gas separation properties lying on or above the upper bound trade-off limit, indicating they possess superior transport properties to most polymers.

In another study Zimmerman and Koros [142] reported activation energies for permeation and diffusion as well as heat of sorption for He, CO_2, O_2, N_2, and CH_4 in the 6FDA/PMDA-TAB copolymer series. These gas transport properties were

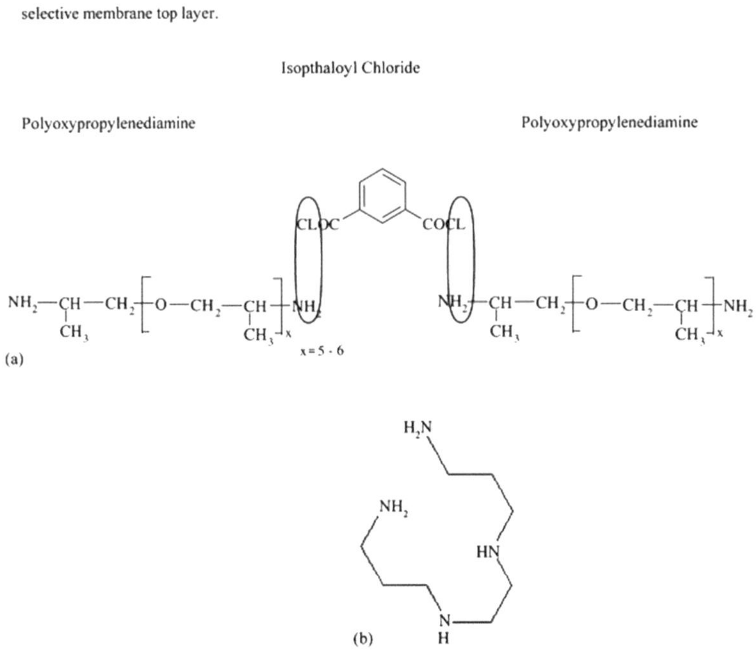

Fig. 3.34 (**a**) Interfacial polymerization reaction scheme for forming the polyamide (IPC) selective membrane top layer. (**b**) Cross-linking agent 1,5,8,12-tetraazdodecane (N_4) [143]

interpreted in terms of their polypyrrolone structures, which possess different average interchain spacing. It was also reported that the activation energy for diffusion was dominant in influencing the activation energy for permeation.

Patil et al. [143] studied the permeation of CO_2 for two types of composite polymeric hollow fiber membranes. The membrane consists of a polyamide copolymer (IPC) layer or a polyvinyl alcohol (PVA) layer on top of a polyethersulfone (PES) support membrane. The reaction scheme of the interfacial polymerization process is shown in Fig. 3.34.

As well, two polyamide top layers were fabricated with a different degree of cross-linking. The crosslinked layers were obtained by the addition of 1,5,8,12-tetraazdodecane (N_4) as the cross-linking agent. It was noted that permeance of CO_2 for both membranes had a maximum as a function of feed pressure at about 8 MPa. Both membranes showed an increase in the carbon dioxide permeance as a function of the exposure time to supercritical carbon dioxide. The stability of the IPC membrane was improved by cross-linking of the selective layer.

Wang et al. [144] developed PEI-PVA blend membranes for the removal of CO_2 from natural gas. The blend of a PEI-PVA composite membrane was used as a separation layer and PS (polysulfone) UF membrane as substrate. The permselectivity of the membrane was measured with CO_2/CH_4 mixed gases. The permeances of both CO_2 and CH_4 decreased with the increase of temperature, and the permeances decreased more quickly under low pressure than those under high pressure.

Semsarzadeh and Ghalei [145] fabricated blend membranes of polyurethane (PU) and polyvinyl acetate (PVAc) in the presence of various polyethylene oxide-polyethylene oxide triblock polymer (Pluronic) contents by solution casting technique. The blends with 5 wt% PVAc showed higher CO_2 permeability (~73 Barrer) compared to the PU membrane. A comparative increase in permselectivity of pair gases was shown with an increase in CO_2/N_2 and O_2/N_2 (in the membrane 15 wt% PVAc) by up to 417 % and 200 %, respectively. CO_2/CH_4 (in the membrane with 5 wt% PVAc) increased by up to 220 %.

The addition of polyethylene glycol (PEG) to the poly(amide–*b*-ethylene oxide) copolymer (Pebax) demonstrated that CO_2 permeability and selectivity over H_2 can be simultaneously increased [146]. The enhancement was attributed to the high CO_2 solubility in PEG, but a free volume increase was also taken into consideration because a decrease in density and glass transition temperature (T_g) was observed. Later, the increase in total free volume was demonstrated for the Pebax/PEG blend [147].

Yave et al. developed a nanostructured and CO_2-philic polymer membrane for CO_2 separation [148]. The authors demonstrated that poly(ethylene oxide)-poly(butylene-terephthalate) (PEO-PBT) is a material with high CO_2 separation. The membranes presented outstanding performance for CO_2 separation, and the measured CO_2 flux was extremely high (>2 $m^3\,m^{-2}\,h^{-1}\,bar^{-1}$) with selectivity over H_2 and N_2 of 10 and 40, respectively, making them attractive for CO_2 capture.

Madaeni et al. [149] fabricated pyromellitic dianhydride-co-4,4′-oxy dianiline (PMDA/ODA) polyimide membranes using polyamic acid (PAA) as precursor materials. On adding PVDF as an additive in the preparation of PMDA/ODA polyimide membrane, the gas separation performance of the membrane was improved for pure N_2 and C_2H_4 gases. The ideal selectivity of 100 % was achieved for ethylene by PMDA/ODA polyimide membranes with 1 wt% PVDF at the feed pressure of 1 bar.

Based on PES and polyimide Matrimid 5218 (PI) blends, hollow fibers were spun for CO_2/N_2 separation by Kapantaidakis et al. [150]. The developed membranes exhibited a typical asymmetric structure and remained miscible for each blend composition. PDMS coated hollow fibers had CO_2 permeances varying from 30 to 60 GPU and CO_2/N_2 selectivities varying from 30 to 40. Hollow fibers, rich in PES, showed a pronounced plasticization behavior but the reduction of CO_2/N_2 selectivity was totally reversible after a short period of time. It was also demonstrated that the air-gap distance in the dry/wet spinning processes affected both membrane structure and permeation properties [151]. In another study for gas separation using PES/PI blend hollow fibers, Kapantaidakis et al. [152] reported that by adjusting major process parameters, such as polymer concentration, air-gap distance, bore liquid composition, and take-up velocity, highly permeable, selective, and ultrathin fibers could be produced. Suitable selection of the spinning conditions resulted in gas separation hollow fibers with a thin skin layer (0.1 μm), macrovoid-free structure, high permeation rate (CO_2: 40–60 GPU) and selectivity coefficient (CO_2/N_2: 40).

Koros and Wood [153] studied the effect of elevated temperatures on three asymmetric hollow fiber membranes (polyaramide, polyimide, and composite polyimide on a polyimide/polyetherimide blend support). Polyaramide membranes were shown to exhibit good stability at elevated temperatures and good separation properties after silicon rubber post-treatment. The hydrogen permeance of 300 GPU at 175 °C is acceptable for industrial application. The polyimide-containing membranes had superior room-temperature properties; however, the thin skin aged at elevated temperatures. This aging effect decreased the permeance of the membranes approximately 40 % at 175 °C and slightly increased the permselectivity; however, the effects of aging leveled out over 200–250 h at 175 °C and the membrane properties became constant. At this level, the polyimide membranes exhibited around 400 GPU of hydrogen permeance with 660 selectivity to *n*-butane.

Seo et al. [154] demonstrated a novel concept of a (universal) "organic molecular sieve" and experimentally proved its possibility by showing that organic polymer molecules at the interface between the permeable phase and the impermeable phase play the role of molecular sieves. The authors prepared polymeric composite film by using a semicrystalline polymer (Nylon 6) as a barrier component dispersed in an amorphous matrix polymer (poly(2,6-dimethyl-1,4-phenylene oxide), PPO) and a compatibilizer (poly(styrene-co-maleic anhydride), PSMA). A mixture of Ny6, PPO, and PSMA was extruded at a process temperature of 240 °C in a Brabender twin screw extruder. A film with an even thickness of 100 μm having 3 % error limits was used for the gas separation. They reported that there was a significantly improved selectivity in gas separation, going over the so called "upper-boundary." The performance of the composite film is shown in Fig. 3.35. This study showed that compatibilizer works like a molecular sieve to separate one gas molecule from

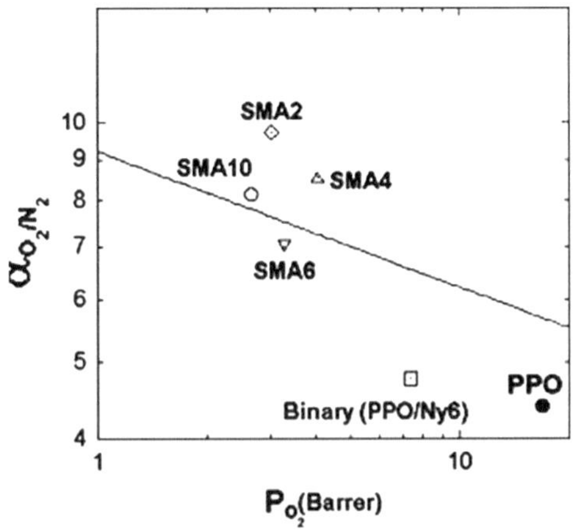

Fig. 3.35 The relationship between the oxygen permeability and the O_2/N_2 selectivity for PPO and Ny6 blended films. (*filled circle*) PPO, (*open square*) a binary blend, (*open diamond*) a ternary blend with 2 wt% PSMA, (*triangle*) a ternary blend with 4 wt% PSMA, (*inverse triangle*) a ternary blend with 6 wt% PSMA, and (*open circle*) a ternary blend with 10 wt% PSMA. The *solid line* is an empirical upper-bound relation. Since the size of error bars was smaller than the size of symbols, the error bars were deleted

the other. Hence, this strategy can be easly used to make extraordinary polymeric gas-separation membranes for all different gas pairs. The film can be used for oxygen gas enrichment or CO_2 gas removal as well as other gas separation. They also claimed that this strategy would be applicable to various separation processes for many chemicals.

3.1.3 Other Polymers

A range of other polymers showed potential for gas separation; however, none has been commercialized. Among others, polybenzoxazoles (PBOs) are a class of polymers with good gas permeation and selectivity properties that show promise [155].

3.1.3.1 Polymers of Intrinsic Microporosity (PIMS)

The idea of PIMs started from the work of McKeown in 1998 [156] Intrinsic microporosity in polymers is defined as "a continuous network of interconnected intermolecular voids, which forms as a direct consequence of the shape and rigidity of the component macromolecules" [157, 158]. Intrinsic microporosity can arise simply from a polymer whose molecular structure is highly rigid and contorted so that space-efficient packing in the solid state is prohibited (Fig. 3.36). The lack of rotational freedom along the polymer backbone ensures that the macromolecules cannot rearrange their conformation to collapse the open structure of the material.

These polymers (PIMs) can exhibit analogous behavior to that of conventional microporous materials, but, in addition, may be processed into convenient forms for use as membranes. These membranes have excellent performance for gas separation and evaporation [159]. In general, due to maximum attractive interactions between the constituent macromolecules, the pack space in the polymer minimizes the void space. Most polymers have sufficient conformational flexibility to allow them to rearrange their shape so as to maximize intermolecular cohesive interactions and pack space efficiently. Due to fused ring structures, PIMS do not possess rotational freedom along the polymer backbone, which ensures that the macromolecular components cannot rearrange their conformation. Thus, during synthesis, their contorted shape does not change [160].

PIMs are prepared by a polymerization reaction based on a double-aromatic nucleophilic substitution mechanism to form the dibenzodioxin linkage. This reaction is one of the few capable of forming two covalent bonds simultaneously, with sufficient efficiency to provide a linking group composed of fused rings and, thus, able to form ladder polymers of high average molecular mass [160]. Generally, aromatic nucleophilic substitutions are known to proceed readily, especially if the halide-containing monomer is activated by an electron-withdrawing substituent (e.g., –CN, F) [161]. This reaction was used previously by Makhseed et al. [168] to prepare phthalocyanine oligomers and extended to planar molecules and oligomers for discotic liquid crystals [162]. Du et al. developed,

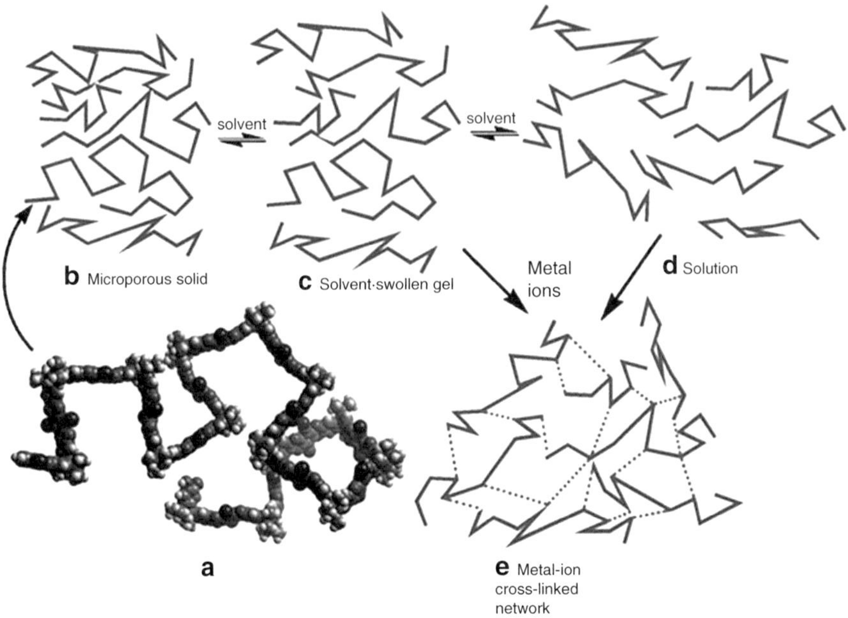

Fig. 3.36 PIMS representation. (**a**) Model of a molecular fragment of PIM-1 showing its randomly contorted structure together with cartoon representation of the various states that can be obtained from a PIM including: (**b**) a microporous solid due to the inability of the polymer molecules to pack efficiently: (**c**) a solvent–swollen gel, (**d**) solution and (**e**) a metal ion cross linked microporous net work [159]

what is called a "high temperature method" [163], in which a high-speed stirring of the mixture in dimethyl acetamide at 155 °C is performed for only 8 min with the addition of toluene to enable the continuation of stirring.

Du et al. introduced azide-based cross-linking of polymers of intrinsic microporosity (PIM) membranes for condensable gas separation [164]. Membranes were prepared by a nitrene reaction from a representative PIM in the presence of two different diazide cross-linkers. These cross-linked polymeric membranes showed excellent gas separation performance and can be used for O_2/N_2 and CO_2/N_2 gas pairs and for the separation of condensable gases such as CO_2/CH_4 and propylene/propane. These membranes were different from typical gas separation membranes derived from glassy polymers as the cross-linked PIMs showed no obvious CO_2 plasticization up to 20 atm pressures of pure CO_2 and CO_2/CH_4 mixtures. Du et al. also discussed these membranes for CO_2 [165].

Incorporation of tetrazoles (TZPIM) into the microporous polymeric framework of PIM, has been shown to create a very high permeability for CO_2 and excellent CO_2/N_2 mixed gas separation, even under plasticization conditions (Fig. 3.37).

The presence of the tetrazole groups leads to favorable sorption and selective pore blocking by presorbed CO_2 molecules, thus limiting access by other light gas molecules such as nitrogen. The introduction of tetrazoles into PIM is the first

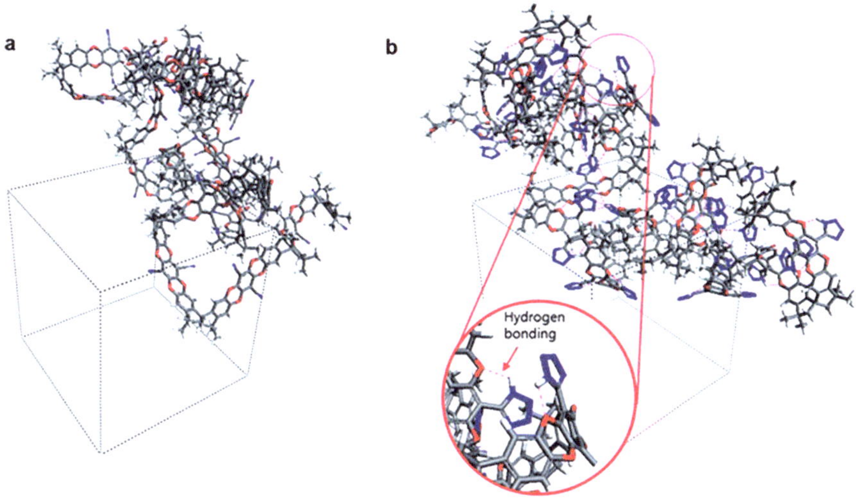

Fig. 3.37 Incorporation of tetrazoles in PIM. (**a**) Three-dimensional view of PIM-1 in an amorphous periodic cell (the number of repeat units in PIM-1 is 20), and (**b**) a three-dimensional view of TZPIM-3 containing tetrazole in an amorphous periodic cell (the number of repeat units in TZPIM is 20; 100 % full conversion from nitrile groups to tetrazole groups; the *blue dotted lines* indicate possible hydrogen bonding modes) [161]

example of a {2+3} cycloaddition of a polymer containing aromatic nitrile groups with an azide. This strategy of incorporating nitrogen heterocycles into PIMs provides new directions in the design of other polymeric membrane materials for important CO_2 separation process. PIMs also undergo some degree of physical aging and plasticization.

3.1.3.2 Cross-linking of Polymers and Other Techniques for Modification

Cross-linking offers the potential to improve the mechanical and thermal properties of membranes. Koros and Mahajan have suggested that cross-linking can be used to increase membrane stability in the presence of aggressive feed gases and to simultaneously reduce plasticization of the membrane [166].

A plasma polymerization process is a technique that allows for obtaining highly cross-linked polymers from nonfunctional monomers that are not utilized in conventional polymer synthesis. Plasma surface modification can improve biocompatibility and biofunctionality. When membrane surfaces are brought into contact with gas plasmas by energetic species such as ions, electrons, radicals, metastables, and photons in the short wave ultraviolet range, their energy is transferred from the plasma to the solid. As a result, the surface of the membrane is etched forming many reactive sites (mostly radicals) on the surface. Polymerization takes place at the reactive sites of the membrane when an organic vapor or a monomer is introduced into the plasma reactor.

Plasma polymers were prepared from three different organosilicon: diethoxydimethyl silane, hexamethyldisiloxane (HMDSO), and octamethyltrisiloxane (OMTSO). Films were deposited upon silicon wafers and on different porous substrates. Silicon-containing polymers are well known as polymers excelling in gas permeation. When they are synthesized by the plasma process, they also exhibit high selectivities because of high cross-linking compared with conventional polymers. Roualdes et al. studied the gas (N_2, H_2, O_2, CO_2, and CH_4) separation properties of organosilicon plasma polymerized membranes [167]. Surfaces of polyimide and PSf membranes were modified by Won et al. [168] by using an ion-beam carbonization technique. To control the structure of membrane skin and to improve gas-transport properties, the irradiation conditions, such as the dosage and the source of ion beam, have been varied. The ideal separation factor of CO_2 over N_2 through the surface modified PI and PSf membranes increased threefold compared with those of the untreated, pristine membranes, whereas the permeability decreased by almost two orders of magnitude. This could be due to the fact the structure of the membrane skin had changed to a barrier layer.

Maya et al. [169] noted that thermally treated copolyimides consisting of flexible PEO segments and rigid polyimide segments are very attractive as CO_2/N_2 separation membranes. After thermal treatment of these membranes under an inert atmosphere, a large improvement in CO_2 permeability was observed, yielding a more productive membrane.

By using a coextrusion and dry jet wet spinning phase-inversion technique with the aid of heat treatment at 75 °C, Li et al. [170] fabricated dual-layer PES hollow fiber membranes with an ultrathin dense-selective layer of 407 Å. The dual-layer hollow fibers had an O_2 permeance of 10.8 GPU and O_2/N_2 selectivity of 6.0 at 25 °C. It was observed that heat treatment at 75 °C improved the gas permeation and ideal selectivity, whereas heat treatment at 150 °C resulted in a significant reduction in both permeation and selectivity due to enhanced substructure resistance. SEM pictures confirmed that higher heat-treatment temperature can significantly reduce pore sizes and the amount of pores in substructure immediately underneath the dense-selective layer.

Castro-Domínguez et al. [171] reported the implementation of perfluorotributylamine (PFTBA) (Fig. 3.38) imbued in porous alumina tubes as a supported liquid membrane to carry out the separation of O_2 and N_2 at 40 °C and 1 atm. The membrane had an average O_2/N_2 separation factor of around 60 with an O_2 permeance of 8×10^{-10} mol m^{-2} s^{-1} Pa^{-1}, an average H_2/N_2 separation factor of 100, and a H_2 permeance of 1×10^{-9} mol m^{-2} s^{-1} Pa^{-1}. The O_2/N_2 selectivity was higher as the temperature increased, but the lifetime of the membrane was reduced.

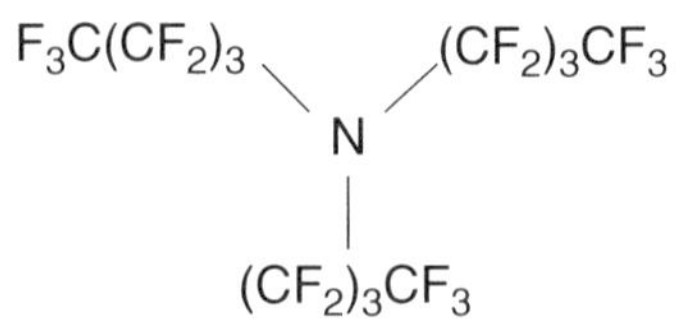

Fig. 3.38 Chemical structure of PFTBA

3.2 Inorganic Membranes

In most cases, gas separation membranes are based on amorphous glassy polymers (polysulfones, polycarbonates, polyimide). Although glassy polymeric membranes exhibit a good combination of gas permeability and selectivity properties, their performance in the separation of gaseous mixtures may decline with time due to aging or plasticization at specific feed conditions. Moreover, the maximum operating conditions of polymeric membranes is about 100 °C, whereas the temperatures encountered in numerous industrial processes are considerably higher. Thus, researchers started to develop other routes or materials for the separation of gases at high temperatures, e.g., inorganic material. Inorganic membranes can be classified in three categories; (1) zeolites, (2) sol–gel based microporous membranes, and (3) Pd-based and Perovskite-like dense membranes. Table 3.17 shows the inorganic materials for gas separation.

There are basically two types of inorganic membranes: (1) dense (nonporous) and (2) porous. Examples of commercial porous inorganic membranes are ceramic membranes, such as alumina, silica, titanium, and glass and porous metals, such as stainless steel and silver. These membranes are characterized by high permeabilities and low selectivities. Dense inorganic membranes are very specific in their separation behaviors; for example, Pd–metal based membranes are hydrogen specific and metal oxide membranes are oxygen specific. Dense membranes prepared from palladium or perovskites only allow certain gases (such as H_2 or O_2) to permeate via mechanisms such as solution-diffusion or solid-state ionic conduction. Such nonporous systems exhibit extremely high selectivities but have limited permeabilities, although substantial research efforts during the last decade have produced fluxes within reach of targets. These membranes further require high capital investment due to the use of precious metals and/or extreme synthesis and operating conditions; the membranes may be mechanically unstable. In contrast, microporous silica membranes have proven to be promising for molecular sieving applications. Precise pore size control (0.3–0.4 nm in diameter) to allow for separation on the basis of size by molecular filtration or "sieving" has, however, not yet been achieved for amorphous inorganic membranes and they are also chemically, mechanically and thermally less robust than zeolite membranes [172].

Table 3.17 Inorganic materials for gas separation membranes

Zeolites/zeolitic materials
Carbon molecular sieves
Nanoporous carbon
Carbon nanotubes
Ultramicroporous amorphous silica or glass
Palladium alloys (metals)
Mixed conducting perovskites
Graphene

Several natural gas resources around the world contain large amounts of CO_2. Economic recovery of the CH_4 from streams having CO_2 as a major component presents numerous technical challenges. One potential separation option is based on inorganic membranes. With the recent advent of commercial ceramic membranes, inorganic membranes are receiving much attention as unique separators and reactors due to their excellent thermal and chemical stabilities. Microporous inorganic membranes ($r_{pore} < 1$ nm) have great potential for gas separation. Compared to polymeric membranes, microporous inorganic membranes with molecular sieve-like properties have relatively high gas permeances and stability in higher temperatures and corrosive atmospheres. Moreover, inorganic membranes can be used in membrane reactors for conversion enhancement such as in dehydrogenation reactions. State-of-the-art microporous silica membranes consist of a microporous top layer on top of a supported mesoporous (1 nm$<r_{pore}<$25 nm) γ-Al_2O_3 membrane. The support of the γ-Al_2O_3 layer provides mechanical strength to the selective silica top layer.

3.2.1 *Ceramic Membranes*

In general, a ceramic membrane can is a permselective barrier or a fine sieve. Ceramic membranes are usually composite, consisting of several layers of one or more different ceramic materials. They generally have a macroporous support, one or two mesoporous intermediate layers and a microporous (or a dense) top layer. The bottom layer provides mechanical support, while the middle layers bridge the pore size differences between the support layer and the top layer where the actual separation takes place. Membrane properties such as permeation and selectivity depend on the microstructures of the membrane/support composite such as pore size and distribution, porosity, and affinity between permeating species and the pore walls. Separation of a gas mixture can take place based on differences in molecular mass, size or shape, or on differences in the affinity of the gas molecules to the membrane material.

Numerous theories for describing transport in microporous media have been presented in the literature [173, 174]. These theories become increasingly complex when the microporous medium is less uniform and when more mobile species are present. For the assessment of membrane quality, a simple phenomenological approach is sufficient [175].

For single gas permeation through amorphous microporous silica membranes, at sufficiently high temperatures and low pressures, transport is activated and permeance is independent of pressure [176–178]. Hence, permeance is described by:

$$P = N / \Delta p = (H_0 D_0) \exp\{(Q - E_D) / RT\} \tag{3.1}$$

where N is the molar flux, H_0 and D_0 are pre-exponential factors related to the Henry and diffusion coefficients, respectively, and R and T have their usual meaning.

The overall thermally activated nature of transport arises from the simultaneous occurrence of diffusion (E_D) and sorption (Q).

Pohl and Heffelfinger [179] simulated pressure-driven gas permeation of gases in a porous silica model using a dual control volume grand canonical molecular dynamics (DCV-GCMD) technique. The molecular sieving nature of microporous zeolites and amorphous silica made by sol–gel methods were discussed and compared. One mesoporous and one microporous membrane model were tested with Lennard–Jones gases corresponding to He, H_2, Ar, and CH_4. The mesoporous membrane model clearly followed a Knudsen diffusion mechanism, while the microporous model, having a hard-sphere cutoff pore diameter of ~3.4 Å, demonstrated molecular sieving of CH_4 ($\sigma = 3.8$ Å) but anomalous behavior for Ar ($\sigma = 3.4$ Å).

Perovskite oxide-type ceramic membranes are used for gas separation etc. The general formula of the perovskite is ABO_3 and the properties are determined by cations occupying its A-site and B-site lattice. The A-site cations are mainly composed of alkaline earth, as well as alkaline and lanthanide ions, while B-site cations are mainly composed of transitional metal ions. These perovskite-type ceramic membranes can be used for oxygen production or gas separation. Oxygen transport through such membranes can occur only via hopping oxygen ions to neighboring vacant sites in the crystal lattice of mixed conductors, whereas the transport of any other species is excluded. Owing to this, gas-tight mixed-conductive membranes possess an infinite permselectivity. Several reviews on mixed conducting membranes (perovskite) for oxygen separation are available, which provide the main understanding on the material composition, structure, preparation as well as the transport mechanism of oxygen permeable membranes [180]. While many perovskite oxide materials have been explored over the past two decades, there are hardly any materials with sufficient practical economic value and performance for large scale applications; the search for new materials is justified and should continue.

Oxygen permeability of a number of dense oxide membranes with perovskite-type structure was studied by Kharton et al. [181]. The cubic perovskite solid membranes derived from $SrCoO_{3-\delta}$ by partial substitution of cobalt with higher valency transition metal cations (Fe, Cr, Ti) exhibited higher permeation fluxes in comparison with other mixed-conducting ceramics. The highest permeation fluxes were observed for the $SrCo_{1-x}Ti_xO_{3-\delta}$ ($x = 0.05$–0.20) and $SrCo_{0:90-x}Fe_{0.10}Cr_xO_{3-\delta}$ ($x = 0.01$–0.20) ceramic membranes.

3.2.2 *Silica Glass Membranes*

Silica (SiO_2) shows unique properties related to the ability of its elemental bricks, i.e., SiO_4 tetrahedra, to be connected together, forming a large numbers of different amorphous or crystallized solids that can be microporous, mesoporous or macroporous. Silica is also known as a chemical compound that contains an oxide of silicon with the chemical formula SiO_2. Silica is most commonly found in nature as sand or quartz. Glass is a hard, brittle substance, typically transparent or translucent, made by fusing sand (silica) with soda lime and some other ingredients.

Homogeneous and defect free amorphous films of silica can be deposited on porous substrate using sol–gel routes or chemical vapor deposition (CVD) methods to produce asymmetric membranes. With sol–gel routes, the starting solutions generally are based on tetramethoxysilane, $Si(OCH_3)_4$, or tetraethoxysilane $Si(OC_2H_5)_4$, diluted in methanol or ethanol, respectively. The formation of the oxide network results from the polymerization of the molecular precursor [182]. The hydrolysis of the alkoxide (a) produces activated species and their condensation by alcoxolation (b) or oxolation (c) leads to a reticulation by formation of siloxane bridge ≡Si–O–Si≡.

- $Si(OR)_4 + H_2O \Rightarrow Si(OR)_3OH + ROH$ (a)
- ≡Si–OH + RO–Si≡ ⇒ ≡Si–O–Si≡ + ROH (b)
- ≡Si–OH + HO–Si≡ ⇒ ≡Si–O–Si≡ + H_2O (c)

The hydrolysis and condensation reactions are catalyzed in acidic and basic media respectively.

CVD Routes: Conventional CVD techniques (atmospheric pressure AP or low pressure LP) at high temperature (400–700 °C) or Plasma Enhanced CVD methods (PECVD) at low temperatures (room temperature—400 °C) can also be easily used to prepare silica layers. The usual silica precursors are SiH_4 or tetraethoxysilane (TEOS). They are mixed with O_2, N_2O, or O_3 as oxidizing reactants. A large number of other organosilanes or alkoxysilanes can also be used.

Highly selective microporous silica membranes with high fluxes can be prepared by sol–gel dip-coating processes. The structure of the thin silica layer mainly depends on the size and shape of the silicalite polymers and their packing behavior during drying and heat treatment. Design of the pore networks is of great importance in deciding the transport properties through the membrane since permeation and permselectivity are mainly determined by the microstructure of the membrane, such as pore size, pore size distribution, porosity, and the interaction of permeating species with the pore walls [178, 183]. The use of mono-dispersed silica spheres of a size smaller than 10 nm, with narrow particle size distribution, make it possible to prepare microporous silica membranes with controllable pore structures for specific applications in gas separation [184].

The gas permeation properties of He, H_2, CO_2, O_2, N_2, and CH_4 in microporous silica membranes were studied as a function of temperature and pressure by Shelekhin et al. [185]. Selectivities were found to be a function of differences in the gas kinetic diameters. The ideal selectivity for He/CH_4 was more than 10,000 at 30 °C. Selectivity decreased with increasing temperature.

Naskar et al. [186] prepared silicate-1 zeolite membranes hydrothermally on the porous ceramic supports, both unmodified and modified with 3-aminopropyl triethoxysilane (APTES) as coupling agents, following ex situ (secondary) crystal growth process. The membrane developed on surface-modified support rendered a lower permeance value, i.e., 9×10^{-7} mol m^{-2} s^{-1} Pa^{-1} of N_2 compared to that formed on the unmodified support, which gave a permeance value of 20×10^{-7} mol m^{-2} s^{-1} Pa^{-1} of N_2.

Glass membranes are actually not very important for hydrogen separation. One of the reasons is their low selectivity. Glass membranes are porous. Depending on the pore size, they can be subdivided into micro porous (pores below 2 nm) and mesoporous (pores 2–5 nm). Microporous membranes have higher selectivity yet lower fluxes. Both membrane types are usually produced from silica using the leaching manufacturing process. The temperature range where they can be used has an upper limit of 400–500 °C. Vycor glass membranes are commercially available. Porous glass membranes are characterized by a wide range of pore sizes and a good accessibility to the active sites. Porous glasses possess, in comparison with other porous inorganic solids, high thermal stability, chemical resistance, high optical transparency, and good accessibility to eventually available active side inside the porous structure. They can be prepared in different geometrical forms. Due to their large surface areas, molecular sieving ability and controlled host–sorbate interactions, microporous and mesoporous glass membranes can be applied as a medium for gas separation.

Marković et al. [187] demonstrated the preparation and quantitative investigation of gas transport and equilibrium properties of porous glass membranes with pore diameters in a relatively narrow range between 2.3 and 4.2 nm. Original glass for the membranes consisted of 70 % SiO_2, 23 % B_2O_3, and 7 % Na_2O. This composition assured the absence of stresses during the cooling process of glass melt. The authors used three mesoporous glass membranes with pore diameters around 2.3 nm (membrane M1), 3.1 nm (membrane M2), and 4.2 nm (membrane M3) as determined by low-temperature nitrogen adsorption. To change surface affinities of the membranes, the surface of membrane M1 was modified with trimethylsilyl groups (membrane M1-mod). As a result, the surface properties of the modified membranes favored interaction with nonpolar gases. During this modification procedure the pore diameter of the membrane was only slightly reduced. Figure 3.39 demonstrates that the permeabilities decrease for all investigated membranes, and for all tested gases, with increasing temperature.

Permeabilities of all examined gases are presented in Fig. 3.40 as a function of the membrane pore diameters at two different temperatures (293 K: Fig. 3.37a, and 433 K: Fig. 3.38b). The same curve shapes were obtained for inert gases for both temperatures. At higher temperatures (Fig. 3.37b) the permeabilities of the adsorbable gases are increasing with increasing pore diameters, but at lower temperature (Fig. 3.38a) they are obviously affected more by the differences in adsorption affinity. The curves of CO_2 and C_3H_8 pass through a maximum observed at 2.3 nm (membrane M1) due to the largest adsorption contribution. Similar trends, as in Fig. 3.38b (where adsorption effects are negligible), were predicted by Bhatia and Nicholson [187] using molecular dynamic simulations for single gas transport in the same range of nanopore sizes.

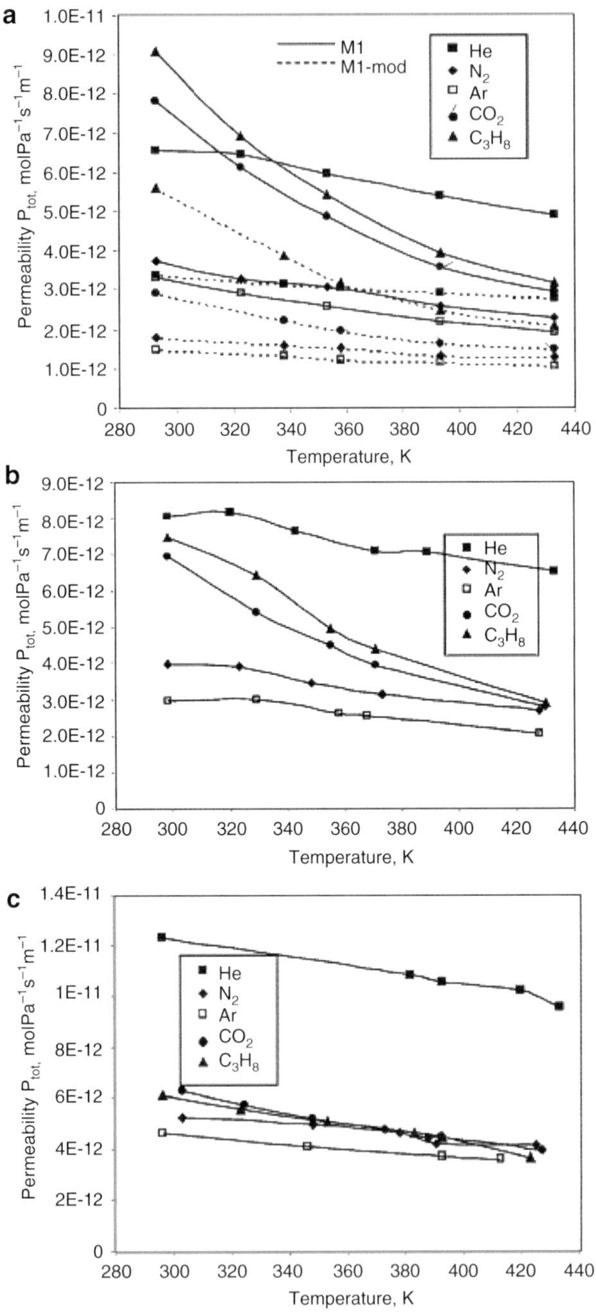

Fig. 3.39 Experimentally observed permeability data of different gases as a function of temperature for: (**a**) membrane M1 and M1-mod, (**b**) membrane M2 and (**c**) membrane M3 [187]

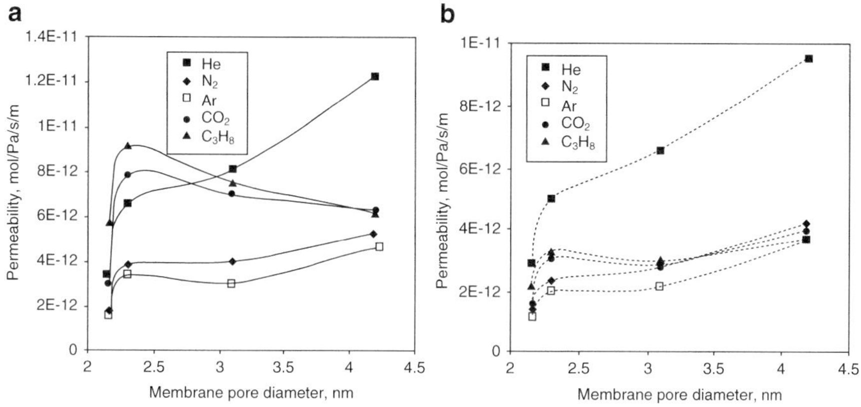

Fig. 3.40 Experimentally observed permeability as a function of pore diameters for all examined membranes presented at two different temperatures: (**a**) 293 K and (**b**) 433 K. M1-mod: 2.18 nm, M1: 2.3 nm, M2: 3.1 nm, and M3: 4.2 nm

3.2.3 Zeolites

Zeolite is a crystalline aluminosilicate made up of a 3D framework that forms uniformly sized pores of molecular dimensions (2–20 Å). Zeotype is any crystalline material (e.g., aluminophosphates, titanosilicates) with a 3D framework in which one of the tetrahedral sites occupied by Si is replaced with another element. Many zeotypes have the same structure as known zeolites. Molecules with different sizes and shapes can be discriminated or separated by zeolites through their channels. Like most silicates, the zeolites are based on TO_4 tetrahedra, where T is an aluminum or silicon atom (phosphorus in aluminophosphates). The vast three-dimensional networks are a result of all four corners for the tetrahedra being shared, producing low density microporous materials [188]. Zeolite structures are made of finite or infinite (chains, layers, etc.) component units. The finite units that have been found to occur are shown in Fig. 3.41.

In Fig. 3.41 the T atom of the TO_4 tetrahedron is located at each of the corners, and the oxygens are located towards the mid-points of the lines joining each T atom (*the oxygens are not shown to aid clarity*). These secondary building units (SBUs)—the primary building units being the TO_4 tetrahedra—can contain up to 16 T atoms. SBUs are non-chiral (neither left nor right "*handed*"). A unit cell always contains the same number of SBUs, and although rare, some materials can have different combinations of SBUs within the zeolite framework [188].

The exact definition of the term "Zeolite" is still the subject of discussion. The naming of zeolites in the literature seldom follows a scientific system such as defined by the *The Atlas of Zeolite Structure Types* published and frequently updated by the IZA (International Zeolite Association) Structure Commission. The maximum size of the molecular or ionic species that can enter the pores of a zeolite is

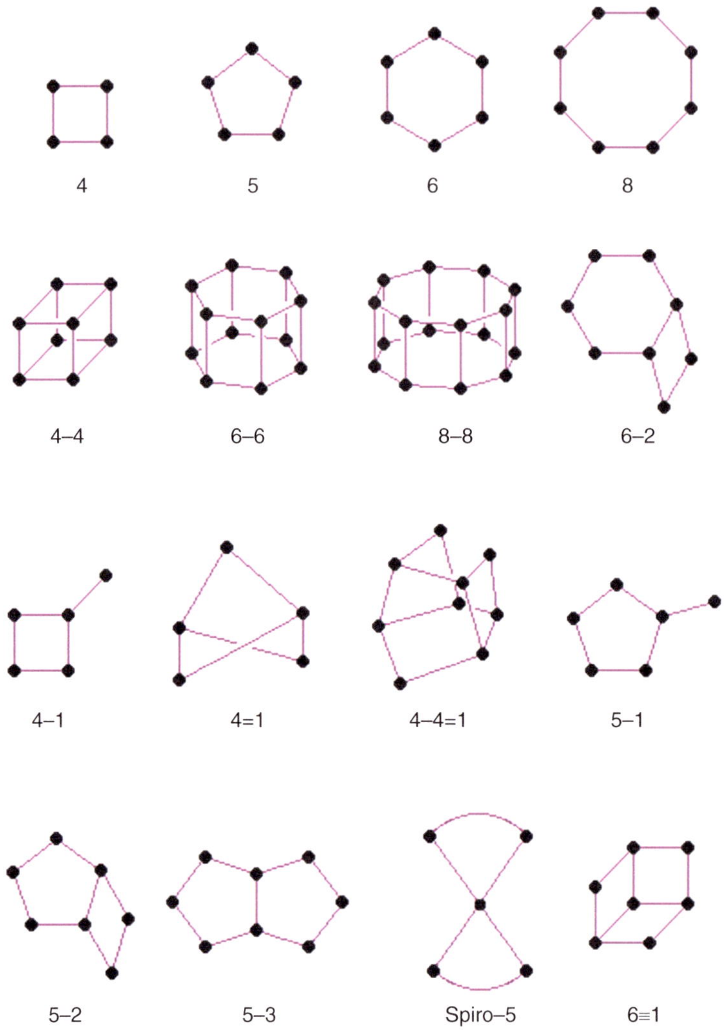

Fig. 3.41 Secondary Building Units (SBUs) in zeolites. (The corners of the polyhedra represent tetrahedral atoms)

controlled by the dimensions of the channels. These are conventionally defined by the ring size of the aperture, where, for example, the term "8-ring" refers to a closed loop that is built from eight tetrahedrally coordinated silicon (or aluminum) atoms and eight oxygen atoms. These rings are not always perfectly symmetrical due to a variety of effects, including strain induced by the bonding between units that are needed to produce the overall structure, or coordination of some of the oxygen atoms of the rings to cations within the structure. Therefore, the pores in many zeolites are not cylindrical.

Zeolites have a porous structure that can accommodate a wide variety of cations, such as Na^+, K^+, Ca^{2+}, Mg^{2+}, and others. These positive ions are rather loosely held and can readily be exchanged for others in contact with solution. Some of the more common mineral (natural) zeolites are analcime, chabazite, clinoptilolite, heulandite, natrolite, phillipsite, and stilbite. An example zeolite mineral formula is $Na_2Al_2Si_3O_{10}{\cdot}2H_2O$, the formula for natrolite. There are two types of zeolites, natural and synthetic and some of the differences between the two include:

1. Synthetics are manufactured from energy-consuming chemicals and naturals are processed from natural ore bodies.
2. Synthetic zeolites have a silica-to-alumina ratio of 1-to-1 and natural clinoptilolite (clino) zeolites have a 5-to-1 ratio.
3. Clino natural zeolites do not break down in a mildly acid environment, whereas synthetic zeolites do. The natural zeolite structure has more acid resistant silica to hold its structure together.

Membranes made from zeolites are promising to achieve high selectivities based on molecular recognition by the membrane pores. As well, zeolite membranes show selective adsorption properties and catalytic abilities.

The preparation of a "defect free" zeolite layer for gas separation was and is still a matter of study. Therefore, the aim of zeolite membrane design is to tune the size of zeolite pores and/or to decrease the number of defects. For example, the intercrystal pores formed inherently in polycrystalline zeolite films should be minimized, because the existence of intercrystal pores is the major cause for decline in molecular separation efficiency [189].

The framework of the zeolite can be modified by synthesizing them with metal cations other than aluminum and silicon in the framework. There are many propriety methods to modify zeolites that impart unique characteristics to them. However, zeolite membranes have been too expensive to replace competing polymeric membranes.

Synthetic zeolites hold some key advantages over their natural analogs. The synthetics can, of course, be manufactured in a uniform, phase-pure state. It is also possible to manufacture desirable zeolite structures which do not appear in nature. Zeolite A is a well-known example. Since the principal raw materials used to manufacture zeolites are silica and alumina, which are among the most abundant mineral components on earth, the potential to supply zeolites is virtually unlimited.

As of October 2012, 206 unique zeolite frameworks have been identified, and over 40 naturally occurring zeolite frameworks are known [188, 190, 191]. There are more than 200 different zeolites with different structures that have been reported by the International Zeolite Association (IZA). Among them, only 15 structures have been tried to fabricate membranes [192, 193]. Table 3.18 provides a brief description of structure and pore sizes for the few different zeolites used for membranes.

3.2.3.1 Preparation of Zeolite Membrane by Crystallization and Seeding

In order to achieve a better separation performance, zeolite membranes should be preferably made of pure zeolite crystals with uniform and small particle sizes. Zeolitic nucleation is affected by the surface composition and chemistry of the support material [194].

Table 3.18 Brief description of structure and pore sizes of different zeolites used for membranes etc

Zeolite/zeotype	Brief description of structure	Nanopore size
T	Intergrowth type zeolite of erionite and offretite	0.36 nm × 0.51 nm
FAU (faujasite), A (LTA)	12-membered ring (NaX, Si/Al:1/1.5), pore 4 Å (NaY, Si/Al:>1.5)	~0.74 nm (pd) (NaA zeolite—pore size 0.41 nm)
MOR	12-membered ring	
MFI ZSM-5	Ten-membered ring	0.56 nm (pd), pore size ~ 6 Å
MEL		
MER (merlinoite)	Comprises double-eight-rings and γ cage	pd. 0.27–0.51 nm
W	Same framework topology as the mineral merlinoite (MER), eight membered ring	Channel dimension 0.31 nm × 0.35 nm
FER	Eight-membered ring	
L	Si/Al:3/1	Cavity, 0.48 × 1.24 × 1.07 nm
DDR	Comprises silicon and oxygen atoms, eight membered ring	0.36 × 0.44 nm
Boron substituted ZSM-5	Boron substituted into the framework for silicon	0.53–0.56 nm pores
SUZ-4	Composition $K_5Al_5Si_{31}O_{72}$; framework topology is related to zeolites ferrierite and ZSM-5 and contains straight channels having apertures defined by rings of ten (Si,Al)–O species. A novel cage may serve as the site for non-exchangeable potassium ions.	97 nm pd
Imidazolate (ZIFs)	A subclass of metal–organic frameworks (MOFs)	ZIF-8, 11.6 Å Cell dimension 16.32 Å
ZIF-8	Sodalite (SOD) type	11.6 Å
AlPOs	3D framework, AlO_4^- and PO_4^- tetrahedral building units	3.8 Å
ITQ	Similar to Zeolite A, but a much higher Si/Al ratio (up to infinity, i.e., pure silica)	Hydrophobic smaller pore size than Zeolite A
SAPO-34	Micropore, similar to chabazite	Internal cages ~ 1.4 nm in diameter and each cage has six pores. XRD pore diameter of 0.38 nm
SAPO-44	Micro pore, similar to chabazite	~0.43 nm
A-Type zeolite	Hydrophilic	0.4–0.43 nm
UZM	16 cages per unit cell	6–8 Å

pd means pore diameter

There are several types of synthetic zeolites that form by a process of slow crystallization of a silica–alumina gel in the presence of alkalis and organic templates. One of the important processes used to carry out zeolite synthesis is sol–gel processing. The product properties depend on reaction mixture composition, pH of the system, operating temperature, pre-reaction "seeding" time, reaction time, and the templates used. In the sol–gel process, other elements (metals, metal oxides) can be easily incorporated. The silicalite sol formed by the hydrothermal method is very stable. The ease of scaling up this process makes it a favorite route for zeolite synthesis.

A number of synthetic strategies have been applied to zeolite membrane synthesis, such as in situ (without seeding) hydrothermal synthesis (in situ crystallization), secondary (seeded) growth and post-treatments of zeolite membranes.

Zeolite membrane synthesis is more commonly carried out by the in situ crystallization technique which involves placing a porous support in contact with a synthesis solution or gel under hydrothermal conditions. For successful membrane formation, proper conditions are necessary to allow for preferential nucleation and growth of zeolite crystals on the support surface (possibly competing with solution events) in an interlocking fashion with minimal non-selective interzeolitic porosity.

The other approach for zeolite membrane formation is a technique called secondary (seeded) growth, which involves attaching a closely packed layer of zeolite seed crystals on the surface of a support. The first seeding was demonstrated by Horii et al. [195]. The use of seed crystals facilitates the formation of zeolite membranes since a seeded support grows to a pure-phase zeolite membrane more easily, even when the crystallization conditions and the chemical batch compositions are not optimum. Seeded growth has significant advantages such as better control over membrane microstructure (thickness, orientation), higher reproducibility, and a wider range of hydrothermal synthesis conditions leading to continuous film formation. Elimination of the constraints imposed by the need for crystal nucleation, due to the preexistence of nuclei on the support surface, renders crystal growth as the main film formation mechanism and, thus, adds improved flexibility in zeolite film and membrane preparation [196].

There are four main ways to attach the seeds to the support [197].

1. Charging the support surface by pH control such that seeds and support have opposite surface charges for an electrostatic attachment.
2. Charging the support surface by adsorption of positively charged cationic polymers like poly-DADMAC (diallyl dimethyl ammonium chloride) or Redifloc (Trade name of EKA Chemicals, a polyamine) to adjust different zeta potentials between the ceramic support and the zeolite nanocrystals to be attached as seeds. The counter ions of the ammonium polymer are usually chlorides which go in the solution, and negatively charged silica nanoparticles are attached. The use of seeded supports usually results in a c-orientation of the MFI (Mordenite Framework Inverted, example ZSM-5, medium pores 5.3 Å) layer but under certain conditions also for secondary growth the desired b-orientation can be obtained.
3. Electrophoretic deposition of nanosized seeds on solid supports.
4. Immersion of the dried support into a seed solution followed by thermal treatment of the seeded support to burn helping organic additives and to fix the seeds via de-hydroxylation to the support.

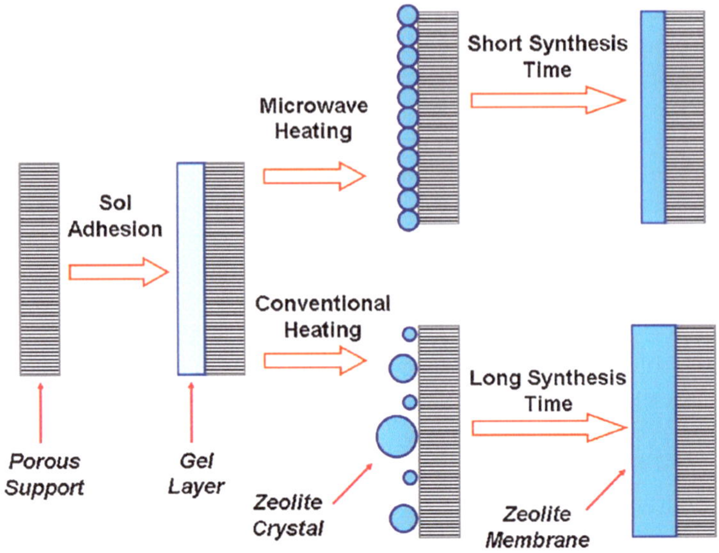

Fig. 3.42 Comparative synthesis model of zeolite membrane by microwave heating and conventional heating

Microwave synthesis of zeolite can be much faster than conventional heating and create a more selective membrane product. Microwave synthesis has often proven to create more uniform (defect-free) products than from conventional hydrothermal synthesis. As compared with conventional hydrothermal synthesis, microwave synthesis of zeolites has the advantages of a very short time, small zeolite particle size, narrow particle size distribution and high purity. All these characteristics make it a promising method for rapid preparation of high performance zeolite membranes. However, microwave synthesis is in the early stages of development, and further research is required [198]. Articles written by Li and Yang [199], Cundy [200], and Li et al. [201] are recommended for more general information on the subject of synthesis of microwave zeolites and microwave chemistry. Figure 3.42 illustrates a comparative synthesis of zeolite membrane by microwave heating and conventional heating [201].

Li et al. [201] developed a method called, "in situ aging–microwave synthesis" of zeolitic membranes without seeding. It decouples two successive steps in the formation of zeolite membranes. The first step is the rearrangements of synthesis mixture and formation of germ nuclei on the support surface obtained by in situ aging. The second step is the nucleation and crystal growth on the support achieved by in situ crystallization under fast and homogeneous heating. Choi et al. [202] demonstrated that rapid thermal processing in the preparation of zeolitic membranes can improve the separation performance of thick columnar films of a certain zeolite (silicate-1) by eliminating grain boundary defects, possibly by strengthening grain bonding at grain boundaries.

Varoon et al. [203] demonstrated the synthesis and structure determination of highly crystalline nanosheets of zeolite frameworks MWW. MWW structured zeolites such as MCM-22 possess two independent pore systems. One system consists of two-dimensional sinusoidal 10-member ring (MR) channels with an elliptical ring cross section of 4.1 Å × 5.1 Å. The other is composed of a large 12-MR supercage connected by 10-MR windows [204] and MFI (mordenite framework inverted). The purity and morphological integrity of those nanosheets allow them to pack well on porous supports, facilitating the fabrication of molecular sieve membranes.

3.2.3.2 LTA Zeolite

Zeolite A exhibits the LTA (Linde Type A) structure. It has a three-dimensional pore structure with pores running perpendicular to each other in the *x*, *y*, and *z* planes, and is made of secondary building units 4, 6, 8, and 4-4. The pore diameter is defined by an eight member oxygen ring and is as small as 4.2 Å. This leads to a larger cavity of minimum free diameter 11.4 Å. The cavity is surrounded by eight sodalite cages (truncated octahedra) connected by their square faces in a cubic structure. The zeolite LTA membrane, with a relatively small pore size of about 0.4 nm, is a candidate for the separation of small sized molecules such as H_2 and CO_2, while zeolite FAU (faujasite) membrane with larger pore size will not reduce the permeation flux like the narrower LTA layer. Further, the composition difference between the zeolite LTA and FAU layers is relatively small and both of them have strong hydrophilicity due to their low Si/Al ratio. Therefore, sandwich-structured composite LTA-FAU membranes will be formed and show higher gas separation performances corresponding to single-phase zeolite membranes.

Huang et al. [205] reported that sandwich-structured zeolite membranes enhanced hydrogen selectivity. Sandwich-structured LTA–FAU composite membranes were prepared by using 3-aminopropyltriethoxysilane (APTES) as an interlayer between the LTA and FAU layers as well as between the alumina support and the zeolite LTA layer, as shown in Fig. 3.43.

Figure 3.44 shows the permeances of the single gases through the LTA–FAU composite membrane as well as the single zeolite LTA and FAU membranes at 100 °C and 1 bar pressure difference as a function of the gas kinetic diameters of the permeating molecules.

H_2 has the highest permeance for all membranes due to its smallest kinetic diameter of 0.29 nm. At 100 °C for the LTA–FAU composite membrane, the ideal separation factors of H_2 from CO_2, N_2, CH_4, and C_3H_8 are 10.6, 8.6, 7.1, and 24.3, respectively, which not only exceed the corresponding Knudsen coefficients (4.7, 3.7, 2.8, and 4.7, respectively) but are also higher than the corresponding ideal separation factors of the single phase zeolite LTA (7.0, 5.8, 4.9, and 6.8) and FAU (8.0, 7.2, 5.6, and 18.6) membranes. This data suggests that the LTA–FAU composite membrane displays higher hydrogen selectivities. The enhancement of the separation performance could be due to the novel sandwich structure of zeolite membranes by using APTES as an interlayer.

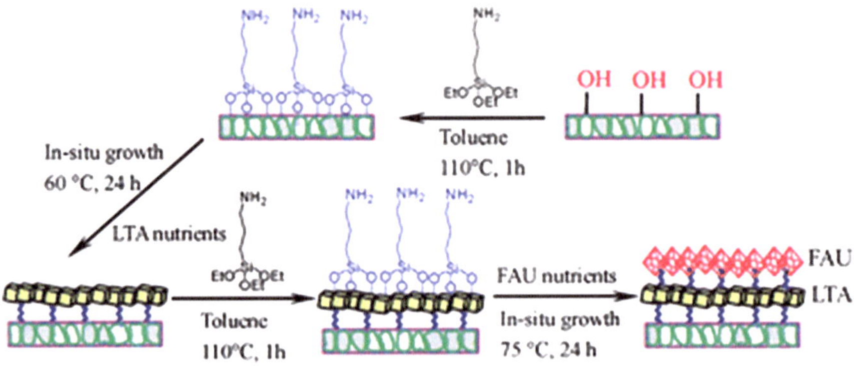

Fig. 3.43 Schematic diagram for stepwise synthesis of a sandwich-structured zeolite LTA–FAU composite membrane by using APTES as an interlayer [205]

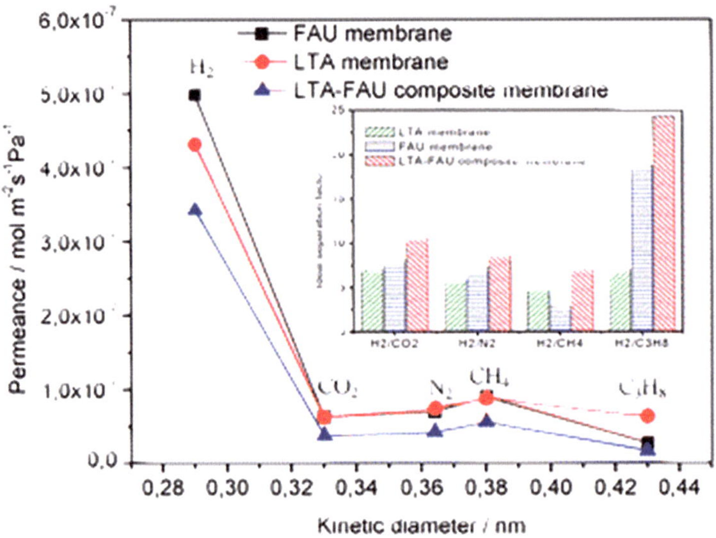

Fig. 3.44 Single gas permeances of different gases through the LTA–FAU composite membrane as well as the single zeolite LTA and FAU membranes, at 100 °C and 1 bar pressure difference as a function of the gas kinetic diameter. (The *inset* shows the ideal separation factors of the membranes for H_2 over other gases [205])

Cheng et al. [206] synthesized NaA zeolite using microwave heating. The method involved two steps: prior seeding of 120 nm of LTA crystals on substrate and then employing a secondary hydrothermal synthesis. The effects of seeding time and synthesis time on performance were studied. The ideal H_2/N_2 selectivity increased from 1.90 of the substrate to 6.37 of the three-stage synthesized membrane, which was distinctly higher than the corresponding Knudsen diffusion selectivity of 3.74.

3.2.3.3 NaA Zeolite

NaA zeolite is one of the microporous crystalline aluminosilicate zeolites, which has a channel opening size of 0.41 nm. The pore size of NaA zeolite is close to many molecular diameters. As it has strong hydrophilicity, the synthesized NaA zeolite membrane has great potential in many fields, such as gas separation. For example, the molecular kinetic diameters of O_2 and N_2 are 0.346 nm and 0.364 nm, respectively. Due to configurational diffusion, the slight difference in molecular diameter between O_2 and N_2 leads to a big difference in the rate of diffusion through the NaA zeolite channels, with the diffusion rate of O_2 being faster than that of N_2. Thus, nitrogen–oxygen mixtures can be separated effectively by the NaA zeolite membrane. Moreover, NaA zeolite membranes can be used for separating many different mixtures of small gases [207]. The permeation of n-C_4H_{10} via a high quality NaA zeolite membrane reveals that the NaA zeolite membrane has intercrystalline pores larger than those of the NaA zeolite channels [208]. Though the NaA type zeolite membrane shows excellent performance, the acid stability of the membrane is poor. The surface of A-type zeolite is hydrophilic due to the low Si/Al ratio in the framework. Few papers have been published on the small gas permeation characteristics for A-type zeolite membrane [209, 210]. Aoki et al. [209] suggested that the behavior of the NaA zeolite ($d_p = 0.41$ nm) membrane is dominated by the molecular sieving mechanism, despite the presence of defects larger than the structural pores. Another consequence of the low Si/Al ratio found in A-type zeolite is a lack of thermal stability, which is why there have been no reports on its gas permeation characteristics at high temperatures (>300 °C).

Dey et al. [211] synthesized NaA-zeolite membrane hydrothermally by a secondary crystallization process at different temperatures (55–75 °C) on porous alpha-alumina support tubes (inner side) precoated with a poly(ethyleneimine) (PEI) buffer layer. The application of PEI as a buffer layer was found to be very effective for proper attachment of NaA crystals with the support. Adsorption/interaction of CO_2 in the NaA zeolite channel was stronger than the other gases, which is reflected in their permeation behaviors. The permeance values of different gases through PEI-modified membranes were in the order of $CO_2 < N_2 < H_2$. The order of permselectivity was $(H_2/CO_2) > (H_2/N_2) > (N_2/CO_2)$. Thus, the PEI-modified NaA membrane could be very effective in separation of CO_2 from other gases.

3.2.3.4 DDR Type Zeolite

Recently, DDR type zeolites have been widely used for gas separation. The all-silica zeolite deca-dodecasil 3R (DD3R) is a clathrasil (8-ring) first synthesized by Gies [212]. Its crystal structure consists of a three-dimensional arrangement of building units. The highly siliceous DDR (Deca-Dodecasil 3R)-type zeolite contains pores formed by a polyhedron with an oxygen eight-membered ring. DDR type zeolite particles have an aperture of 0.36×0.44 nm. This zeolite has a high thermal stability allowing for the study of gas permeation (or diffusion) at high temperatures where adsorption is negligible. This stability avoids the necessity to measure the equilibrium of gas adsorption in supported, thin zeolite film. Furthermore, this zeolite has small pore opening (0.36×0.44 nm^2) making it an ideal candidate to study the effects of size or molecular weight of gases on permeation or diffusion properties for zeolites and zeolite membranes. The DDR critical diameter with eight-membered-ring windows closely matches the diameters of light hydrocarbons and carbon dioxide.

Nakayama et al. [213] patented the method for the preparation of DDR zeolites. They reported that the DDR type zeolite membrane separates at least one type of a gas component from a mixed gas containing at least *t* from a group consisting of carbon dioxide, hydrogen, oxygen, nitrogen, methane, propane, propylene, carbon monooxide, and nitrogen oxide. Single gas permeances were different both at room temperature and 100 °C, enabling the separation of at least one selected gas component from the mixed gas.

Bergh et al. [214] measured the equilibrium adsorption data of pure CO_2, N_2, and CH_4 and their mixtures in pure silica DDR membranes at a temperature range of 200–400 K and at pressures up to 1,500 kPa. From the measured equilibrium data of these gases, the membrane's transport parameters were derived by using the unary permeation. A model based on the generalized Maxwell–Stefan equations was used to simulate the component fluxes and separation factors of the binary mixtures as a function of temperature and total feed pressure. It was concluded that DDR membranes have very high selectivities for CO_2/CH_4 separations with good permeances. At a total pressure of 101 kPa and at 225 K, the CO_2 selectivity of an equimolecular CO_2/CH_4 mixture was found to be >3,000 and for N_2 over CH_4 selectivity was 40 with a 50/50 feed. The N_2/CH_4 selectivity was constant with pressure, while the selectivity for CO_2/CH_4 decreased. Figure 3.45 shows the selectivities of equimolar mixtures through the DDR membrane as a function of the temperature at constant total feed pressure of 101 kPa, using sweep gas He at 101 kPa [215].

Figure 3.46 shows the CO_2 selectivity of a DDR membrane for an equimolar CO_2/CH_4 mixture as a function of permeation temperature.

Tomita et al. [216] coated molecular-sieve type zeolite (DDR) with an aperture of 0.36×0.44 nm on a porous alumina substrate using a hydrothermal process. The permeation through the membrane in the single gas feed of helium, hydrogen, carbon dioxide, oxygen, nitrogen, methane, *n*-butane, *i*-butane, and sulfur hexafluoride was measured at 301 and 373 K up to 0.5 MPa. Figure 3.47 shows the permeance plotted against the kinetic diameter of various gases. The permeance decreased by more than three orders of magnitude between 0.35 and 0.40 nm of the kinetic diameter of permeated gas at both 301 and 373 K. The separation factor of CO_2 to CH_4 in 50 % CO_2 and 50 % CH_4 mixed gas feed was 220 and 100, at 301 and 373 K, respectively. Total gas feed pressure was 0.5 MPa. The DDR type membranes were found to have few defects and worked as molecular-sieving membranes. It was suggested that DDR type zeolite membranes are not perfectly hydrophobic.

Kanezashi et al. [217] prepared DDR type zeolite membranes by the secondary growth method on a porous α-Al_2O_3 disc. To eliminate the crystalline micropores, the surface was modified by an on-stream counter diffusion chemical vapor deposition (CVD) technique. Table 3.19 shows the permeance for four gases for the

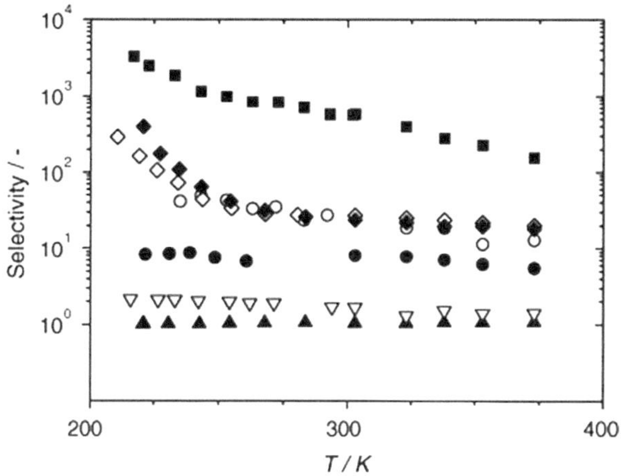

Fig. 3.45 Selectivities of equimolar mixtures through the DDR membrane as a function of the temperature constant total feed pressure of 101 kPa, sweep gas He at 101 kPa. CO_2/CH_4 (*filled square*), N_2/CH_4 (*open circle*), CO_2/Air (*filled diamond*), N_2O/Air (*open diamond*), Air/Kr (*filled circle*), O_2/N_2 (*inverse triangle*), N_2O/CO_2 (*filled triangle*) [215]

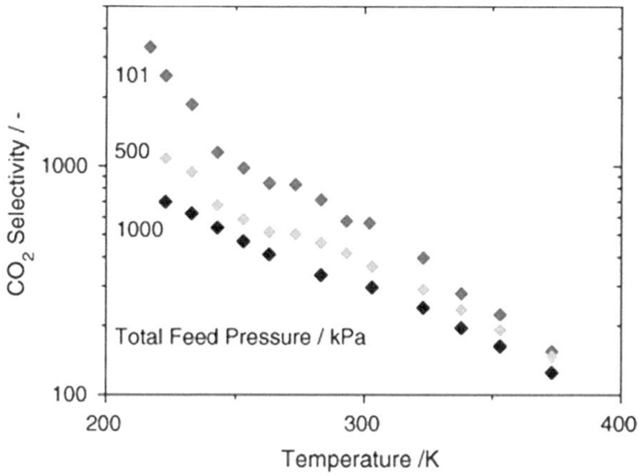

Fig. 3.46 CO_2 selectivity of a DDR membrane for an equimolar CO_2/CH_4 mixture as function of permeation temperature [214]

CVD-modified DDR type zeolite membranes before and after exposure to steam at 500 °C. The table clearly shows that there is negligible change in gas permeation before and after exposure to steam. Similar results were also observed for activation energy. It proved that CVD-modified DDR type membranes are hydrothermally stable, and that single gas permeance decreases as H_2 > He > CO. This order is determined by both molecular size and weight of permeating gases. This permeance indicates the presence of intercrystalline pores in the as-synthesized DDR-type zeolite membranes.

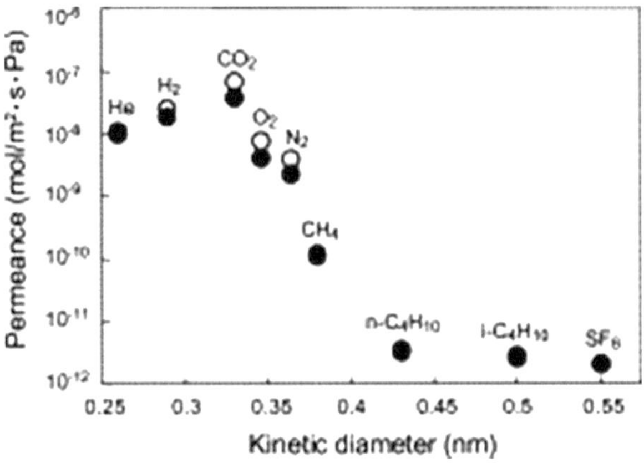

Fig. 3.47 Single gas permeance versus kinetic diameter of gas for DDR type zeolite membrane at 301 K (*open circle*) and 373 K (*filled circle*) [216]

Table 3.19 Permeance of gases for CVD-modified DDR-type zeolite membrane before and after exposure to steam at 500 °C (partial pressure of steam: 50 kPa)

	Permeance at 500 °C (10^{-8} mol m^{-2} s^{-1} Pa^{-1})	
Gases	Before exposure to steam	After exposure to steam
He	2.79	2.74
H_2	2.34	2.18
CO_2	0.38	0.39
CO	0.20	0.18

Kanezashi et al. [217] suggest that high temperature diffusion data for the small gases in the DDR type zeolites measured by the macroscopic membrane permeation method are consistent with the theory of translation gas diffusion in zeolites as proposed by Xiao and Wei [218].

Himeno et al. [219] coated the outer surface of a porous α-alumina tube with highly hydrophobic DDR zeolite membrane. Single gas permeance for CO_2, CH_4, He, H_2, O_2, and N_2, and CO_2/CH_4 binary gas were measured. The permeances were in the following order: $CO_2>H_2>He>O_2>N_2>CH_4$. Single-gas permeance was dependent on the relative molecular size of the DDR to the pore diameter; however, CO_2 permeance was dominated by the adsorption affinity to the pore wall of DDR zeolite. The respective single-gas permeances of CO_2 and CH_4 at 298 K at a feed pressure of 0.2 MPa and a permeate pressure of 0.1 MPa were 4.2×10^{-7} and 1.2×10^{-9} mol m^{-2} s^{-1} Pa^{-1}; the ideal selectivity for CO_2/CH_4 was 340. These CO_2/CH_4 selectivities and CO_2 permeance are better than other zeolite membranes. Himeno et al. [220] developed a membrane separation process for biogas using a DDR-type zeolite membrane with high CO_2/CH_4 selectivities. Biogas produced in a sewage plant was separated by using a DDR-type zeolite membrane, and the performance and durability of the membrane was estimated. Himeno et al. reported that the developed membrane separation process was fully applicable to the separation and purification of biogas; however, the membrane performance was reduced by long periods of ventilation and by exposure to compounds of high boiling points such as higher hydrocarbons and siloxane, which are present as impurities in the biogas.

3.2.3.5 SAPO-34

The molecular sieve SAPO-34 has the composition $(Si_xAl_yP_z)O_2$, where $x=0.01$–0.98, $y=0.01$–0.60, and $z=0.01$–0.52. A SAPO-34 membrane removed CO_2 from CO_2/H_2 mixtures because CO_2 adsorbed more strongly than H_2; the fluxes of H_2 at low temperatures were orders of magnitude lower in the presence of CO_2. At low temperatures and high pressures, the CO_2/H_2 selectivity was greater than 100, and thus, SAPO-34 membranes may have a significant potential for application in separation of CO_2 from CO_2/H_2 mixtures for the purification of hydrogen [221].

Li et al. [222] synthesized powerful SAPO-34 membranes by in situ crystallization on a porous tubular stainless steel support. For a SAPO-34 membrane with a Si/Al gel ratio of 0.1, a CO_2/CH_4 mixture selectivity of $\alpha=170$ with a CO_2 permeance of $P=1.2\times10^{-7}$ mol m^{-2} s^{-1} Pa^{-1} was achieved at 22 °C. With decreasing temperature the selectivity increases, and at −21 °C a CO_2/CH_4 separation factor $\alpha=560$ was achieved. Membranes of SAPO-34 prepared with the higher Si/Al ratio of 0.15 showed slightly lower selectivity ($\alpha=115$), but a higher permeance for CO_2 (4×10^{-7} mol m^{-2} s^{-1} Pa^{-1}) at 35 °C. At 7 MPa, the SAPO-34 membrane had a $\alpha=100$ for a 50/50 feed of CO_2/CH_4 mixture at room temperature.

In another study, Li et al. [223] also reported that SAPO-34 membrane can separate CO_2 from CO_2/CH_4 mixtures best at low temperatures with a selectivity of $\alpha=270$ at −20 °C. Another study noted that SAPO-34 on porous alumina support can be used for the separation of light gases at both low and high temperature [224]. Zhou et al. [225] made a SAPO-34 membrane which showed high permeances (maximum of 1.2×10^{-6} mol m^{-2} s^{-1} Pa^{-1}) and high CO_2/CH_4 selectivities (70) at 46 bar of feed pressure.

In an additional study, Li et al. [226] fabricated high flux SAPO-34 membranes on porous, tubular stainless steel supports with CO_2/CH_4 separation selectivities greater than 200. Monolith-supported SAPO-34 also was used for the separation of a CO_2/CH_4 mixture. Monoliths increased the membrane surface area per volume and, thus, had the potential to decrease membrane module cost; however, CO_2 permeances and CO_2/CH_4 selectivities at 4.6 MPa feed pressure were similar to SAPO-34 membranes on single channel supports [227]. Humidity has a strong effect on permeation of gases through SAPO-34 membranes, and the effect depends on the fraction of permeation through non-SAPO pores in the membrane. For high quality membranes, water almost completely blocks the SAPO pores and dramatically decreases the permeances of gases that can enter the pores. The degradation accelerates with prolonged exposure until the membranes have low CO_2/CH_4 selectivities and exhibit viscous flow [228].

3.2.3.6 AlPO-18

Microporous aluminophosphates (AlPOs) are a class of zeolites with framework structures built of AlO_4^- and PO_4^- tetrahedral building units. In particular, AlPO-18 is an appealing zeolite composition for membrane preparation. The AEI framework topology of this aluminophosphate is characterized by a three-dimensional framework possessing eight-membered intersecting channels with a diameter of 3.8 Å. Due to its pore size, this membrane was used for the separation of CO_2 (kinetic diameter 3.3 Å). The membrane can be synthesized mainly via a hydrothermal approach and under microwave irradiation [237]. The separation performances of these membranes for equimolar CO_2/CH_4 gas mixtures was studied by Carreon et al. [229] and they reported that the AlPO-18 membranes displayed CO_2 permeances as high as ~6.6×10^{-8} mol m^{-2} s^{-1} Pa^{-1} with CO_2/CH_4 selectivities in the ~52–60 range at 295 K and 138 kPa.

3.2.3.7 Beta Zeolite or ZSM Zeolite (MFI Zeolite Membranes (ZSM-5))

Beta zeolite is an old zeolite discovered before Mobil began the "ZSM" naming sequence. As the name implies, it was the second in an earlier sequence. The very complex structure of beta zeolite was only recently determined because interest was not high enough until the material became important for some dewaxing operations. Beta zeolite consists of an intergrowth of two distinct structures termed Polymorphs A and B. The polymorphs grow as two-dimensional sheets and the sheets randomly alternate between the two. Both polymorphs have a three-dimensional network of 12-ring pores.

Among the different types of zeolites available, zeolite MFI (ZSM-5 and its Al-free analog, silicate-1) has been more commonly used in zeolite membrane synthesis because of its pore size (~5.5 Å) suitable for several industrially important separations, and the relatively easy synthesis from a variety of silica sources and structure directing agents. ZSM-5 is composed of several pentasil units linked together by oxygen bridges to form pentasil chains. A pentasil unit consists of eight 5-membered rings. In these rings, the vertices are Al or Si and an O is assumed to be bonded between the vertices. The pentasil chains are interconnected by oxygen bridges corrugated with 10-ring holes. Its chemical formula is $Na_nAl_nSi_{96-n}O_{192}\cdot16H_2O$ $(0<n<27)$ [230].

Poshusta et al. [231] measured the effect of pressure on the CO_2 permeance of three different MFI membranes chosen to represent membranes with small, medium and large amounts of non-zeolite pores. A model expressing the flux as the sum of surface diffusion, Knudsen diffusion, and viscous flow showed that the differences in the pressure behavior of the membranes were due to different relative amounts of zeolite and non-zeolite pores.

Membranes with the largest permeation through non-zeolite pores had the lowest CO_2/CH_4 mixture selectivity. The highest CO_2/CH_4 mixture selectivity was 5.5 at room temperature, and for all membranes, the selectivity decreased with temperature because of a decrease in competitive adsorption. The separation performance decreased with pressure despite an increase in the selectivity.

MFI zeolite membranes were prepared by Takata et al. [232] via secondary growth on α-alumina microfiltration membranes. Colloidal silicate (size around 100 nm) was used as a seed crystal. An MFI membrane in which the zeolite layer was oriented to the (1 0 1) plane, showed a n-C_4H_{10} permeance of 1.5×10^{-5} m^3 (STP) m^{-2} s^{-1} kPa^{-1} and a n-C_4H_{10}/*iso*-C_4H_{10} selectivity of 15 at 150 °C. N_2 permeated faster than He at temperatures lower than 150 °C. The authors reported that at high temperatures (300 °C) the permeation mechanism obeys the Knudsen mechanism, irrespective of molecular size in the experimental range.

ZSM-5 membrane was prepared by Kwon et al. [233] on the porous alumina support using in situ seeding techniques in hydrothermal conditions. The packed density of zeolite membrane was controlled by the hydrothermal time and temperature. The sample packed at 100 °C was densely packed with ZSM-5 seeds on the top of the alumina substrate. The prepared zeolite films were characterized with SEM and thin film XRD. The hydrogen permeance and selectivity toward carbon dioxide gas were 0.6×10^{-6} mol m^{-2} s^{-1} Pa^{-1} and 3.16, respectively. The hydrogen selective zeolite membranes show promising application in hydrogen separation from coal gasification such as Integrated Gasification Combined Cycle (IGCC).

Welk and Nenoff [234] studied the permeance and selectivity of both ZSM-5 and silicate-1 zeolite membranes under the flow of gas mixtures, two chosen as benchmarks (50/50 mol% H_2/CH_4 and 50/50 mol% H_2/CO_2) and one chosen to simulate an industrial methane reformate stream. Permeation experiments of mixed gases through both ZSM-5 and silicate-1 zeolite membranes revealed the extraordinary selectivities of these membranes for H_2. The ZSM-5 membrane had the following H_2 selectivities for the 50/50 H_2/CH_4 mixture, the 50/50 H_2/CO_2 mixture and the reformate mixture: 39.4, 60.1 and 58.8, respectively.

Richter et al. [235] coated ZSM-5 zeolite membrane on the inner surface of ceramic tubes and capillaries to get useful membrane shapes with a high membrane area/module volume ratio for industrial application. On flat discs and by a two-step-crystallization of ZSM-5 membranes of 20 μm thickness, a very low H_2 permeance of only 30 l/(m^2 h bar) and a high H_2/SF_6 single gas permselectivity of 51 were achieved (Fig. 3.48).

The coated membranes were tested by H_2 and SF_6 single gas permeation at 110 °C. The results are given in Table 3.20.

In Table 3.20, all tubular membranes had a much higher H_2 permeance and a lower H_2/SF_6 permselectivity compared with flat discs because of the simpler one-step crystallization. The preparation in a resting synthesis solution causes very low H_2/SF_6 permselectivities of 10–14. By using nano-sized MFI seeds and by pumping the synthesis solution through the support tubes, a homogeneous ZSM-5 membranes of 30 μm thickness, a high permeance of 4,500 l/(m^2 h bar), and an H_2/SF_6 single gas permselectivity above the Knudsen factor were achieved.

Bernal et al. [196] have grown zeolite MFI membranes (thickness 15–20 μm) on the surface of macroporous α-alumina and stainless steel support tubes (pore size 200 and 500 nm respectively) by the secondary (seeded) growth technique.

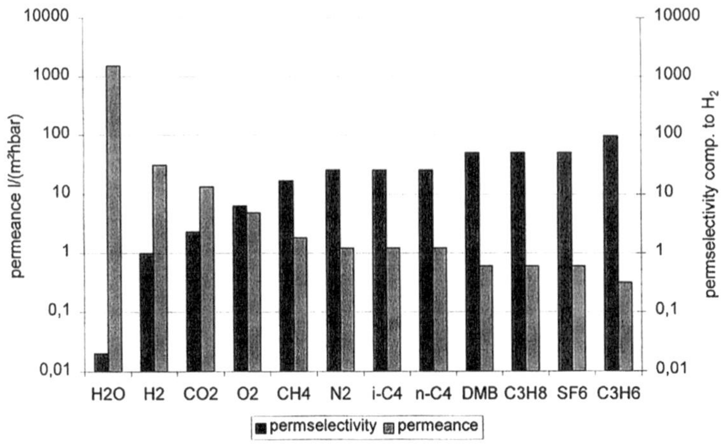

Fig. 3.48 Single gas permeance of ZSM-5 membrane on flat support at 105 °C, prepared by two-step-crystallization

Table 3.20 Single gas permeation of ZSM-5 membranes at 110 °C

Support	Seeding	H_2 permeance (l/m² h bar)	SF_6 permeance (l/m² h bar)	H_2/SF_6 permselectivity
Two-step-crystallization, resting synthesis solution				
Flat disc	MFI powder	30[a]	0.6[a]	51[a]
One-step-crystallization, resting synthesis solution				
Tube	MFI powder	6,300	460	14
Small tube	MFI powder	–	–	–
Capillary	MFI powder	–	–	–
Tube	Nano seeds	18,040	1,520	12
Small tube	Nano seeds	8,580	610	14
Capillary	Nano seeds	3,140	660	5
One step-crystallization, moved synthesis solution				
Tube	Nano seeds	9,540	480	20
Small tube	Nano seeds	5,310	170	31
Capillary	Nano seeds	4,550	210	22

[a]Measurement at 105 °C

The tubular supports were dipped vertically in an aqueous suspension of colloidal silicate-1 seed crystals (particle size 100 nm) and withdrawn at a speed of 1–2 cm/h to allow the uniform formation of seed layers on the outer (for stainless steel) or inner (for α-alumina) cylindrical surface of the support tubes. The seeded tubes were treated hydrothermally with clear solution of different compositions given in Table 3.21.

Aoki et al. [236] hydrothermally synthesized ZSM-5 zeolite (Si/Al = 2,525 and 600). These membranes were ion exchanged with H^+, Na^+, K^+, Cs^+, Ca^{2+}, and Ba^{2+}.

Table 3.21 Molar compositions of clear solutions for secondary growth

Code	KOH	TPABr	TPAOH	SiO_2	H_2O
A[a]	1.0	1.0	–	4.5	1,000
B[a]	–	–	3.0	25	1,500
C[a]	1.0	1.0	–	9.0	1,000
D[b]	1.6	2.0	–	40	1,000

[a]Silica source is TEOS
[b]Silica source is Ludox AS-40

Their gas permeation properties were measured over the temperature range of 323–523 K. Both the Si/Al ratio and the exchanged ion size affected the separation performance. For the membrane with a Si/Al ratio of 600, ion exchange only changed the single gas permeation of i-C_4H_{10}, because the number of exchange sites was small. For the membrane with Si/Al = 25, single gas permeances increased for the exchanged forms in the order: $K^+ < Ba^{2+} \sim Ca^{2+} < Cs^+ < Na^+ \sim H^+$, which coincides with the decrease in ion size and only the Cs ion does not fit this trend. It was concluded that zeolite membranes can be ion exchanged without irreversibly damaging the membrane performance.

Tuan et al. [237] prepared boron-substituted ZSM-5 membranes on porous stainless-steel and α-alumina support. These membranes had higher n-C_4H_{10}/i-C4H10 separation selectivities, and effectively separated these isomer mixtures at higher temperatures than membranes with aluminum substituted into the framework. Membranes were prepared with Si/B ratios as low as 12, and the best membranes were prepared from alkali-free gels. The highest n-C_4H_{10}/i-C_4H_{10} permselectivity was 60 at 473 K, and 24 at 527 K. It was also reported that B-ZSM-5 membrane preparation was reproducible and the membranes were stable at elevated temperatures.

Wang and Lin [238] fabricated a thin, high quality ZSM-5 top layer on a thick silicalite layer (bottom) to form a ZSM-5/silicalite bilayer membrane. The ZSM-5/silicalite bilayer membrane was supported on a porous alpha alumina support coated with an yttria stabilized zirconia (YSZ) intermediate barrier layer for membrane stability improvement. Both membranes were CCD (catalytic cracking deposition) modified using methyldiethoxysilane (MDES) as the precursor to reduce the zeolitic pore size for improving the H_2/CO_2 separation factor. Compared with the ZSM-5 zeolite membrane, the ZSM-5/silicalite bilayer membrane showed more improvement in H_2/CO_2 separation and less reduction in H_2 permeance by CCD modification. The CCD modified ZSM-5/silicalite bilayer zeolite membrane exhibited good stability and hydrogen separation performance.

Cheng et al. [239] studied the effects of synthesis parameters on the properties of a Ce-ZSM-5 zeolite membrane grown onto porous α-alumina supports. Under optimum synthesis conditions, the permeances for H_2, N_2, CH_4, and CO_2 were 4.73×10^{-6}, 1.56×10^{-6}, 1.51×10^{-6}, and 1.36×10^{-6} mol m^{-2} s^{-1} Pa^{-1}, respectively; the ideal selectivity of the membrane at room temperature for H_2/CO_2, H_2/N_2, and H_2/CH_4 were 3.13, 3.03, 3.48, respectively.

3.2.3.8 FAU-Type Zeolite

The FAU-type zeolite has relatively large pores, which are composed of 12-membered oxygen rings of approximately 0.74 nm in diameter. Thus, FAU-type zeolite membranes do not show strict molecular-sieving properties unlike the MFI-types, and separation by FAU-type zeolite membranes is usually achieved by differences in the adsorptivities of permeates. The larger pores of FAU-type zeolite membranes are, however, beneficial for higher permeation rates compared to the MFI- and LTA-type zeolite.

The FAU-type zeolite includes X- and Y-types, Si/Al ratios of which are 1–1.15 and 1.5–3.0, respectively; the number of cations that can be coordinated in the X-type structure is larger than that in the Y-type structure. It has been reported that the CO_2/N_2 reaches a maximum of approximately 100 for the NaY-type zeolite membrane on porous alumina at a permeation temperature of 308 K [240]. Hasegawa et al. [240] fabricated hydrothermally FAU-type zeolite membranes with different Si/Al ratios on the outer surface of a porous α-alumina support tube. The membranes were ion-exchanged with Rb^+ and K^+ ions, and their permeation properties were investigated by using equimolar mixtures of CO_2/CH_4 and CO_2/N_2 at 308 K. Permeances for single component system at a temperature of 308 K were in the order of $CO_2 > N_2 > CH_4$, and decreased with decreasing Si/Al ratio. Permeances and selectivities of CO_2 for the CO_2/CH_4 system were approximately half the values for the CO_2/N_2 system. The NaX-type zeolite containing Si/Al ratio 1.26 showed the maximum CO_2 selectivities, which were 28 for the CO_2/CH_4 system and 78 for CO_2/N_2 system, respectively. The effect of the ion-exchange was the highest for the NaY-type zeolite membrane. The CO_2 separation ability of the NaY-type zeolite membrane was further improved by ion exchange with K^+. The CO_2/CH_4 selectivity of the KY-type zeolite membrane was in the range of 25–40, and CO_2 permeance was in the range of $(7.5–9.0) \times 10^{-7}$ mol m^{-2} s^{-1} Pa^{-1}.

Gu et al. [241] studied the effects of seeding methods, synthesis gel chemistry, and conditions of hydrothermal treatment on the FAU membrane quality in terms of the crystal phase purity and CO_2/N_2 separation performance. The membrane exhibited a CO_2 separation factor of 31.2 at room temperature for the equimolecular CO_2/N_2 dry mixture. The addition of water vapor to the mixture was found to significantly increase the selectivity for CO_2 above 110 °C, but decreased the CO_2 permeance. Kumar et al. [242] reported the removal of H_2S from mixtures containing N_2, CO_2, and CO by using zeolite X and Y in sodium form and ion-exchanged with silver and copper cations. Ag and Cu exchanged zeolite demonstrated very high H_2S adsorption capacities at room temperatures and in H_2S–He mixtures at 150 °C.

3.2.3.9 Hydroxy-Sodalite Zeolite Membrane (HDS-zeolite)

Hydroxy-sodalite has the same framework structure as sodalite and consists of the cubic array of beta-cages. It has a six-membered ring aperture with a pore size of 2.8 Å. Its pore size is smaller than that of the zeolites with an eight-membered ring aperture, e.g., NaA zeolite. Only small molecules, such as helium and hydrogen, can enter into the pore of HDS-zeolite. Thus, HDS-zeolite membrane has a better performance on the separation of small molecules from gas mixtures than the zeolite membranes with a bigger pore size. Julbe et al. [243] synthesized a sodalite/ α-Al_2O_3 composite membrane by microwave heating (MAHS), which has a low He/ N_2 permselectivity of 6.2 at 115 °C. Xu et al. [244] further explored the MAHS method for the synthesis of HDS membranes. They synthesized a high-quality pure HDS-zeolite membrane on an alpha-alumina support by a novel MAHS method. The pure HDS zeolite membrane was found to be well inter-grown and the thickness of the membrane was 6–7 μm. Gas permeation results showed that the hydrogen/*n*-butane permselectivity of the HDS membrane was larger than 1,000. The authors claimed that the HDS zeolite membrane is a promising candidate for the separation of hydrogen from gas mixtures and important for the emerging hydrogen energy fuel system.

3.2.3.10 Zeolite T

Zeolite T with a Si/Al ratio of 3–4 is less hydrophilic than zeolite Y with a Si/Al ratio of less than 3. Zeolite T is an intergrowth-type zeolite of erionite and offretite, of which the pore sizes are 0.36×0.51 nm and 0.67×0.68 nm, respectively [245]. Zeolite T has both hydrophilic and fairly high acid resistant properties because of their proper Si/Al ratio. Chen et al. [246] described a new seeding method, namely, varying temperature hot-dip coating (VTHDC) for synthesis of zeolite T membrane. The authors revealed that the method was flexible and effective for combined control over the seed suspension concentration, seed size, and coating temperature, leading to better control of the seed layer over the seed size, thickness, and coverage defect.

Cui et al. [245] synthesized the T-zeolite membrane by hydrothermal synthesis on porous mullite tubes seeded with zeolite T crystals. Single gas and mixed-gas permeation experiments through zeolite-T membranes were carried out by a vacuum method at 303–473 K using He, H_2, CO_2, O_2, N_2, CH_4, C_2H_6, and C_3H_8 single component gases and CO_2/N_2, CO_2/CH_4, and other CO_2/hydrocarbon mixtures, respectively. In single-gas permeation experiments, with increasing kinetic diameters from 0.33 nm of CO_2 to 0.43 nm of C_3H_8, the gas permeation rate decreased by four orders. Permeance of CO_2 was much higher than those of N_2 and CH_4 and the ideal selectivities for CO_2/N_2 and CO_2/CH_4 were 31 and 266 at 343 K, respectively. In mixed-gas permeation, zeolite T membranes showed high selectivities for CO_2/ N_2 and CO_2/CH_4 pairs of 107 and 400, respectively, at 308 K. The selectivity α decreased with an increase in temperature, but even at 473 K w still at high levels of 20 and 52 for CO_2/N_2 and CO_2/CH_4, respectively. This is due to the synergetic effects of competitive adsorption of CO_2 and molecular sieving of zeolitic pores. Because of the increasing effect of single file diffusion, the selectivities for CO_2/ C_2H_6 (α=61) and CO_2/C_3H_8 (α=17) were rather low. Zeolite T showed excellent CO_2 separation performance for CO_2/N_2, and CO_2/CH_4 systems even at a high temperature of 473 K, mainly due to the molecular sieving effect.

3.2.3.11 Zeolite L

The crystal structure of the synthetic zeolite, Linde L, $K_6Na_3Al_9Si_{27}O_{72}\cdot 21H_2O$, was determined by Barrer and Villger [247]. The zeolite L is hexagonal with uni-cell dimensions $a = 18.4$ Å and $c = 7.5$ Å. The minimum constricting aperture is defined by a ring of 12 tetrahedral atoms (T atoms, e.g., Si, Al) that forms an opening of 0.71 nm [248]. Zeolite L crystals contain a one-dimensional pore with an opening of 0.71 nm, which runs along its *c*-axis. The channel contains cationic sites, which can strongly interact with negatively charged or polarized molecules. Zeolite L can separate gas species by selective adsorption in addition to size discrimination. Zeolite L can be synthesized without the aid of an organic template, eliminating the strong agglomeration of the nano-sized zeolite particles at high temperature calcinations, which is beneficial to the homogeneous distribution of zeolite particles in the carbon matrix [249]. Yin et al. [249] fabricated thin zeolite L/carbon nanocomposite membranes. The results of CO_2 adsorption isotherms indicated that the zeolite L/carbon composite materials had greater CO_2 sorption ability than the pure carbon materials; therefore, the composite membranes can exhibit higher gas permeance and selectivities of CO_2/CH_4 and CO_2/N_2 than the pure carbon membranes. At 298 K, CO_2/CH_4 and CO_2/N_2 separation factors reached 43.59 and 27.21, respectively in experiments with gas mixtures; these results are higher than the 35.75 and 20.43 attained in the single gas permeation experiments.

3.2.3.12 ITQ-29 zeolite

Corma et al. [250] first introduced ITQ-29 as a zeolite with the same topological structure as zeolite A, but a much higher Si/Al ratio (up to infinity, i.e., pure silica), using a bulky organic template obtained from the self-assembly of two identical organic cationic moieties through π–π type interactions. Casado-Coterillo et al. [251] prepared ITQ-29 crystals and used in ITQ-29/polysulfone mixed-matrix-membranes for gas separation. The molar composition of the synthesis gels was:

$(1-x)SiO_2{:}xGeO_2{:}0.25$ ROH (4-methyl-2,3,6,7-tetrahydro-1H,5H-pyrido(3.2.1-ij) quinolium hydroxide):0.25 TMAOH (tetramethylammonium hydroxide):0.5 HF:yH_2O, where $x = 0$–0.05 and $y = 3$–12.

These membranes were used for the gas separation of H_2/CH_4 mixtures and showed promising results (highest H_2 permeability of 21.9 Barrer and a separation factor of 118 for the 4 wt% ITQ-29/polysulfone membrane) were obtained.

Pure-silica ITQ-29 is a hydrophobic small pore zeolite, which gives ITQ-29 the possibility of sieving and processing small organic molecules with high precision, even in the presence of water or other polar molecules. Al-free ITQ-29 was tested for N_2, CH_4, and propane, but the selectivity and the permeability were not very high [252]. ITQ-29 membranes prepared by using Kryptofix 222 as SDA (structure directing agent) and activated in situ in presence of oxygen at 300 °C, gave a separation factor of 127 for H_2/C_3H_8 separation [253].

3.2.3.13 UZM Zeolites

Moiscoso et al. [254] and Blackwell et al. [255] introduced a new family of crystalline alumino-silicate zeolites, named UZM. These zeolites are represented by the following empirical formula:

$$\mathbf{M}_m^{n+}\mathbf{R}_r^{p+}\mathbf{Al}_{(1-x)}\mathbf{E}_x\mathbf{Si}_y\mathbf{O}_z$$

where M is an alkali or alkaline earth metal such as lithium and strontium, R is nitrogen containing organic cation such as tetramethyl ammonium, and E is a framework element such as gallium.

Recently Liu et al. [256] patented two methods for the preparation of small pore microporous UZM-5 zeolite membrane: (1) in situ crystallization of one or more layers of UZM-5 zeolite crystals on a porous membrane support; and (2) a seeding method by in situ crystallization of a continuous second layer of UZM-5 zeolite crystals on a seed layer supported on a porous membrane support. The membrane in the form of discs, tubes, or hollow fibers had superior thermal and chemical stability, good erosion resistance, high CO_2 plasticization resistance, and significantly improved selectivity over polymer membranes for gas, vapor, and liquid separation. Liu et al. [256] claimed that the microporous UZM-5 zeolite membranes are useful for liquid separation such as deep desulfurization of gasoline and diesel fuels, ethanol/water separation, and pervaporation dehydration of aqueous/organic mixtures; these membranes are also useful for a variety of gas and vapor separations such as CO_2/CH_4, CO_2/N_2, H_2/CH_4, and O_2/N_2. They can also be used for separating olefin/paraffins such as propylene/propane and iso/normal paraffins, polar molecules such as H_2O, and H_2S, and NH_3 mixtures with CH_4, N_2, H_2, and other light gases.

3.2.3.14 Zeolite W

Zeolite W is a synthetic zeolite that has the same framework topology as the mineral merlinoite (MER). It has an eight-membered ring (8MR); the channel dimensions are 0.31×0.35 nm, 0.27×0.36 nm, and 0.51×0.34 nm [257]. Membranes consisting of W-type zeolite have a unique nanopore structure which is stable at low pH. Mohammdi and Maghsoodloorad [258] fabricated W-type zeolite membranes on the surface of α-alumina (porous) via grown hydrothermally. Studies showed that maximum gas permeation flux (minimum ideal selectivity for O_2/SF_6 gas separation) was obtained for single layer membranes synthesized at 200 °C for 6 h. With changing synthesis temperature, synthesis time and number of layers, and keeping other effective parameters constant, maximum ideal selectivity of 20.1 for O_2/SF_6 gas separation was achieved using double layer zeolite membrane synthesized via $Al_2O_3:SiO_2:K_2O:H_2O = 1.0:6.4:5.6:164.6$ gel formula over the flat support at 185 °C for 18 h.

3.2.3.15 Zeolitic Imidazole Frameworks (ZIFs)

Zeolitic imidazolate frameworks (ZIFs) are three-dimensional structures consisting of rigid MN_4 tetrahedra (M = metal ion), linked through bridging imidazolate (Im) $C_3H_3N_2^-$ anions. Among numerous MOFs, zeolitic imidazole frameworks have drawn significant attention due to their superior chemical and thermal stability. They are very promising material to be used in gas storage and separation processes. This zeolite has sodalite topology with a six-member ring and is constructed by connecting zinc metal clusters with benzimidazole linker. The size of ZIF-7 pore openings is estimated to be 0.30 nm. Several ZIFs can be synthesized easily and fast and at low cost, such as ZIF-8, ZIF-90, and ZIF-7.

Several studies have investigated ZIFs, such as ZIF-8, ZIF-90, and ZIF-7, as porous fillers with various polymers, including polysulfone, Matrimid, PBI, and polyimide for MMMs for gas separation and pervaporation. Almost every report showed improved permeability, sometimes coupled with enhanced selectivity for gas separation [259]. Bae et al. [260] synthesized ZIF-90 particles and used them to fabricate nanocomposite membranes with three different poly(imide)s (Ultem, Matrimid, and 6FDA-DAM). The imidazole linker in ZIF-90 contains a carbonyl group which has a favorable chemical noncovalent interaction with CO_2. The ZIF-90 crystals showed excellent adhesion with the poly(imide)s without any surface-compatibilization procedures. Interfacial voids were absent, and the MOF crystals were well dispersed. Figure 3.46 shows the pure-component CO_2 and CH_4 gas-transport properties of MMMs containing 15 wt% of ZIF-90 crystals. Ultem and Matrimid MMMs showed significantly enhanced CO_2 permeability without any loss of CO_2–CH_4 selectivity. MMMs fabricated with 6FDA-DAM (a highly permeable polymer) showed substantial enhancement in both CO_2 permeability and CO_2/CH_4 selectivity, suggesting that the membrane is defect free and that permeabilities of the MOF and the polymer are well matched (Fig. 3.49).

Bux et al. [261] fabricated gas separating ZIF membranes, selective for H_2 over other gases via a novel microwave-assisted solvothermal process. The membrane achieved a fine balance between flux and selectivity compared to other MOF membranes. The hydrogen permeance of the novel MOF membrane with a relatively thick ZIF-8 layer was around 50 % of the hydrogen permeances of zeolite membranes of the same selectivity. Caro's group [262] fabricated an ultramicroporous imidazolate framework, ZIF-7 molecular sieve membrane on porous alumina support and tested for H_2 gas separation. ZIF-7 has many advantages for hydrogen separation: (1) its pore dimension approaches the size of H_2, and therefore, a high H_2 selectivity could be obtained without any sophisticated pore-size engineering as is essential for zeolite membranes targeting H_2/CO_2 separation; (2) it is thermally stable for use at elevated temperatures (ZIF-7 is stable at least up to 500 °C in air); and (3) its hydrophobic property endows it with very high hydrothermal stability.

ZIF-69 membranes were also used for the separation of gases. Liu et al. [263] studied single gas permeation through ZIF-69 membranes by a vacuum method at room temperature using H_2, CH_4, CO, CO_2, and SF_6. The permeances were in the order of $H_2 > CO_2 > CH_4 > CO > SF_6$. The permselectivity of the CO_2/CO gas mixture

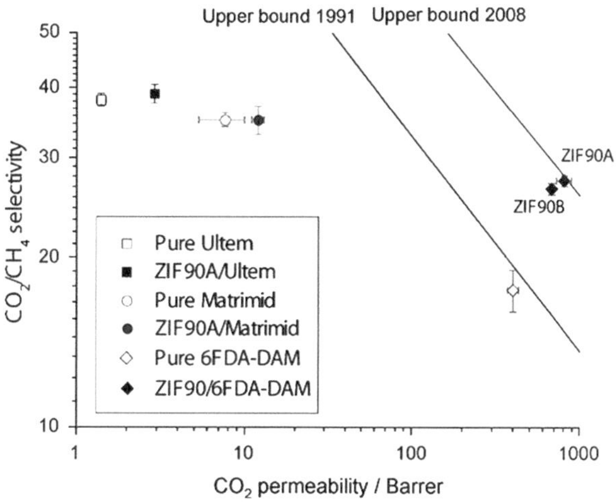

Fig. 3.49 Gas-permeation properties of mixed-matrix membranes containing 15 wt% of ZIF-90 crystals obtained from experiments with pure gases. (Measurements were performed at 35 °C and 4.5 atm upstream pressure for Ultem and Matrimid membranes, and at 25 °C and 2 atm upstream pressure for 6FDA-DAM membranes. The data for pure Ultem and Matrimid are averaged values from the literature. The upper bounds for polymer membrane performance as defined in 1991 and 2008 are shown [260])

at room temperature was 3.5 ± 0.1 with CO_2 permeance of $3.6 \pm 0.3 \times 10^{-8}$ mol m^{-2} s^{-1} Pa^{-1}. Venna and Carreon [264] synthesized thin ZIF-8 membranes with around 5–9 μm thicknesses by secondary seeded growth on tubular alpha-Al_2O_3 porous supports. The separation performance of these membranes for equimolecular CO_2/CH_4 gas mixtures was demonstrated. The membrane displayed unprecedented CO_2 permeance as high as ~2.4×10^{-5} mol m^{-2} s^{-1} Pa^{-1} with CO_2/CH_4 selectivities of ~4–7 and separation indexes π $\left\{\pi = \left(P_{CO_2} \times (\text{selectivity} - 1)\right) \times \text{Permeation pressure}\right\}$ in the ~6.5–10 range at 295 K, and a feed pressure of 139.5 KPa.

3.2.3.16 Other Zeolitic Type or Ceramic/Inorganic Membranes

Sandström et al. [265] used an MFI membrane comprising an approximately 0.7 μm silicate-film on a fully open and graded porous alumina support to separate CO_2/H_2, CO_2/CO/H_2, and CO_2/CH_4 mixtures at high pressures. CO_2 fluxes up to 657 kg m^{-2} h^{-1} were observed, which are many times larger than those previously reported for any zeolite membrane. The very high fluxes were a result of low film thickness, open support, high pressure, relatively high diffusivity, molecular weight and pressure drop. The maximum CO_2/H_2 ideal selectivity was 32.1, which was observed at 1,000 kPa feed pressure, a permeate pressure of 200 kPa and a temperature of 275 K. The highest measured CO_2 permeance for the binary CO_2/H_2 mixture was 93×10^{-7} mol m^{-2} s^{-1} Pa^{-1}. The membrane was also CO_2 selective for a CO_2/CO/H_2 mixture. However, both the

CO_2 flux and the CO_2/H_2 separation factor were reduced slightly in the presence of CO, probably as a result of competing adsorption between CO and CO_2. The highest measured CO_2/CH_4 separation factor was 4.5.

Nair et al. [266] discussed the results on the separation of close boiling point hydrocarbon mixtures by means of zeolite membranes. In the case of silicate membranes (MFI), the selectivity was found to depend on the microstructure. Permeation of xylene isomers through the silicate membranes occurred through both zeolitic and non-zeolitic (intercrystalline) nanopores. The faujasite membranes were found to have high selectivities (40–150) for the separation of binary mixtures containing one aromatic component, and modest selectivities (4–9) for the separation of unsaturates from saturated low-molecular-weight hydrocarbons.

Choi et al. [116] synthesized composites of polybenzimidazole (PBI) with proton-exchanged AMH-3 (a silicate with three-dimensional microporous layers) and swollen AMH-3. Proton-exchanged AMH-3 was prepared under mild conditions by the ion exchange of Sr and Na cations in the original AMH-3 using aqueous solution of DL-histidine. Swollen AMH-3 was prepared by sequential intercalation of dodecylamine following the ion exchange in the presence of DL-histidine. Both silicate materials were introduced into a continuous phase of PBI as a selective phase. Mixed matrix nano-composite membranes, prepared under certain casting conditions, with only 3 wt% of swollen AMH-3 presented a substantial increase of hydrogen/carbon dioxide ideal selectivity at 35 °C, i.e., by a factor of more than 2 compared to pure PBI membranes (40 vs. 15). Similar ideal selectivity was observed using higher loadings (e.g., 14 %) of proton-exchanged AMH-3 particles, suggesting that transport of hydrogen is faster than carbon dioxide in AMH-3-derived silicates. However, the ideal selectivity of MMMs approaches that of a pure polymer as the operating temperature increases to 100 °C and 200 °C.

By using the stepwise deposition method, Bétard et al. [267] fabricated membranes from MOF **1** and **2**, using macroporous alumina and titania as supports. SEM images (Fig. 3.50) reveal that the MOF crystallites have grown on the support surface and also up to 30 μm deep inside the support pores, where they form a foam-like, lamellar structure.

Figure 3.51 shows permeances from MOF **1** in the range of 10^{-8} mol m^{-2} s^{-1} Pa^{-1} for both CO_2 and CH_4 which moderately increase with increasing pressure. Particularly at lower pressure, obvious reductions of mixed gas permeances occur compared to the pure gas.

Figure 3.52 shows the permeances of pure and equimolar mixed CO_2 and CH_4 measured for the $[Cu_2(BME\text{-}bdc)_2(dabco)]_n$.

The CO_2 and CH_4 permeance of the membrane **2** (Fig. 3.52) are of the same order of magnitude as the permeances of the membrane **1**. It should be noted that the selectivity is less than unity for membrane 1, while it is more than unity for membrane 2. Bennett et al. [268] studied the ZIF-4, a metal organic framework (MOF) with a zeolitic structure, and suggested an avenue for designing broad new families of amorphous and glasslike materials that exploit the chemical and structural diversity of MOFs. Microporous metal organic framework (MMOF) membranes on porous alumina supports, synthesized by the seeded growth method, showed moderate ideal selectivity for H_2/N_2 [269].

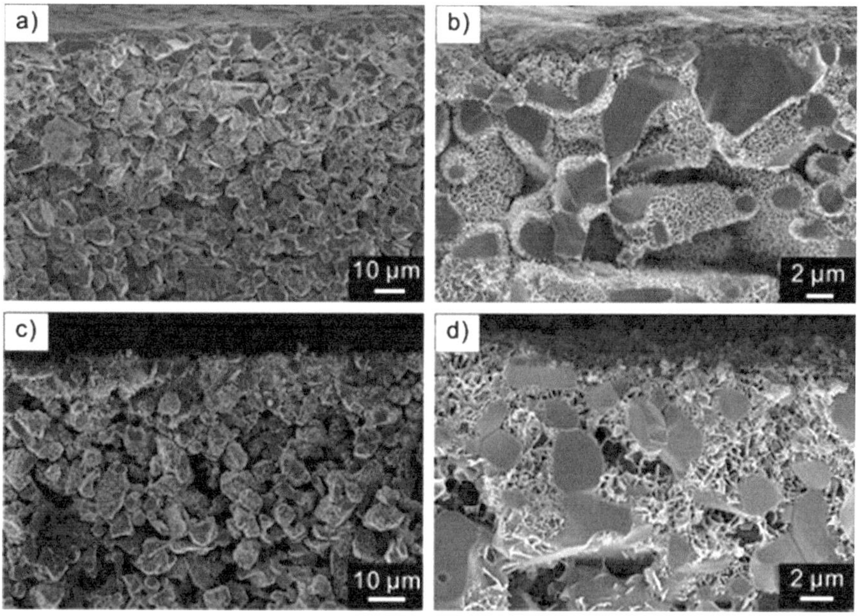

Fig. 3.50 Representative examples of SEM micrographs of the cross-sections of $[Cu_2(ndc)_2(dabco)]_n$. (**1**) On alumina (**a**, **b**) and $[Cu_2(BME\text{-}bdc)_2(dabco)]_n$ (**2**) on titania substrates (**c**, **d**) at different magnifications. (The morphology of the resulting MOF crystals seems to be independent of the support type)

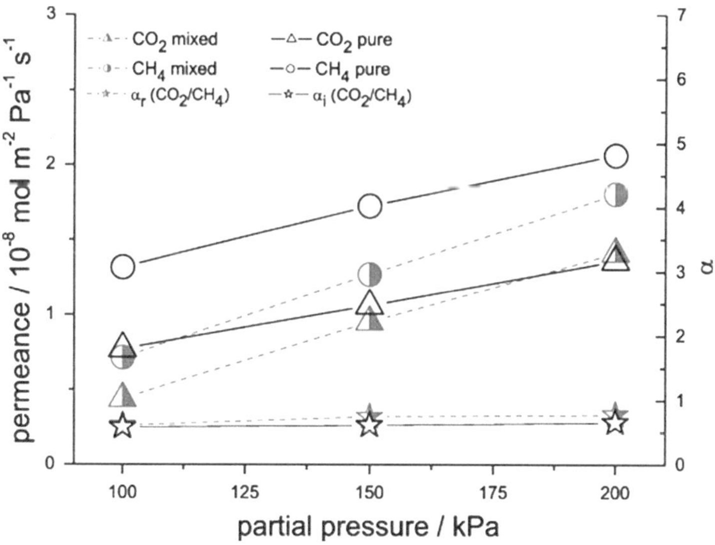

Fig. 3.51 Permeances of pure and equimolar mixed CO_2 and CH_4 measured for the $[Cu_2(ndc)_2(dabco)]_n$. (**1**) Membrane at room temperature ($T = 298$ K) as function of pressures at the feed side (total pressures for pure gases, partial pressures for the gas mixture). The ideal and mixed gas separation factor α_i and α_r were calculated from the corresponding ratio of the CO_2/CH_4 permeances

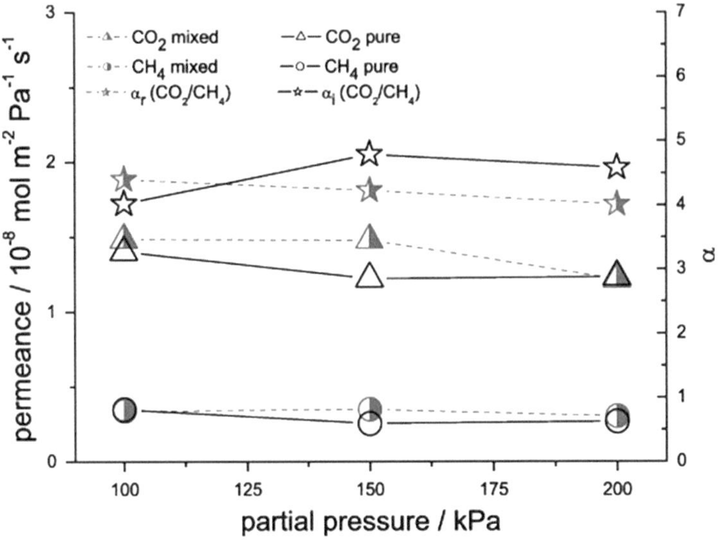

Fig. 3.52 Permeances of pure and equimolar mixed CO_2 and CH_4 measured for the $[Cu_2(BME\text{-}bdc)_2(dabco)]_n$. (**2**) Membrane at room temperature ($T = 298$ K) as function of pressures at the feed side (total pressures for pure gases, partial pressures for the gas mixture). (The ideal and mixed gas separation factor α_i and α_r were calculated from the corresponding ratio of the CO_2/CH_4 permeances)

Bohrman and Carreon [270] used metal–adeninate biometal organic frameworks (Bio-MOFs) for the separation of a CO_2/CH_4 mixture. Bio-MOFs were introduced by Rosi's group [271, 272]. Bio-MOFs have permanent microporosity, high surface areas, chemical stability, and exceptional CO_2 adsorption capacities due to the presence of basic bio-molecule building units. In particular, $Zn_8(ad)_4(BPDC)_6O{\cdot}2Me_2NH_2$ (where ad = adeninate, BPDC = biphenyldicarboxylate, denoted as Bio-MOF-1), a three-dimensional metal organic framework with infinite-adeninate columnar secondary building units (SBUs) which are interconnected via biphenyldicarboxylate linkers, is an attractive material with great potential for CO_2 separation if prepared in a membrane form. These membranes displayed high CO_2 permeances and separation ability for CO_2/CH_4 gas mixtures. The observed CO_2/CH_4 selectivities above the Knudsen selectivity indicated the separation was promoted by preferential CO_2 adsorption over CH_4. The preferential CO_2 adsorption was attributed to the presence of adeninate amino basic sites present in the Bio-MOF-1 structure. However, selectivity was slightly below in Robeson plot for CO_2/CH_4 mixtures [3]. Table 3.22 shows the CO_2/CH_4 separation performance of the stainless-supported Bio-MOF-1 membranes.

Xomeritakis et al. [273] demonstrated a novel and efficient method for molecular engineering of the pore size and porosity of microporous sol–gel silica membranes. The addition of a suitable organic template (e.g., tetraethyl- or tetrapropylammonium bromide) in polymeric silica sols resulted in pores in the range of 5–6 Å.

Table 3.22 CO_2/CH_4 separation performance of Bio-MOF-1 membranes at a pressure drop of 138 kPa and 298 K [271, 272]

Permeance mol m^{-2} s^{-1} Pa^{-1} ($\times10^{-7}$)			
Membrane[a]	CO_2	CH_4	CO_2/CH_4 selectivity
M1 (3)	11.5	4.6	2.5
M2 (3)	11.9	4.6	2.1
M3 (4)	10.5	4.8	2.6
M4 (7)	5.8	4.7	–

[a]Number in parentheses indicate number of layers

In general, without any modification, the pore size of the synthetic zeolites will be in the range of 3–4 Å. With pore sizes of 5–6 Å, sol–gel silica membranes are useful for hydrocarbon isomer separations. The templated membranes exhibit as high as 10^{-7}–10^{-6} mol m^{-2} s^{-1} Pa^{-1} for molecules with $d_k < 4.0$ Å (e.g., CO_2, N_2, CH_4), coupled with single-component selectivities of 100–1,800 for N_2/SF_6, 20–40 for *n*-butane/*iso*-butane, and 10–20 for *para*-xylene/*ortho*-xylene. The unique features of this new approach include simple, fast and scaleable processing under ambient conditions, and demonstrate the possibility to tune membrane pore size and porosity by proper choice of the type and amount of template.

Dual-layer polyethersulfone (PES)-zeolite beta/BTDA-TDI/MDI co-polyimide (P84) composite hollow fibers, were applied to fabricate dual layer nanocomposite hollow fiber membranes through pyrolysis by Li and Chung [274]. After pyrolysis at 800 °C, these nanocomposite hollow fibers exhibited a significantly enhanced O_2/N_2 and CO_2/CH_4 selectivity of 11.3 and 152, respectively in the pure gas measurement. It was also noticed a comparable CO_2/CH_4 selectivity of 140 in the mixed gas measurement. The authors claimed that these dual-layer hollow fibers are a potential type of excellent membrane material for oxygen enrichment and natural gas separation in industrial applications.

Zeolite membranes were synthesized by the dry gel method, using a tubular support of stainless steel by Alfaro and Valenzuala [275]. The composition of the precursor gel was $0.22Na_2O$:$10SiO_2$:$280H_2O$:0.5TPABr (tetrapropylammonium bromide). It was observed that the amount of each hydrocarbon that permeated from the mixture of hydrocarbons was as follows: $n\text{-}C_4H_{10} > i\text{-}C_4H_{10} > C_3H_8 > C_2H_6 > CH_4$. The separation factor measured for the N_2/SF_6 ratio was five times higher than the theoretical one. This improved selectivity to N_2 was explained in terms of properties such as the pore size of the membrane, controlling Knudsen-type diffusion. In hydrocarbon separation of natural gas, *n*-butane showed the higher concentration in the permeated side of the membrane.

Kuznicki [276] synthesized a new class of molecular sieve materials, which are generally denoted as Engelhard titano-silicates (ETS). Unlike zeolites their frameworks are built by corner sharing SiO_4 tetrahedra and TiO_6 octahedra, resulting in new structures that cannot be built by connecting only tetrahedral units, the framework units of zeolites. The ETS materials contain a channel system like zeolites which, in principle, enables them to be used in similar molecular sieve and/or adsorptive separations. Membranes consisting of thin intergrown layer of Na-ETS-4 on porous titania supports were highly water permselective with selectivities as high as 400 and a corresponding water flux of 0.01 mol m^{-2} s^{-1} in room temperature pervaporation experiments using 1:1 water/ethanol mixtures. With increasing water content in the feed solution (in the range of 10–90 %) the water flux was increasing linearly, while the selectivity did not vary significantly. The selectivity of the ETS-4 membranes was similar to the highest reported for Na-X and Na-Y membranes. These Na-ETS-4 membranes may find applications in pervaporation as well as separation of permanent gases.

Stoeger et al. [277] fabricated highly c-oriented, intergrown, continuous crystalline alumino-phosphate $AlPO_4$-5 and CoAlPO-5 (both of the AFI framework type) films, grown on porous α-alumina by the seeded method. The membrane quality was inspected through pervaporation measurements consisting of a liquid hydrocarbon feed of *n*-heptane and 1,3,5-triisopropylbenzene. It was observed that the separation factor was 2.8 with a corresponding flux of 1.2 kg m^{-2} h^{-1}. However, further investigation is needed, focusing on growth, calcination, and microstructure optimization.

MER type zeolite was investigated by Nagase et al. [278]. The micropore structure of MER-type zeolite comprises double-eight rings and γ-cage, and pore diameter is 0.27–0.51 nm. The MER (merlinoite) zeolite membranes are relatively acid tolerant in the low-silica 8MR zeolite group, and the low-SAR type. MER membranes exhibit high water selectivity for pervaporation of 90 % acetic acid solution. The separation factor of the membrane is as high as that of the LTA-type zeolite membrane (more than 5,000). However, the dehydration performance and permeation mechanism of the membrane are not clear. Hasegawa et al. [279] determined the stabilities and dehydration performances of MER-type zeolite membranes prepared on porous α-alumina tubes by the secondary growth of seed crystals ($SiO_2/Al_2O_3=4.7$). The membranes showed relatively high stability, permeability, and separation performances. The permeation flux and separation factors were 1.9 kg m^{-2} h^{-1} and 9,300, respectively, for an equimolar mixture of ethanol and water at 350 K. Membranes were also used for the dehydration of several organic solvents (methanol, *n*-propanol, *i*-propanol, and acetone) containing water. It was noticed that the separation factor increased with the molecular diameter of the organic solvents.

Kim et al. [280] investigated modified MFI-type zeolite membranes as high-temperature water-gas shift (WGS) membrane reactors (MRs) using nanocrystalline Fe/Ce WGS catalyst. The effects of the MR operating conditions and the membrane separation performance on the CO conversion (χ_{co}) were studied experimentally and by calculations using an ample one-dimensional plug-flow reactor (PFR) model. The model calculations indicated that the membrane had the potential to achieve high CO conversion of $\chi_{co}>99$ % under practical operating conditions. Due to its excellent hydrothermal stability and chemical resistance, the modified MFI-type zeolite membranes are potentially useful for constructing MR for high temperature WGS reaction of coal-derived syngas.

Recently, UK researchers have developed a porous material named NOTT-202 that can soak up CO_2 from the atmosphere. NOTT-202 is a "metal–organic framework" that works like a sponge, absorbing a number of gases at high pressures. But as the pressure is reduced, CO_2 is retained as other gases are released [281]. Solvothermal reaction of H_4L (biphenyl-3,3′,5,5′-tetra(phenyl-4-carboxylic acid)) (Fig. 3.50) with $In(NO)_3$ in an acidic (HNO_3) mixture of CH_3CN/DMF(dimethylformamide) (v/v = 1:2) at 90 °C affords the solvated framework complex $Me_2(NH_2)_{1.75}$ $[In(L)]_{1.75}(DMF)_{12}(H_2O)_{10}$ (NOTT-202). The counter-cation $Me_2NH_2^+$ is generated by in situ decomposition of the DMF solvent during the reaction.

The unique partially interpenetrated metal–organic framework (MOF) NOTT-202 represents a new class of dynamic material that undergoes pronounced framework phase transition on desolvation. NOTT-202 consists of two MOF networks attached to a central indium metal atom and overlaid in such a way as to leave gaps where the carbon dioxide is stored. This discovery holds promise for carbon dioxide capture and storage, or even for removing CO_2 from the exhaust gases of power plants and factories. However, there are some drawbacks to the material, i.e., its stability in the presence of high-temperature water vapor is questionable. Large-scale production of this type of material has long been considered a major challenge [282, 283]. Table 3.23 shows the gas permeation/separations by using different types of zeolites.

3.3 Metal–Organic Framework Membranes for Gas Separations

Metal–organic frameworks (MOFs) are a relatively new class of hybrid materials consisting of organic and inorganic moieties in crystalline lattice. Pore size tailorability combined with tunable sorption behavior provides promising avenues for applications of MOFs as membranes for gas separation applications. Synthesis conditions are less energy intensive as compared to zeolites. For instance, most MOFs do not require high-temperature/pressure conditions for their fabrications and can be synthesized using click chemistry. Also, unlike zeolites, structure-directing agents are not required; therefore, a subsequent calcination step is not necessary [284]. This new class of porous material is attracting attention due to demonstration of their large pore sizes, high apparent surface areas, and selective uptake of small molecules. Most important is that their synthesis from molecular building blocks holds the potential for directed tailoring of these properties [285]. MOFs consist of metal–oxygen polyhedral, interconnected with a variety of organic linker molecules, resulting in tailored nanoporous materials. With a judicious choice of organic linker groups, it is possible to fine-tune size, shape, and chemical functionality of the cavities and the internal surfaces. This unique structural feature offers unprecedented opportunities in small-molecule separation as well as chiral separations [286]. MOF membranes are polycrystalline in nature just like zeolite membranes.

Table 3.23 Gas permeation/separation by using different types of zeolites

Permeation/separation of gases	Reference
LTA zeolite	
H_2, CO_2, N_2, CH_4, C_3H_8, H_2/CO_2	[205]
H_2, CO_2, N_2, CH_4, C_3H_8	[206]
Zeolite NaA, Faujasite	
H_2, O_2, N_2, $H_2/n\text{-}C_4H_{10}$	[207, 208]
He, H_2, CO_2, N_2, CH_4, O_2, C_3H_8, H_2/N_2	[209, 210]
CO_2, N_2, H_2	[211]
DDR type zeolite	
H_2, CO_2, N_2, CH_4, O_2, CO, propane, propylene	[213]
CO_2, N_2, and CH_4 and their mixtures	[214, 215]
He, H_2, CO_2, N_2, CH_4, O_2, C_3H_8, *n*-butane, *i*-butane, and sulfur hexafluoride	[216]
He, H_2, CO, CO_2	[217]
CO_2, CH_4, He, H_2, O_2, and N_2, and CO_2/CH_4	[219]
SAPO-34	
CO_2/H_2, H_2/CH_4	[223–228]
AlPO-18	
CO_2/CH_4	[229]
Beta zeolite or ZSM Zeolite (MFI Zeolite Membranes (ZSM-5)	
CO_2	[231]
He, N_2, $n\text{-}C_4H_{10}/iso\text{-}C_4H_{10}$	[232]
H_2, CO_2	[233]
H_2/CH_4, H_2/CO_2	[234]
H_2/SF_6	[236]
$i\text{-}C_4H_{10}$	[236]
$n\text{-}C_4H_{10}/i\text{-}C_4H_{10}$	[237]
H_2/CO_2	[238]
H_2, N_2, CH_4, CO_2	[239]
FAU-type zeolite	
CO_2/N_2, CO_2/CH_4	[240]
CO_2/N_2	[241]
N_2, CO_2, CO, H_2S/He	[242]
Hydroxy-sodalite zeolite membrane (HDS-zeolite)	
He, H_2, N_2	[243]
H_2/*n*-butane	[244]
T-Zeolite	
He, H_2, CO_2, O_2, N_2, CH_4, C_2H_6, C_3H_8, CO_2/N_2, CO_2/CH_4	[245]
L Zeolite	
CO_2/CH_4 and CO_2/N_2	[249]
ITQ-29 Zeolite	
H_2/CH_4	[251]
N_2, CH_4,propane	[252]
H_2/C_3H_8	[253]

(continued)

Table 3.23 (continued)

Permeation/separation of gases	Reference
UZM Zeolites	
CO_2/CH_4, CO_2/N_2, H_2/CH_4, O_2/N_2, olefin/paraffin such as propylene/propane, iso/normal paraffin, polar molecules such as H_2O, H_2S, and NH_3 mixtures with CH_4, N_2, H_2, and other light gases	[256]
W-Type zeolite	
O_2/SF_6	[258]
Zeolitic imidazole frameworks (ZIFs)	
CO_2, CO_2/CH_4	[260]
H_2 over other gases	[261, 262]
H_2, CH_4, CO, CO_2, SF_6	[263]
CO_2/CH_4	[264]
Other zeolitic type or ceramic/inorganic membranes	
CO_2/H_2, $CO_2/CO/H_2$, CO_2/CH_4	[265, 267]
CO_2/CH_4	[270]
CO_2, N_2, CH_4	[273]
CO_2/CH_4	[274]
CH_4, C_2H_6, C_3H_8, n-C_4H_{10}	[275]

The first MOF membranes were reported in 2009 by the Lai and Jeong groups [286, 287]. Like zeolite, fabrication of thin films of crystalline framework materials follows one of two approaches, in situ growth and secondary or seeded growth [197]:

1. In situ growth: substrate is immersed in the growth solution without any crystals previously attached to the surface, and during the fabrication, nucleation, growth, and intergrowth of crystals on the substrate will happen.
2. Secondary or seeded growth: refers to film growth from preattached seed crystals. The advantage of this method is to make tailored membrane.

Other methods for the fabrication of MOF membranes have been reported, such as chemical modification of the support surfaces with self-assembled monolayers [197]. Klinowski et al. [288] discussed microwave heating for the preparation of MOF membranes. Macroscopic or microscopic cracks in polycrystalline films can form for a number of reasons and will likely ruin membrane performance for gas separation. In the preparation of MOF membranes the prevention of cracks is a subject of importance. Mixed matrix membranes (MMMs) with MOFs are a new class of membranes.

Hu et al. [289] developed a facile reactive seeding (RS) method for the preparation of continuous MOF membranes on alumina porous supports, in which the porous supports acted as the inorganic source reacting with the organic precursor to grow a seeding layer. Figure 3.53 shows the schematic of the RS method with the MIL-53 membrane as an example. In this technique, α-Al_2O_3 support itself acts as the aluminum precursor in place of $Al(NO_3)_3{\cdot}9H_2O$, which reacts with 1,4-benzenedicarboxylic acid (H_2BDC) under mild hydrothermal conditions to

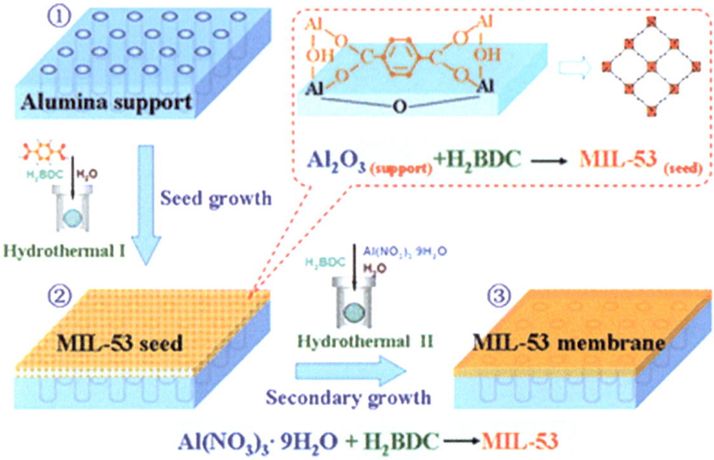

Fig. 3.53 Schematic diagram of preparation of the MIL-53 membrane on alumina support via the RS method

produce a seed layer. This is followed by a secondary growth process when $Al(NO_3)_3{\cdot}9H_2O$ and H_2BDC form the MIL-53 membrane under hydrothermal conditions (typical synthesis condition: 220 °C for 12 h).

Schoedel et al. [290] demonstrated that the gel-layer approach enables the synthesis of differently oriented MOF structures on functionalized gold surfaces at room temperature. Lu and Zhu [291] developed a method for MOF membrane preparation based on a liquid–liquid interfacial coordination mechanism. For example, MOF precursors, zinc nitrate and terephthalic acid (TPA or H_2BDC) as well as catalyst triethylamine (TEA), were dissolved in two immiscible solvents, dimethylforamide (DMF) and hexane. The reaction of $Zn(NO_3)_2$ and TPA in DMF was catalyzed by TEA in hexane at the solvent interface, thus forming a free-standing membrane. Ben et al. [292] reported a convenient and universal method to prepare MOF membranes. The polymer-supported and free-standing MOF membrane preparation is illustrated in Fig. 3.54.

In a typical procedure, preparations of the HKUST-1 membrane, polymethyl methacrylate (PMMA) are dissolved in chloroform and then spin-coated onto a silica wafer, which serves as a shape template substrate. When the solvent is evaporated, the PMMA-coated substrate is immersed into 12 M sulfuric acid for 20 s to hydrolyze the external PMMA into polymethacrylic acid (PMAA). After careful washing with deionized water, the PMMA–PMAA–silica substrate is introduced into a water/ethanol solution (1:1, v/v) of $Cu(NO_3)_2{\cdot}3H_2O$ and trimesic acid in a Teflon-lined autoclave where crystal intergrowth takes place at 120 °C for 3 days. Then, the membrane is washed several times with ethanol and dried at room temperature.

As shown in Fig. 3.55 polymer-inorganic nanocomposite membranes can be classified into two types according to their structure: (a) polymer and inorganic phases connected by covalent bonds and (b) polymer and inorganic phases connected by van der Waals force or hydrogen bonds [293].

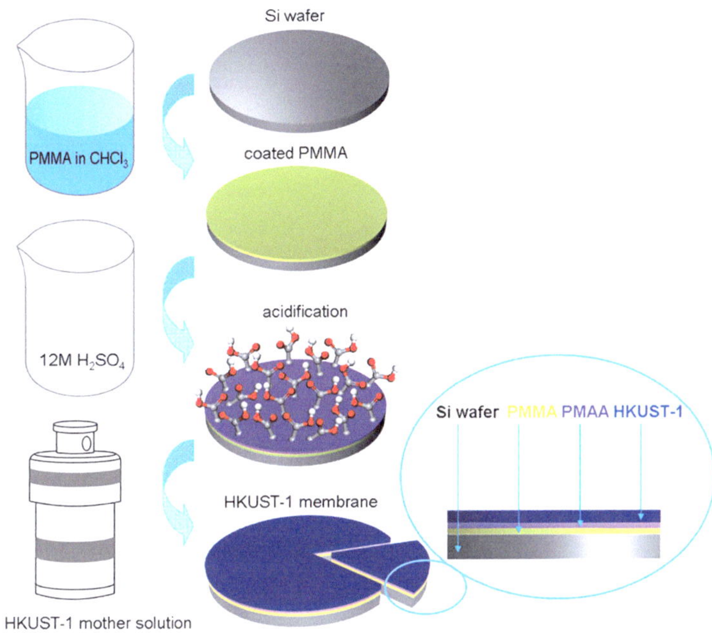

Fig. 3.54 Schematic illustration of the preparation procedure for the free-standing HKUST membrane

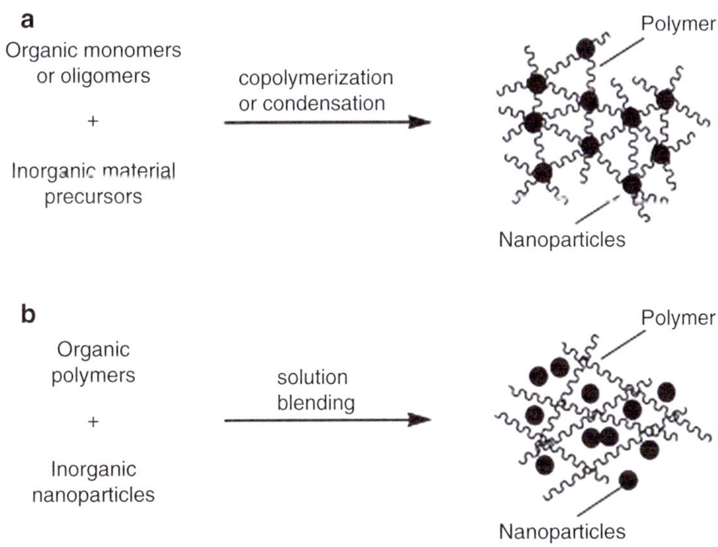

Fig. 3.55 Illustration of different types of polymer–inorganic nanocomposite membranes. (**a**) Polymer and inorganic phases connected by covalent bonds and (**b**) polymer and inorganic phases connected by van der Waals force or hydrogen bonds

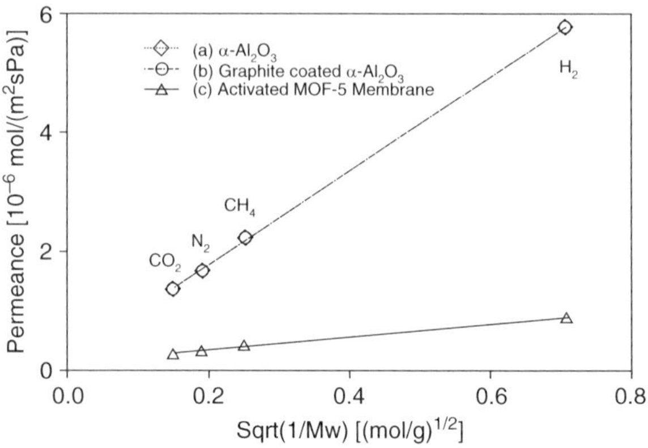

Fig. 3.56 Permeation of various gas molecules. (**a**) α-Alumina support, (**b**) graphite-coated α-alumina support, and (**c**) activated randomly oriented MOF-5 membrane (note that three membrane samples were prepared under the same condition and their performance was tested and plotted) [286]

Yoo et al. [286] fabricated MOF-5, or IRMOF-1 (isoreticular metal–organic framework 1, consisting of four Zn_4O clusters in octahedral subunits interconnected with benzene dicarboxylate linkers to form a three-dimensionally porous open framework structure) membranes on porous α-alumina substrate by secondary growth method. Figure 3.56 [288] shows the permeance of gas molecules as a function of their molecular weight. First, the permeation through graphite coated α-alumina substrate is compared with that of bare α-alumina, confirming that there is no resistance to permeation of gas molecule due to the presence of thin graphite powders on the substrate surface. Both substrates show Knudsen diffusion behavior (i.e., the permeance is proportional to $1/\sqrt{\text{Mwt}}$). Permeation of gas molecules through the activated MOF-5 membranes indicates the behavior of the Knudsen diffusion process.

Bétard et al. [267] fabricated a metal–organic framework (MOF) membrane by stepwise deposition of reactants. Two pillared layered MOFs with the general formula $[Cu_2L_2P]_n$ (L = dicarboxylate linker, P = pillaring ligand) were selected. For this demonstration, they selected the nonpolar $[Cu_2(ndc)_2(dabco)]_n$ (**1**: ndc = 1,4-naphthalene dicarboxylate; dabco = 1,4-diazabicyclo(2,2,2)octane) and the polar $[Cu_2(BME\text{-}bdc)_2(dabco)]_n$ (**2**: BME-bdc = 2,5-bis(2-methoxyethoxy)-1,4-benzene dicarboxylate). The framework structure of **1** is shown in Fig. 3.57.

The performances of both membranes were evaluated in gas separation experiments of CO_2/CH_4 (50:50) mixtures using a modified Wicke–Kallenbach technique. The separation-active MOF layer was located inside the macroporous support in a depth ranging in the μm scale. The microstructures of the MOF-based membranes resemble a foam with the inter-grown lamellae as transport-selective membrane. Proof of principle that the functionalization of linker can induce CO_2 membrane selectivity was found.

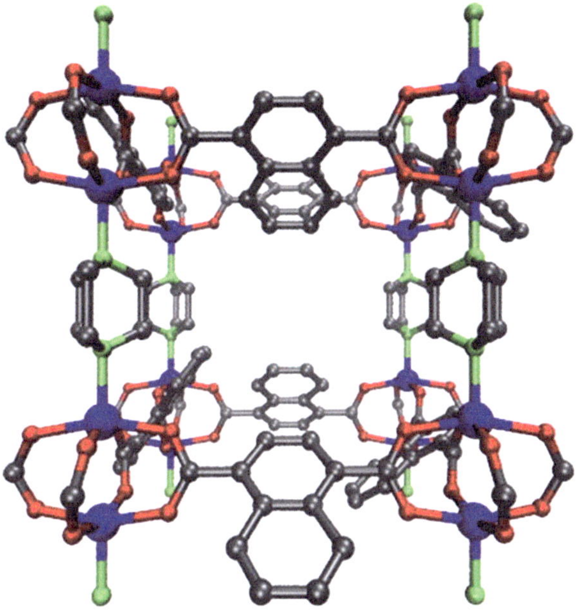

Fig. 3.57 Structure of $[Cu_2L_2P]_n$ MOFs (here $\{Cu_2(ndc)_2(dabco)]_n$). (**1**) With linker L=ndc(1,4-naphthalenedicarboxylate) and ligand P=dabco(1,4-diazabicyclo(2,2,2)octane seen in [1 0 0] direction. (Structure **2** is quite similar to **1** by just replacing ndc by BME-bdc (see Fig. 3.58))

3.4 Mixed Matrix Membranes (MMMs)

Gas separation by selective transport through polymeric membranes is one of the fastest growing branches of membrane technology. However, the existing polymeric membrane materials are inadequate to fully exploit the application opportunities on an industrial scale: the improvement in permeability is at the expense of selectivity, and vice versa. New types of membrane material emerging, with the potential for future applications, are mixed-matrix materials composed of homogeneously interpenetrating polymeric and inorganic particle matrices.

Robeson [3] predicted the upper limits for the performance of polymeric membranes in gas separation. The performance of various materials available for the separation of O_2/N_2 is depicted in Fig. 3.59. The figure presents both the permeability of the fast gas oxygen on the abscissa and O_2/N_2 selectivity on the ordinate, on a logarithmic scale. For the polymeric materials, either rubbery or glassy, a rather general trade-off exists between permeability and selectivity, with an *upper bound line shown* in Fig. 3.59.

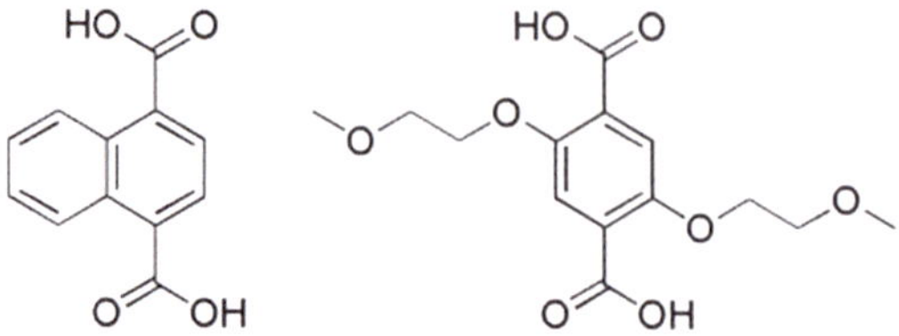

Fig. 3.58 Dicarboxylic acid linkers used in this work. *Left*, H_2ndc (ndc: 1,4-naphthalene dicarboxylate) and *right*, H_2BME-bdc (BME-bdc: 2,5-bis(2-methoxyethoxy)-1,4-benzene dicarboxylate)

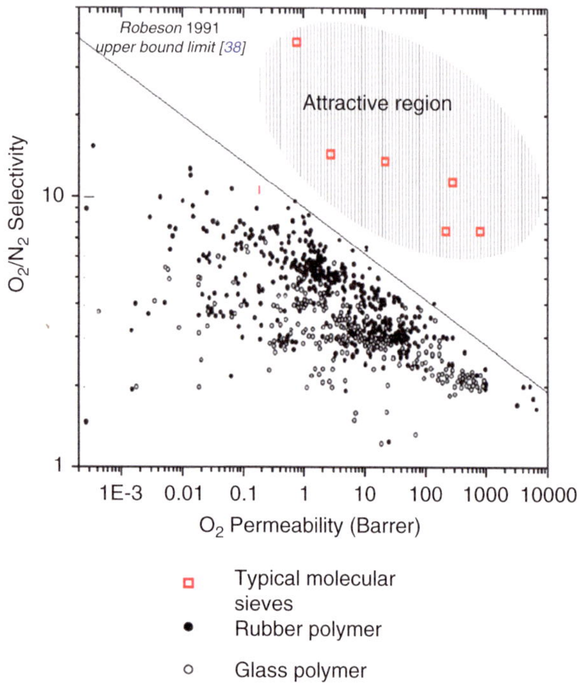

Fig. 3.59 Relationship between the O_2/N_2 selectivity and O_2 permeability for polymeric membranes. (The *dots* indicates the performance of polymeric materials [3])

When the materials with separation properties near this limit were modified on the traditional structure–property relation, the points specific to the modified materials tracked along this line but did not exceed it. The inorganic materials have the properties lying far beyond the upper limit for the organic membranes, as shown by the square symbols in the desired region [294]. Although tremendous developments have occurred in tailoring polymer structures to enhance separation properties, further progress exceeding the trade-off line seems to present a challenge in the near future. The application of inorganic membranes is still hindered by the lack of technology to form continuous and defect free-membranes, the cost of membrane production, and handling problems (e.g., brittleness). Thus, a new approach is needed to provide an alternative membrane with separation properties well above the upper-bound limit between permeability and selectivity.

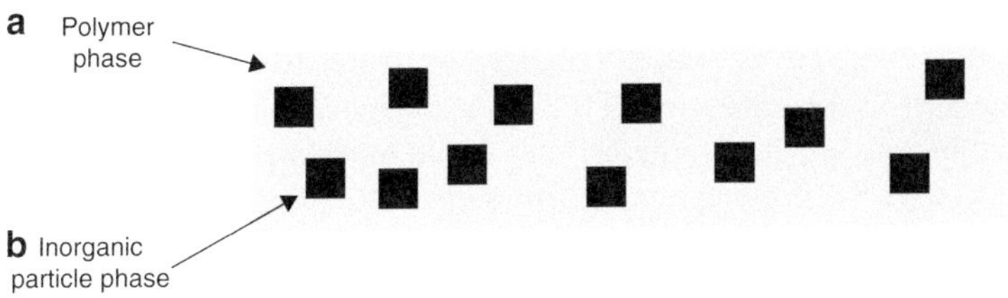

Fig. 3.60 Schematic of a mixed matrix membrane (MMM)

The inclusion of dispersed particles can have three possible effects on the permeability of gases: the discrete particles can act as molecular sieves, altering permeability in relation to molecular size, the particles can disrupt the polymeric matrix resulting in increased micro cavities and hence increase permeability, or they can act as a barrier to the gas transport and reduce permeability [19]. The mixed matrix membranes provide the opportunity to overcome the individual deficiencies of molecular sieves and polymers, and achieve gas separation performance well above famous Robeson's upper bound (again, see Fig. 3.57).

The schematic illustration of MMMs is shown in Fig. 3.60.

The bulk phase (phase A) is typically a polymer; the dispersed phase (phase B) represents the inorganic particles, which may be zeolite, carbon molecular sieves, or nano-size particles. Thus, MMMs have the potential to achieve higher selectivity, permeability, or both relative to the existing polymeric membranes, resulting from the addition of the inorganic particles with their superior inherent separation characteristics.

The review written by Chung et al. [294] gives an outline of the concept and the key advances in MMMs. The research conducted to date on MMMs has focused on the addition of porous inorganic filler to a polymer matrix. It is necessary that both materials should be selective for the same gas pairs. However, in most cases, the inorganic fillers have selectivity far superior to the neat polymer. Ideally, the incorporation of a small volume fraction of inorganic fillers into the polymer matrix can result in a significant increase in overall separation efficiency, as predicted by the so-called Maxwell model [295]. The Maxwell model equation for MMMs with dilute suspension of spherical particles can be written as follows:

$$P_{eff} = P_c\left[\left(P_d + 2P_c - 2\Phi_d\left(P_c - P_d\right)\right) / \left(P_d + 2P_c - \Phi_d\left(P_c - P_d\right)\right)\right]$$

where P_{eff} is the effective composite membrane permeability, Φ the volume fraction, P the single component permeability, and the subscripts d and c refer to the dispersed and continuous phase, respectively. To properly choose the dispersed and continuous phases, one must take into consideration the transport mechanisms and the gas component preferentially transporting through the membrane [294].

Funk and Lloyd [296] introduced the concept of zeolite-filled mixed matrix membranes, referred to as ZeoTIPS membranes, which were formed using the thermally induced phase separation (TIPS) process and consisted of zeolite particles supported in a microporous polymer matrix. On modeling, it was reported that the performance of these membranes could surpass Robeson's upper bound for gas separation polymers [3]. Paul and Kemps reported the delay in diffusion time lag effect for CO_2 and CH_4 when adding 5A zeolite into rubbery polymer PDMS [297]. On adding 5A zeolite into a polymer matrix, a very large increase in the diffusion time lag was observed but it had only minor effects on the steady-state permeation. Kulprathipanja et al. [298] reported that mixed matrix systems of polymer/adsorbent might yield superior separation performance to that of a pure polymeric system. They observed an enhanced O_2/N_2 selectivity from 3.0 to 4.3 when increasing silicate content in the cellulose acetate (CA) matrix. Kulprathipanja et al. [299] also observed that the calculated separation factor for CO_2/H_2 (50/50 mol%) was 5.15 ± 2.2 in a MMMs membrane in comparison with 0.77 ± 2.2 in a CA membrane, under similar conditions. This indicated that the presence of silicate in the membrane phase alternates the selectivity of H_2 over CO_2. MMMs comprising PDMS as continuous phase and zeolite (ZSM-5) as dispersed phase were prepared by Hussain and König [300] and used for the separation of CO_2 from gas mixtures. ZSM-5 incorporation in PDMS significantly increased the permeability of single gases and a similar effect was observed for gas mixtures. Membrane performance was evaluated using the Maxwell model and as a result an interphase gap between the filler and the polymeric phase was identified.

The gas separation performance of hollow fiber MMMs (PES-zeolite 4A) was enhanced through coating by silicon rubber solution, which could be due to sealing of the defect of the outermost skin of the membrane fiber [301]. Widjojo et al. [302] fabricated polyethersulfone (PES)-beta zeolite/PES-Al_2O_3 dual-layer mixed-matrix hollow-fiber membranes. The incorporation of 20 wt% beta zeolite in the outer selective layer and 60 wt% Al_2O_3 in the inner layer, coupled with spinning at high elongation draw ratios, yielded membranes with an O_2/N_2 selectivity of 6.89. Chaidou et al. [303] fabricated MMMs of polyimide (PI) with different types of zeolites via a solution-casting procedure. The effect of zeolite loading, pore size, and hydrophilicity/hydrophobicity of zeolite on gas separation properties of MMMs was studied. It was observed that the permeability of studied gases (He, H_2, CO_2, and N_2) for Matrimid-zeolite membranes increased with an increase in zeolite loading. The MMMs with zeolite ZSM-5 exhibited the highest permeabilities of all gases, for the same concentrations of zeolite, which could be due to the specific structure and properties of this particular zeolite. Boroglu and Gurkayank [304] fabricated new monomers having silica groups as an intermediate for the preparation of poly(imide-siloxane)-zeolite 4A and 13X MMMs. The addition of particles improved the thermal strength of the polymer to be sufficient for gas separation applications. Zeolites were well distributed throughout the membrane and the zeolites and polymers had good contact at the interface. The transport parameters for all the membranes were determined for N_2 and O_2. However, the permeability of all gases for the poly(imide siloxane)-zeolite 4A membrane decreased with an increase in zeolite loading.

Table 3.24 Averaged gas permeabilities ±1 standard deviation (in Barrers) of pure PVAc and 15 % CuTPA PVAc MMMs (1 Barrer = 7.5×10^{-8} m^3(STP) m m^{-2} s^{-1} Pa^{-1})

Membrane	P_{He}	P_{O_2}	P_{N_2}	P_{CH_4}	P_{CO_2}
PVAc	15.1 ± 0.8	0.514 ± 0.034	0.0783 ± 0.0064	0.0697 ± 0.0034	2.44 ± 0.32
MMMs	18.0 ± 0.5	0.624 ± 0.026	0.0912 ± 0.0032	0.0806 ± 0.0035	3.26 ± 0.23

Interfacial void-free MMMs of polyimide (PI)/zeolite were developed by Karkhanechi et al. [305] using 13X and Linde type A nano-zeolites for gas separation. Fabrication of a void-free polymer-zeolite interface was verified by the decreasing permeability developed by the MMMs for the examined gases, in comparison to the pure PI membrane. The molecular sieving effect introduced by zeolite 13X improved the CO_2/N_2 and CO_2/CH_4 selectivity of the MMMs. The Koros group [306] synthesized a MOF of copper and terephthalic acid (CuTPA) and used it for MMMs (using polymer PVAc). The gas transport properties of these CuTPA MMMs showed improvements over the pure polymer transport properties. Table 3.24 summarizes the average pure gas permeabilities for pure PVAc and the MMMs.

Nik et al. [307] synthesized a glassy polyimide, 6FDA-ODA (diamine), and mixed it with several as-synthesized MOF fillers at 25 % content for CO_2/CH_4 gas separation MMMs. The gas separation properties improved. The data revealed that the presence of $-NH_2$ functional groups in the MOF structure could lead to creating a rigidified polymer at the interface of the filler and polymer matrix and therefore decrease the permeability while increasing the selectivity. Tanh Jeazet et al. [308] studied MOFs for MMMs including:

- [$Cu(SiF_6)$(4,4′-BIPY)$_2$].
- [$Cu_3(BTC)_2(H_2O)_3$] (HKUST-1, Cu-BTC).
- [Cu(BDC)(DMF)].
- [$Zn_4O(BDC)_3$] (MOF-5).
- [Zn(2-methylimidazolate)$_2$] (ZIF-8).
- [Zn(purinate)$_2$] (ZIF-20).
- [Zn(2-carboxyaldehyde imidazolate)$_2$] (ZIF-90).
- $Mn(HCOO)_2$.
- [Al(BDC)(μ-OH)] (MIL-53(Al)).
- [Al(NH_2-BDC)(μ-OH)] (NH_2-MIL-53(Al)).
- [$Cr_3O(BDC)_3(F,OH)(H_2O)_2$] (MIL-101) (4,4′-BIPY = 4,4′-bipyridine,
- BTC = benzene-1,3,5-tricarboxylate,
- BDC = benzene-1,4-dicarboxylate, terephthalate).

MOF-polymer MMMs were investigated for the permeability of the single gases H_2, N_2, O_2, CH_4, and CO_2 and the gas mixtures O_2/N_2, H_2/CH_4, CO_2/CH_4, H_2/CO_2, CH_4/N_2, and CO_2/N_2. Results showed that MOF-MMMs had higher separation performance than pure polymer membranes for gas separation.

MOF crystals of $Cu_2(BTC)_2$ (surface area 1,396 m^2 g^{-1}) were mixed with polyimide (PI) to prepare hollow MMMs and the permeation of gases was studied [309]. The H_2 permeance and the selectivity of H_2 with respect to other gases such as N_2, O_2, CO_2, and CH_4 both increased markedly with increased $Cu_2(BTC)_2$ loading. At a loading of 6 wt% $Cu_2(BTC)_2$, the permeance of H_2 increased by 45 %, and the ideal selectivity increased by a factor of 2–3 compared to the corresponding data for the pure PI. The influence of three different MOFs in MMMs for binary gas mixtures was reported by preparing dense and asymmetric Matrimid® membranes filled with $Cu_3(BTC)_2$, ZIF-8, and MIL-53(Al). Dense membranes and asymmetric membranes for all three studied MOFs showed improvement in CO_2/CH_4 and CO_2/N_2 selectivity, and permeance, as compared to the unfilled reference membrane [310]. Li et al. [311] reported that a newly developed dual layer PES/P84 (copolyimide) hollow fiber with a PES-zeolite betamixed-matrix outer layer showed comparable permeance and selectivity of O_2/N_2 and CO_2/CH_4 in both single and mixed gas tests. Heating the membranes also improved the performances.

Li et al. [259] reported the first mixed matrix composite membrane made of commercially available poly(amide-b-ethylene oxide) (Pebax®1657, Arkema), mixed with the nano-sized zeolite imidazole framework ZIF-7. ZIF-7 was successfully deposited as a thin layer (less than 1 μm) on a porous polyacrylonitrile (PAN) support. An intermediate gutter layer of PTMSP was applied to serve as a flat and smooth surface for coating to avoid polymer penetration into the porous support. The performance of the composite membrane was characterized by single gas permeation measurements of CO_2, N_2, and CH_4. Both permeability (P_{CO_2} up to 145 Barrer) and gas selectivity (CO_2/N_2 up to 97 and CO_2/CH_4 up to 30) could be increased at low ZIF-loading. The CO_2/CH_4 selectivity could be further increased to 44 with a filler loading of 34 %, but the permeability was reduced compared to the pure Pebax® 1657 membrane.

Jiang et al. [312] reported that during the spinning of polymer-zeolite mixed matrix hollow fiber the particles located near the outer surface of the hollow fibers may form a defect free mixed-matrix structure with the surrounding polymer, which arises from the instantaneously solidification of the polymer phase. However, a continuous defect-free skin cannot be obtained because of the defects existing in the polymer matrix and the detachment of polymer chains from the particle surface during post treatment. More uniform and less defective mixed-matrix structures in the outer mixed-matrix layer of the dual-layer hollow fibers can be obtained by thermal treatment with a *p*-xylenediamine/methanol soaking method. Some fibers could obtain selectivity much higher than Knudsen diffusion even without silicon rubber coating, which indicates the significantly reduced amount of defects. Zeolite/carbon composite membranes represent another type of MMMS. A polyimide precursor containing MFI crystals was cast onto a stainless steel support and calcined at 580 °C in nitrogen. A medium O_2/N_2 selectivity of 4–6 with relatively high oxygen permeances of about 10^{-8} mol m^{-2} s^{-1} Pa^{-1} was found [197].

3.4.1 Preparation of MMMs

Polymer-inorganic nanocomposite membranes present an interesting approach to improve the separation properties of polymer membranes because they possess the properties of both organic and inorganic membranes—good permeability, selectivity, mechanical strength, and thermal/chemical stability. The methods for the fabrication of mixed matrix membranes are very similar to ordinary polymer membrane fabrication. The most commonly used for the preparation of nanocomposite membranes can be divided into the following three types [313].

1. Solution blending: In this technique, the inorganic nanoparticles are mixed with the polymer solution and dispersed by stirring. The nanocomposite membrane is cast by removing the solvent through conventional means. The solution blending method is easy to operate and suitable for all kinds of inorganic materials, and the concentration of the polymer and inorganic components are easy to control; however, the inorganic ingredients are liable to aggregate in the membranes.
2. In situ polymerization: In this method, the nanoparticles are mixed well with organic monomers, and then the monomers are polymerized. There are some functional groups such as hydroxyl and carboxyl on the surface of inorganic particles, which can generate initiating radicals, cations or anions under high energy radiation, plasma or other circumstances to initiate the polymerization of the monomers on the surface. For instance, nanocomposite membranes of poly(methacrylic acid) (PMA)/TiO_2 were synthesized from TiO_2 nanopowder/methacrylic acid dispersions under microwave radiation [314]. In the in situ polymerization method, inorganic nanoparticles with functional groups can be connected with polymer chains by covalent bonds. However, it is still difficult to avoid the aggregation of inorganic nanoparticles in the final product (membrane).
3. Sol–gel method: In this method, organic monomers, oligomers, or polymers and inorganic precursors are mixed together in the solution. The inorganic precursors hydrolyze and condense into well-dispersed nanoparticles in the polymer matrix [313]. The advantages of this method include:
   (a) The reaction conditions are moderate—usually room temperature and ambient pressure.
   (b) The concentration of organic and inorganic components is easy to control in solution.
   (c) The organic and inorganic ingredients are dispersed at the molecular or nanometer level in the membranes, and thus, membranes are homogeneous.

Iwata et al. [315] reported that by using the sol–gel method, a nanocomposite membrane of polyacrylonitrile (PAN) with hydrolysate of tetraethoxysilane (TEOS) as the inorganic phase showed a significant performance in O_2/N_2 separation. Ahmad et al. [316] studied the chemical, mechanical, and gas separation properties of PVA/TiO_2 nanocomposite membranes. The membrane was prepared using a polymer and TiO_2 (AEROXIDE hydrophilic fumed TiO_2P25) via solution blending (in water). It was reported that the addition of TiO_2 (up to 20 wt%) to PVA increased the selectivity of gas pairs O_2/N_2, H_2/N_2, H_2/CO_2, and CO_2/N_2 by 60 %, 55 %, 23 %, and 26 % respectively, with corresponding decreases in permeability. At higher loading of TiO_2, a reverse trend was noticed. MOF materials are also used in making MMMs. MMMs fabricated from MOF-5 nanocrystals with a high surface area (3,000 m^2/g) and high thermal stability (up to 400 °C), along with Matrmid® were used for gas separation. Despite the high surface area of the MOF-5, no increase in ideal selectivity for any gas pairs was observed. However, up to a 120 % increase in permeability was achieved due to the porosity of the MOF-5 nanocrystals. Gas mixtures (CO_2/CH_4, N_2/CH_4) showed a marked increase in selectivity for CH_4 due to the larger solubility of CO_2 and N_2 in the polymer matrix [317].

3.5 Other Materials

3.5.1 Metallic Membranes

Gas separation membranes based on Pd/Pt alloys can be used either independently or in conjunction with porous ceramic supports. Pd/Pt alloys have the ability to dissolve considerable amounts of hydrogen and to demonstrate increasing permeability. The major drawbacks to their industrial use are the high cost for Pd, the relatively low flux, and the irreversible change that takes place in the palladium lattice structure during cycling above and below a critical temperature, resulting in significant damage to the membrane.

Palladium thin films are known to selectivity transmit hydrogen via an adsorption-desorption mechanism. Permeability of hydrogen as high as 10^{-6} mol m^{-2} s^{-1} Pa^{-1} with H_2/N_2 permselectivity higher than 10,000 has been achieved [318]. As well, γ-alumina membranes modified by the deposition of metals such as Ru, Pd, Rh, and Pt possess hydrogen separation values that exceed limitation of Knudsen diffusion. Mixed proton- and electron-conducting materials consisting of barium cerate doped with rare-earth ions, that is, $BaCe_{1-x}M_xO_{3-\delta}$, where $M = Nd^{3+}$, La^{3+}, Y^{3+}, or Gd^{3+}, have been found to be of potential interest for hydrogen separation [319]. The presence of H_2S (low concentration) in the feed decreased the permeation of hydrogen through Pd and Pd–Cu alloy membranes by blocking H_2 dissociation sites. At high H_2S concentrations, a sulfur (due to decomposition of H_2S) surface layer did not allow H_2 to penetrate into the Pd–Cu surface [320].

If very pure hydrogen is required, dense metallic membranes may be a good option. Palladium and palladium alloys (practically the only types of hydrogen selective metallic membranes used) are extremely selective because only hydrogen can permeate through them [321]. Hydrogen transport through the membrane can best be described by the solution/diffusion mechanism. Hydrogen is adsorbed on one side of the membrane, splits into two atoms, diffuses through the metal matrix, and recombines and desorbs at the permeate side. To improve fluxes and reduce membrane costs (material cost of palladium is very high), usually thin layers deposited on a porous ceramic or metallic support are used. If palladium membranes are exposed to hydrogen at lower temperatures, they can be seriously damaged, because hydrogen can become locked inside the palladium lattice. This will cause stresses in the membrane, increasing the likelihood of membrane failure. A solution to this problem is to dope the palladium with other elements such as silver or copper. Operating temperatures of today's palladium alloy membranes are in the range 300–600 °C. A major technical disadvantage of palladium membranes in most applications is their high sensitivity to chemicals such as sulfur, chlorine, and even CO. These chemicals can poison the membrane surface reducing the effective hydrogen fluxes by 20 to even 100 %. Although much attention is focused on development of palladium membranes, their commercial availability is still limited. Johnson Matthey produces palladium-silver alloy membranes up to 60 cm in size commercially for the production of ultra pure hydrogen in the electronics industry. However, there are some drawbacks to the material, i.e., its stability in the presence of high-temperature water vapor is questionable. Large-scale production of this type of material has long been considered a major challenge [282, 283].

3.5.2 *Carbon-Based Membranes*

Carbon-based membranes can be classified into three categories.

1. Carbon molecular sieve membranes (CMSMs) and adsorption selective carbon membranes (ASCMs).
2. Carbon nanotube (CNTs) membranes.
3. Graphene.

3.5.2.1 Carbon Molecular Sieve Membranes (CMSMs) and Adsorption Selective Carbon Membranes (ASCMs)

Depending on the separation mechanism, two types of carbon membranes can be distinguished: molecular sieve carbon membranes (MSCMs) and adsorption-selective carbon membranes (ASCMs). The separation of gas molecules by means of MSCM takes place via a molecular sieving mechanism. Since MSCMs have micropores with sizes close to the dimensions of permanent gases (<4 Å), the diffusivity of these gases through the membrane changes abruptly with the molecular size and shape. This allows the separation of gases with similar molecular sizes. These membranes have been demonstrated to be effective at separating gas mixtures, such as O_2/N_2, CO_2/N_2, and CO_2/CH_4. Gas separation by MSCM is limited to gases with molecular sizes smaller than 4.0–4.5 Å. However, MSCMs are not suitable to separate gas mixtures, such as *iso*-butane/*n*-butane or gas–vapor mixtures (i.e., air/hydrocarbons, H_2/hydrocarbons, etc.) [322].

The separation of gas molecules by means of ASCM takes place as a consequence of their adsorption properties. The more strongly condensable components are preferentially adsorbed on the micropores of the membrane. This reduces the open porosity and consequently limits the diffusion in the micropores of the less adsorbable gases. As a consequence, the more strongly adsorbed components permeate preferentially through the ASCM membrane, being separated as the permeate stream (low-pressure side); whereas the less adsorbed components of the feed gas mixture are mainly recovered at the high-pressure side (retentate stream). Thus, ASCMs are effective at separating non-adsorbable or weakly adsorbable gases (i.e., He, H_2, air, O_2, N_2, CH_4, etc.) from adsorbable gases, such as hydrocarbons (C_{2+}), NH_3, SO_2, H_2S, and CFCs. From a structural point of view, ASCMs are constituted by a carbon film with micropores slightly wider than those characteristic of CMSMs, probably in the range 5–7 Å.

Molecular sieving membranes are identified as promising, both in terms of separation properties (including achievable fluxes) and stabilities, but are not yet commercially available at a sufficiently large scale. The pore sizes are in the order of the size of H_2-molecules. Reported selectivities are in the range of 4–20. Adsorption selective carbon membranes separate non- (or weakly) adsorbable gases from adsorbable gases (such as H_2S, NH_3, and CFCs). The performance of these membranes will deteriorate severely if feed streams contain organic traces or other strongly adsorbing vapors. Carbon membranes can be used in non-oxidizing environments with temperatures in the range of 500–900 °C. A disadvantage of carbon membranes is that they are brittle and therefore difficult to package if the membrane surfaces become larger. Furthermore, the price of carbon membranes is still high and optimum manufacturing conditions still need to be determined.

The configuration of carbon membranes can be divided into two categories: supported and unsupported. Unsupported membranes have three different configurations: flat (film), hollow fiber, and capillary, while supported membranes consist of two configurations: flat and tube. Figure 3.61 shows the configurations of carbon membranes.

3.5.2.2 Carbon Molecular Sieve Membranes (CMSMs)

Carbon molecular sieve membranes have been seen as a very promising candidate for gas separations, both in terms of separation properties and stability. Carbon molecular sieves are porous solids that contain constricted apertures that approach the molecular dimensions of diffusing gas molecules. At this constriction the interaction energy between the molecule and the carbon comprises both dispersive and repulsive interactions. When the opening becomes sufficiently small relative to the size of the diffusing molecule, the repulsive forces dominate and the molecule requires activation energy to pass through the constrictions. In this region of activated diffusion, molecules with only slight differences in size can be effectively separated through molecular sieving [323]. The mechanism of gas permeation uptake through porous solid is thus closely related to the internal surface area and

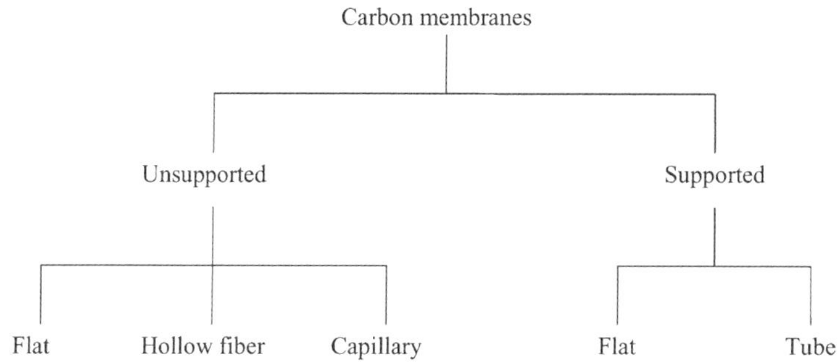

Fig. 3.61 Configurations of carbon membranes

dimensions of the pores and to the surface properties of the solid, rather than to the bulk properties of the solid (as in the case with polymers). CMSMs show excellent intrinsic performances for gas separation applications. This micro- to nanosize materials are obtained through the pyrolysis (at high temperature in an inert atmosphere) of polymeric precursors already processed in the form of membranes, and may be considered as just a type of "very high free-volume" material.

There are four different mechanisms for separation of a gas mixture through a porous membrane as discussed earlier (Chap. 2). The predominant transport mechanism of most carbon membranes is molecular sieving as shown in Fig. 3.60. The carbon membrane contains constrictions in the carbon matrix that approach the molecular dimensions of the absorbing species. In this manner, they are able to separate the gas molecules with similar size effectively. According to this mechanism, the separation is caused by passage of smaller molecules of a gas mixture through the pores while the larger molecules are obstructed. This mechanism exhibits high selectivity and permeability for the smaller components of a gas mixture. The carbon matrix is assumed to be impervious, and permeation through carbon membranes is attributed entirely to the pore system. The pore system consists of relatively wide openings with narrow constrictions. The openings contribute the major part of the pore volume and are thus responsible for the adsorption capacity, while the constrictions are responsible for the stereoselectivity of pore penetration by host molecules and for the kinetics of penetration. Hence, the diffusivity of gases in carbon molecular sieves changes abruptly depending on the size and shape of molecules because the carbon molecular sieve has pore size close to dimension of gas molecules (Fig. 3.62) [324].

The size of pores along the carbon fiber membrane can be controlled during the production process; it is possible to "tailor" the pore size distribution so that the diameter of virtually all pores will fall between the size of the large and small molecules of the gas mixture to be separated.

The fabrication of carbon membranes involves six important steps as shown in Fig. 3.63.

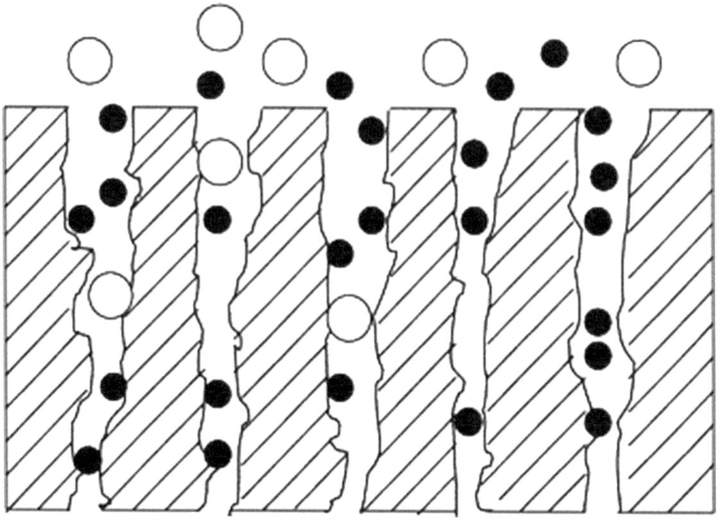

Fig. 3.62 Typical molecular sieving transport mechanism

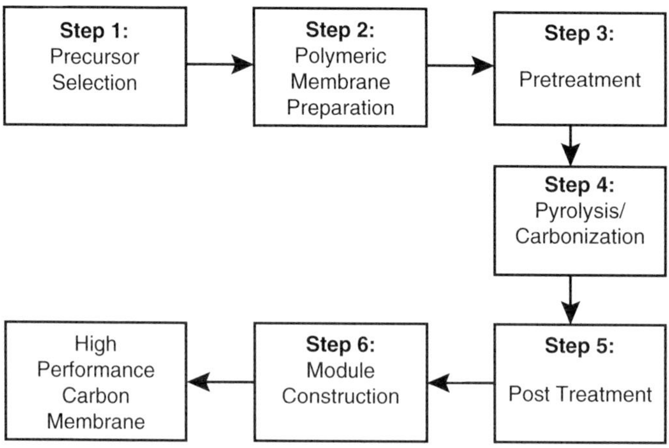

Fig. 3.63 Carbon membrane fabrication process [324]

At the same time, several other factors need to be considered, such as temperature and environment of carbonization, and polymeric precursors. The pyrolysis process is the most important step and can be regarded as the heart of the carbon membrane production process. During this stage, the pore structure of the carbon membrane is formed, and this determines the ability of a carbon membrane to separate gases.

The choice of the polymeric precursor is the first important factor. Pyrolysis of different precursors may produce different kinds of carbon membranes. Carbon membranes can be produced through the carbonization or pyrolysis process of suitable carbon containing materials such as resin, graphite, coal, pitch and plants, under inert atmosphere or in a vacuum. Lately, numerous synthetic precursors have been used to form carbon membranes, including polyimide and its derivatives, polyacrylonitrile (PAN), phenolic resin, polyfurfuryl alcohol (PFA), polyvinylidene chloride-acrylate terpolymer (PVDC-AC), phenol formaldehyde, and cellulose.

A polymeric membrane can be produced in two main configurations as precursor for carbon membranes, namely an unsupported membrane (flat, hollow fiber, capillary) and a supported membrane (flat, tube). For making the supported carbon membranes, various options are available for coating the supports with thin polymeric films, such as ultrasonic deposition, dip coating, vapor deposition, spin coating, and spray coating [325]. Because of the shrinkage of the polymer material during pyrolysis, the coating procedure has to be repeated until a defect-free carbon molecular sieve is obtained. The coating of the support surface with an intermediate layer reduces the number of defects existing on the original substrate.

Polymeric membranes are often subjected to pretreatments before they undergo a pyrolysis process. This step can ensure the stability of the polymeric precursor and the preservation of its structure during pyrolysis. In fact, carbon membranes of good quality, in terms of stability and separation performance, can be produced using specific pretreatments for a given precursor. Pretreatment methods can be divided into physical and chemical methods. Physical pretreatment consists of stretching or drawing hollow fiber membranes prior to pyrolysis. Chemical pretreatments involve some chemical reagents which are applied to the polymeric precursor.

Pyrolysis (sometimes referred to as carbonization) is a process in which a suitable carbon precursor is heated in a controlled atmosphere (vacuum or inert) to the pyrolysis temperature at a specific heating rate for a sufficiently long thermal soak time. During pyrolysis of a polymer, byproducts of different volatilities are produced resulting in a large weight loss. Depending on the polymer, typical volatile byproducts such as ammonia, hydrogen cyanide, methane, hydrogen, nitrogen, carbon monoxide, carbon dioxide and others may be produced. The polymer precursors are initially cross-linked or become cross-linked during pyrolysis. Pyrolysis creates an amorphous carbon material that exhibits a distribution of micropore dimensions with only short-range orders of specific pore sizes and also pores larger than the ultramicropores required to exhibit molecular sieving properties. One can determine which pyrolysis parameters are important and contribute most significantly to the structural changes of the material, where it would be possible to predict the trends of transport properties for a given carbon material more effectively.

Pyrolysis temperature was found to significantly change the structure and properties of carbon membranes based on PAN [326]. A similarly carbonization atmosphere was found to have a profound effect on the PAN-based carbon membranes [327]. As mentioned above, the MSCMs can separate gas components by means of their microporosity, which discriminates between molecules according to their size, shape and strength of interaction with the pore surfaces. Excellent performance with respect to hydrogen permeability and selectivity in the separation of hydrogen from light hydrocarbons such as methane has been shown with MSCMs. MSCMs derived from the carbonization of cellulosic films were developed for the purpose of hydrogen recovery, by evaluating the effect of copper (II) nitrate addition to the cellulose precursor, carbonization temperature and environment on MSCMs performance [328]. The performance of MSCMs synthesized in this way was better than polymeric membranes for hydrogen/methane separation in terms of the Robeson tradeoff curve, which plots membrane productivity (often represented by permeability against selectivity) [3].

Favvas et al. [329] prepared and characterized three different MSCMs from Matrimid® 5218 polyimide hollow fiber precursor. The formation of the selective layer on the fibers was independent from the initial orientation of the asymmetric polyimide precursor. The size of the pores was influenced by the temperature of the carbonization process whereas the pore volume was influence by the pyrolysis environment conditions. Carbon dioxide seems to be a more effective oxidizing agent than water, one reason being the hydrophobicity of the carbon surface. Figure 3.64 shows schematically this oxidization process. The developed carbon fibers exhibited H_2 permeances varying from 20 to 52 GPU with a highest permselectivity coefficient of 137. Permeation rates of He, H_2, Ar, CH_4, CO_2, CO, O_2, and N_2 at various pressures were measured too. In most cases, permeation properties were independent of feed pressure indicating the absence of compaction. The size of pores was found to be mainly dependent of the carbonizing process rather than the pyrolytic environment, which does not play an important role.

Carbon hollow fiber membranes derived from a polymer blend of polyetherimide and polyvinylpyrrolidone (PVP) were prepared through stabilization under air atmosphere followed by carbonization under N_2 atmosphere by Salleh et al. [330]. The polymer blends with 6 wt% PVP showed the best composition in the preparation of PEI/PVP-based carbon hollow fiber membranes with CO_2/CH_4 and CO_2/N_2 selectivities of 55.53 and 41.5, respectively. These results (data) were superior in comparison with other published data as shown in Tables 3.25 and 3.26.

Yoshimune et al. [337] prepared MSCMs as hollow fibers using PPO and its functionalized derivatives (R-PPO) as a precursor, and gas transport properties were measured for He, H_2, CO_2, O_2, and N_2. PPO MSCMs exhibited higher performances than those polymeric precursors. The highest performance was attained by trimethylsilyl-PPO (TMSPPO) MSCM pyrolyzed at 923 K, of which O_2 permeability was 125 Barrer and O_2/N_2 permselectivity was 10.0 at 298 K.

Amongst the polymer precursors applied for the preparation of carbon membranes, the most frequently used are polyfurfuryl alcohol (PFA), polyvinylidene chloride (PVDC), cellulose, phenolic resins, polyacrylonitrile (PAN), polyetherimides, and polyimides [331]. Polyimides are categorized as the most stable classes of polymer and can be used at temperatures higher than 573 K. They usually decompose before reaching their melting point. They are considered to be excellent precursors for glassy carbon as they do not go through a melting phase transition and thus do not lose their shape [331]. Matrimid®, Kapton®, and P84 polyimides are widely used for the preparation of CMS membranes.

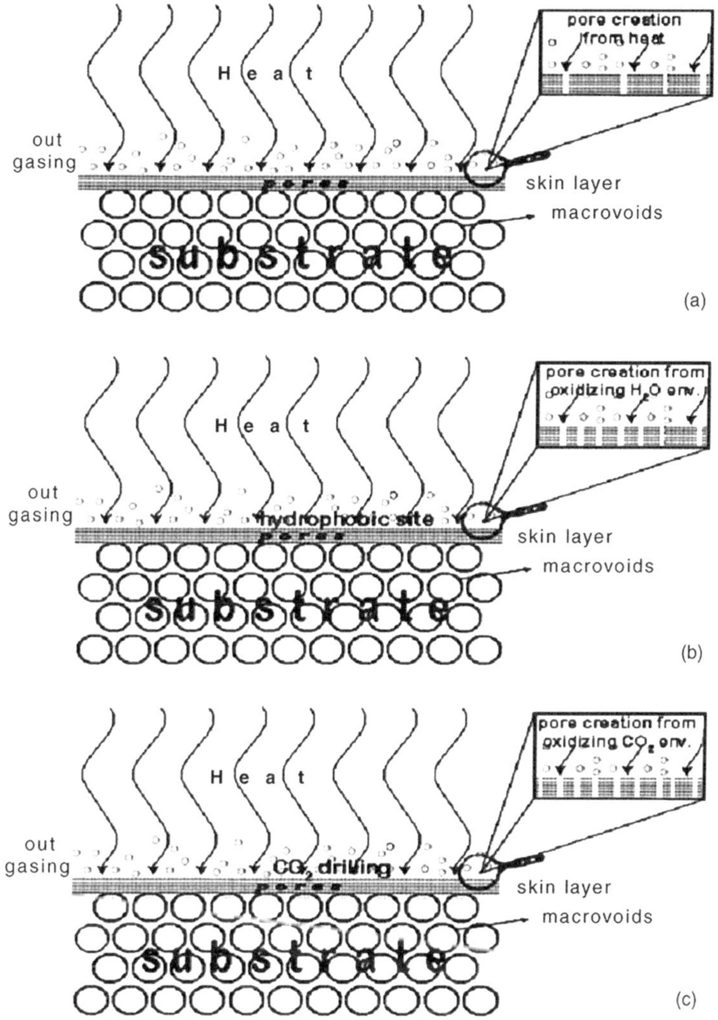

Fig. 3.64 Model of carbonization/activation process. (**a**) Carbonization in inert environment; (**b**) carbonization/activation in oxidizing H_2O environment; and (**c**) carbonization/activation in high oxidizing CO_2 environment [329]

Hattori and coworkers have prepared carbon films, including supported and unsupported, from Kapton-type polyimide [338, 339]. It was reported that the carbon molecular sieve film used for gas separation should be as thin as possible in order to enhance the separation efficiency; however, the thin film should be supported with a porous plate for handling convenience. The flat homogeneous carbon film prepared by pyrolysis at 800 °C had O_2/N_2 selectivities of 4.2.

Table 3.25 Permeation performance of the derived carbon membrane [330]

Membrane from (Precursor)	Configuration	Selectivity	
		CO_2/CH_4	CO_2/N_2
P84 copolyimide [331]	Hollow fiber	38.9	42.8
Matrimid [329]	Hollow fiber	20.86	23.6
PEI/PVP [332]	Flat sheet		13.70
PI/PVP [333]	Flat sheet		~40.00
PI/PVP [334]	Flat sheet		30 -38
PPO/PVP [335]	Tubular	~10.00	~20.00
PPESK/PVP [336]	Flat sheet		25-70

Table 3.26 Comparison of permeance on carbon hollow fiber

Precursor	Permeance (GPU)		
	N_2	CO_2	CH_4
Matrimid [331]	0.270	6.300	0.300
P84 copolyimide [332]	0.006	0.276	0.007
PEI/PVP [329]	0.04	1.66	0.03

Rao and Sirkar [340–342] prepared nanoporous-supported carbon membranes by pyrolysis of the polyvinylidene chloride layer coated on a macroporous graphite disc support. By heat treatment at 800 °C for 3 h, the diameters of macropores were reduced to the order of a nanometer. These membranes were used to separate hydrogen–hydrocarbon mixtures and the results are discussed on the basis of the surface diffusion mechanism, in which selectivity depends on the adsorption of gas molecules on the pore wall. This transport mechanism differs from the molecular sieving mechanism; therefore, these membranes are called selective surface flow (SSF™) membranes. They possess a thin layer (e.g., 2–5 μm) of nanoporous carbon (effective pore diameter in the range of 5–6 Å) supported on a mesoporous inert support, such as graphite or alumina (effective pore diameter in the range of 0.3–1.0 μm).

Centeno et al. [343] demonstrated that the gas separation performance of phenolic resin-based carbon membranes can be adjusted by pyrolysis processing variables (heat treatment temperature, heating rate, soaking time, and atmosphere). A large variety of carbon membranes for gas separation have been developed by simple carbonization of a phenolic resin film deposited on a ceramic tubular support. Thus, molecular sieve carbon membranes (MSCMs) with good capabilities towards the separation of O_2–N_2, CO_2–CH_4, CO_2–N_2, and olefin–paraffin mixtures, as well as adsorption-selective carbon membranes (ASCMS) effective in the recovery of hydrocarbons from hydrocarbon–N_2 mixtures, have been obtained.

A solution to overcome reproducibility problems of nanoporous carbon (NPC) membranes has been introduced by Acharya and Foley [344]. They used spray coating system for the production of thin layers of nanoporous carbon on the surface of a porous stainless steel support. A solution of poly(furfuryl)alcohol (PFA) in acetone was sprayed onto the support in the form of a fine mist using an external mixer brush with nitrogen gas. The advantage of this technique is reproducibility, simplicity and good performance for O_2/N_2 separation. The resulting membranes were found to have oxygen over nitrogen selectivities up to 4 and oxygen fluxes on the order of 10^{-9} mol m^{-2} s^{-1} Pa^{-1}.

Favvas et al. [331] fabricated gas separation carbon hollow fiber membranes based on a 3,3′,4,4′-benzophenone tetracarboxylic dianhydride and 80 % methylphenylene-diamine + 20 % methylene diamine co-polyimide precursor (BTDA-TDI/MDI, P84 Lenzing GmbH). Hollow fibers were initially prepared by the dry/wet phase inversion process in a spinning setup, while the spinning dope consisted of P84 as polymer and NMP as solvent. The developed polymer hollow fibers were further carbonized in nitrogen at temperatures up to 1,173 K. Permeability (Barrer) of He, H_2, CH_4, CO_2, O_2, and N_2 were measured at atmospheric pressure and temperatures of 313, 333, and 373 K and were found to be higher than those of the precursor. Moreover, the calculated permselectivity values were significantly improved. The developed carbon fibers exhibited rather low H_2 permeance values (8.2 GPU or 2.74×10^{-9} mol m^{-2} s^{-1} Pa^{-1} with a highest H_2/CH_4 selectivity coefficient of 843 at 373 K).

Jones and Koros [323] prepared carbon membranes from the pyrolysis of several different hollow fiber polymeric materials, including cellulose acetate, polyaramides, and polyimides. Pyrolysis was done in a vacuum. The selectivities obtained with these membranes were much higher than those found with conventional polymeric materials, and the high selectivities were achieved without the loss of membrane productivity. Membranes were produced by two different temperature protocols, and were evaluated with mixed gas feeds at pressures ranging up to 200 psig (1.48 MPa). The lower temperature protocol yielded membranes with O_2/N_2 selectivities ranging from 8.5 to 11.5, and a higher temperature pyrolysis yielded membranes with selectivities ranging from 11.0 to 14.0. These membranes were found to be quite stable over time periods of several days with high-purity, dry feeds. Limited studies also showed that these membranes were highly effective for the separation of other mixed gas pairs, including CO_2/N_2, CO_2/CH_4, and H_2/CH_4.

Centeno et al. [345] described a method for the preparation of a composite carbon membrane from poly(vinylidene chloride-co-vinyl chloride) for gas separation. The membrane was formed by a thin microporous carbon layer (thickness, 0.8 μm) obtained by pyrolysis of a polymeric film supported over a macroporous carbon substrate (pore size, 1 μm; porosity, 30 %). In a few cases polymeric film was oxidized in presence of air (at 200 °C) before carbonization. An almost defect-free carbon membrane was obtained in only one casting step. This carbon film exhibited molecular sieving properties and allowed the separation of gases depending on their molecular sizes. Single gas permeation experiments with pure gases of different molecular size (He, CO_2, O_2, N_2, and CH_4) were performed at different temperatures between 25 °C and 150 °C. It was revealed that the microporous carbon layers had molecular sieving properties. The carbon membrane showed high selectivities for the separation of permanent gases like the O_2/N_2 system (selectivity ~ 14 at 25 °C).

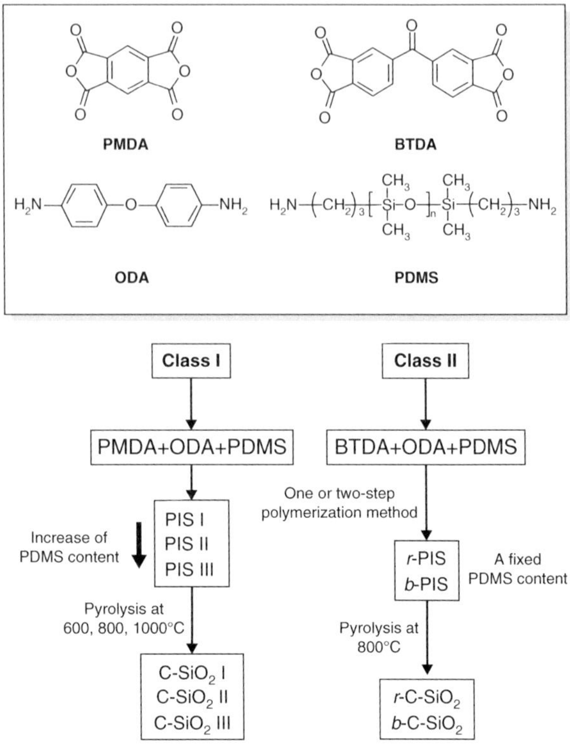

Fig. 3.65 The overall scheme of the Park and Lee study [346]

Air preoxidation at 200 °C for 6 h improved the permselectivity but with a loss in gas permeance. It was also reported that the carbonization temperature had a marked effect on gas permeance.

Park and Lee [346] fabricated carbon–silica membranes, by pyrolysis of the imide siloxane copolymers as the precursor of C-SiO_2. This was the first reported case of a polymeric precursor containing two thermo-stable phases being used for the preparation of carbon membrane implanted SiO_2. Figure 3.65 shows the overall scheme of their study.

The change in morphology in polymeric nanomaterial (block or random copolymer consisting of two phases in nanoscale) was found to affect the permeation properties to a large extent. In the case of O_2/N_2 separation, the O_2/N_2 selectivity versus O_2 permeability for the C-SiO_2 membrane was higher than the values obtained with other gas separation membranes (Fig. 3.66). The authors concluded that the combination of two building blocks with different carbon densities on the nanoscale can provide a hint about a new type of template carbonization, which differs from the conventional method using thermally stable and thermally unstable phases.

Hosseini and Chung [347] studied H_2/CO_2 separation by using carbon membranes derived from a PBI/polyimide blend. The selectivity of H_2/CO_2 was increased as the PBI (polybenzimidazole) content in the blend increased due to the rigidity and high packing density of PBI. PBI/Matrimid (75/25 wt%) blend membranes

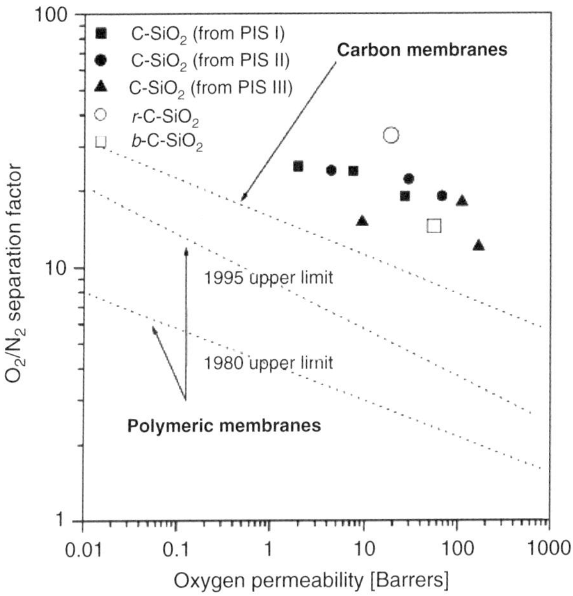

Fig. 3.66 O_2/N_2 selectivity vs. O_2 permeability (Barrer). (*Filled square*), (*filled circle*), (*filled triangle*): C-SiO_2 membrane derived from class I and (*open circle*), (*open square*): C-SiO_2 film derived from class II. 1 Barrer $= 10^{-10} \times cm^3(STP) cm\ cm^{-2}\ s^{-1}\ cmHg^{-1}$

cross-linked with *p*-xylenediamine showed a H_2/CO_2 selectivity of 26 with H_2 flux of 3.6 Barrer. Fuertes [322] fabricated carbon membranes from deposition of a thin film of a phenolic resin on the inner surface of an alumina tube. After carbonization (under vacuum at 700 °C) and air oxidation (300–400 °C), carbon membrane was obtained. The prepared membrane showed high permeabilities and selectivities towards separation of gas mixtures formed by hydrocarbons and N_2. As an example, the values of permeability and selectivity (hydrocarbons/N_2) for the separation of a complex gas mixture formed by 16.3 % CH_4, 16.1 % C_2H_6, 16.2 % C_3H_8, 20 % C_3H_6, and 31.4 % N_2 are: CH_4, 320 Barrer ($\alpha = 2.6$); C_2H_6, 1,104 Barrer ($\alpha = 9.1$); C_3H_6, 2,930 Barrer ($\alpha = 23.4$); C_3H_8, 2,850 Barrer ($\alpha = 22.8$).

Polymeric precursor—poly(phthalazinone ether sulfone) (PPES)—was used as a precursor for the preparation of carbon membranes via stabilization and pyrolysis by Zhang et al. [348]. The evolution of functional groups of membrane was monitored by ATR-FTIR during the formation process of carbon membranes. It was noticed that PPES is a highly thermally stable polymer with the char yield of 38.2 wt% at 700 °C in nitrogen. The functional groups of PPES disappeared by forming graphite-like structures in the membrane matrix during pyrolysis. At the test condition of 0.1 MPa and 30 °C, the gas permeabilities of H_2, CO_2, O_2, and N_2 for carbon membranes prepared at the stabilization and pyrolytic temperature of 240 and 650 °C are 610.13, 439.9, 146.98, and 28.95 Barrer, together with the selectivities of gas pairs H_2/N_2, CO_2/N_2, and O_2/N_2 of 22.6, 16.3, and 5.5, respectively.

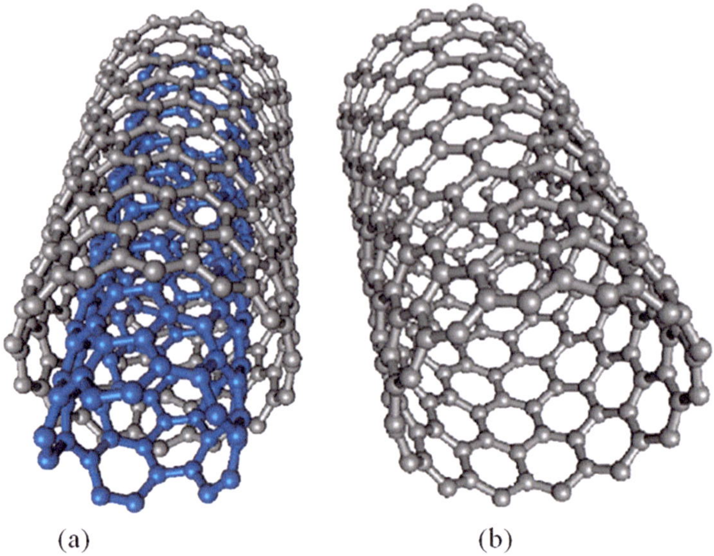

Fig. 3.67 (**a**) Structure of a multi-wall carbon nanotube (MWCNT) and (**b**) structure of a single-wall carbon nanotube (SWCNT) [353]

3.5.2.3 Carbon Nanotubes

Iijima in 1991 [349] reported the first detailed transmission electron microscope images of an arc-grown multiwalled carbon nanotube (MWCNT). The single walled carbon nanotubes (SWCNTs) were reported later [350]. Carbon nanotubes (CNTs) are allotropes of carbon with a cylindrical nanostructure. Nanotubes have been constructed with length to diameter ratios up to 132,000,000:1, significantly larger than for any other material [351, 352]. These cylindrical carbon molecules have unusual properties, which are valuable for nanotechnology, electronics, optics, and other fields of materials science and technology. Carbon nanotubes belong to a family of fullerenes. Nanotubes are categorized as single-walled nanotubes (SWNTs) and multi-walled nanotubes (MWNTs).

Most single-walled nanotubes (SWNT) have a diameter of close to 1 nm, with a tube length that can be many millions of times longer. The structure of a SWNT can be conceptualized by wrapping a one-atom-thick layer of graphite called graphene into a seamless cylinder (Fig. 3.67b). Multi-walled nanotubes (MWNT) consist of multiple rolled layers (concentric tubes) of graphene (Fig. 3.67a). Double-walled carbon nanotubes (DWNT) form a special class of nanotubes because their morphology and properties are similar to those of SWNT but their resistance to chemicals is significantly improved.

Individual nanotubes naturally align themselves into "ropes" held together by van der Waals forces. Other materials are also used as nanotubes composed of metal oxides, and their morphology is always very similar to carbon nanotubes.

All nanotubes have an extremely high aspect ratio in common, which makes them to molecular-level needles. Nanotubes are the strongest materials known, but the ultimate limits of their strength have yet to be reached experimentally [354]. The studies made by Sholl and Johnson [355] suggested that carbon nanotube membranes can have spectacularly high fluxes, must also high selectivity, and can be used in gas separation.

Carbon nanotubes represent a rare experimental realization of a nanofluidic cannel, which has a molecularly smooth wall, and a nanometer-scale inner diameter. This unique combination of properties gives the carbon nanotube channel an ability to support enhanced transport of gases with flows often exceeding those of conventional channels by several order of magnitudes [356]. Noy [356] presented a simplified analytical model that uses classic theory formalism to describe gas transport in carbon nanotube channels and to highlight the role of surface defects and adsorbates in determining transport efficiency, including the possibility of gas molecule diffusion along the nanotube walls. They also mentioned that in all conditions the nanotube channel walls play a critical role in determining transport efficiency and that in some cases obtaining efficient transport has to involve optimization of flows from diffusion through the gas phase and along the nanotube surface.

In general, there are four routes for the preparation of membranes based on CNTs [357].

1. Deposition of carbonaceous materials inside preexisting ordered porous membranes, such as anodized alumina, also known as template-synthesized membranes.
2. Membranes based on the interstice between nanotubes in a vertical array of CNTs, subsequently referred to as the dense-array outer-wall CNT membranes.
3. Encapsulation of as-grown vertically aligned CNTs by a space-filling inert polymer or ceramic matrix followed by opening up to CNT tips using plasma chemistry, or open-ended CNT membranes.
4. Membranes composed of nanotubes as fillers in a polymer matrix, also known as mixed-matrix membranes.

In MMMs for gas separation, carbon nanotubes are among three emerging fillers for membranes (the two others are metal–organic frameworks and clay-layered silicate) [358]. The development of MMMs gas separation membranes started on the basis of simulations, proving the excellent factors of CNTs for gas separation. In MMMs for gas separation carbon nanotubes are among three emerging fillers for membranes (the two others are metal=organic frameworks and clay layered silicate) [358]. Arora and Sandler calculated the kinetic and ideal separation factors for a carbon nanotube membrane, and proved that for single wall carbon nanotubes, high permeance can be obtained along with good kinetic selectivities [359]. Chen and Sholl [360] made atomic calculations to predict the separation between CH_4 and H_2 for SWCNTs and found a remarkable selectivity for CH_4.

Simulation predicting the diffusivity properties of simple gases in CNTs shows that these materials may be suitable filler in a polymer matrix to make MMMs. Atomistic simulations of diffusion of pure Ar and Ne through SWNTs were discussed by Ackerman et al. [361]. They also predicted the diffusion of these gases through the zeolite silicate, a commonly used zeolite for industrial applications with pores of about the same size as the nanotubes (0.81 and 1.36 nm). Their results predicted that the self diffusivity of Ar in SWNTs was orders of magnitude larger than silicate. Theoretical work of Skoulidas et al. [362] has reported atomic simulation results for both self and transport diffusivities of light gases such as H_2 and CH_4 in carbon nanotubes and in zeolites. They reported transport rates in CNTs to be orders of magnitude faster than zeolites and the exceptionally high transport rates in nanotubes.

Kim et al. [363] tried to verify this hypothesis by developing and characterizing novel nanocomposite membranes based on carbon nanotubes dispersed inside a polymer (imide siloxane) matrix. It was observed that the permeability of He dropped with the addition of closed ended CNTs. This large drop in He permeability indicated that the copolymer adhered well to the CNTs and that the prepared CNT MMMs were defect free. The permeability of O_2, N_2, and CH_4 increased in proportion to the open-ended CNTs in the polymer matrix. The permeability of He, H_2, and CO_2 increased after the addition of 2 wt% CNTs. The increase in the diffusion coefficients for O_2, N_2, and CH_4 in MMMs based on open-ended CNTs indicated the presence of high diffusivity CNT tunnels within the poly(imide siloxane) matrix. This diffusion suggested that CNT is an attractive additive for universally enhancing the gas permeability of polymers.

Several articles are reported in literature for gas separation using CNT MMMs, for hydrogen/methane, carbon dioxide/methane, hydrogen/carbon dioxide, oxygen/nitrogen, and carbon dioxide/nitrogen separation [358]. Tseng et al. [364] introduced a new class of multi-wall carbon nanotubes (MWCNTs)/carbon nanocomposite thin films, which were prepared by incorporating (MWCNTs) into polyimide (PI) precursor solution. The carbon films were obtained in only one coating step by a spin-coating technique on a microporous alumina substrate and were carbonized at 773 K. The MWCNTs/carbon nanocomposite thin film exhibited the ideal carbon dioxide flux of 8,656.6 Barrer and the separation factor of CO_2/N_2 was 4.1 at room temperature and 1 atm. This result was 2–4 times of magnitude higher than that of pure carbon membrane prepared by the same procedure and conditions.

Kusworo et al. [365] synthesized a new type of MMM consisting of polyethersulfone (PES) and carbon nanotubes (CNTs) and applied for biogas purification application. PES chains were grafted on the carbon nanotube surface. The modified carbon nanotubes MMMs increase the mechanical properties and permeability of all gases. For PES-modified carbon nanotubes selectivity achieved for CO_2/CH_4 was 23.4.

3.5.2.4 Graphene Membranes

Graphene is a substance composed of pure carbon, with atoms arranged in a regular hexagonal pattern similar to graphite, but in a one-atom thick sheet. It is very light, with a square meter sheet weighing only 0.77 mg. It is an allotrope of carbon whose structure is a single planar sheet of sp^2-bonded carbon atoms that are densely packed in a honey comb crystal lattice (Fig. 3.68) [366].

Fig. 3.68 Graphene is an atomic-scale honeycomb lattice made of carbon atoms

Graphene is most easily visualized as an atomic scale chicken wire made of carbon atoms and their bonds. In other words, graphene—a single layer of graphite—is the ultimate limit of a chemically stable and electrically conducting membrane one-atom in thickness [367]. The crystalline or "flake" form of graphite consists of many graphene sheets stacked together. The carbon-carbon bond length in graphene is about 0.142 nm. Graphene sheets stack to form graphite with an interplanar spacing of 0.335 nm. Graphene represents the first truly two-dimensional atomic crystal. It consists of a single layer of carbon atoms chemically bonded in a hexagonal "chicken wire" lattice. It has a unique atomic structure that gives it remarkable mechanical and thermal properties. Graphene is the basic structural element of some carbon allotropes including graphite, charcoal, carbon nanotubes and fullerene. It can also be considered as an indefinitely large aromatic molecule, the limiting case of the family of flat polycyclic aromatic hydrocarbons. Geim and Novoselov received the Noble Prize for groundbreaking experiments regarding this two-dimensional material [368].

Bunch et al. [369] demonstrated that a monolayer graphene membrane is impermeable to standard gases including helium. Bunch et al. measured both the elastic constant and the mass of a single layer of graphene by applying a pressure difference across the membrane. They claimed that this pressurized membrane is the world's thinnest balloon and provides a unique separation barrier between two distinct regions that is only one atom thick. Bunch et al. also suggested that graphene drumheads offer the opportunity to probe the permeability of gases through atomic vacancies in a single layer of atoms. The authors reported that small molecules like salts passed easily through a graphene membrane's tiny pores, while larger molecules were unable to penetrate. They suggested that the graphene has promising applications, such as membranes that filter microscopic contaminants from water, or that separate specific types of molecules from biological samples.

In the MIT news [370], Professor Rohit Karnik (mechanical engineering at MIT) suggested that a lot of chemical methods can be used to modify the pores in graphene membranes, thus creating a new technology for a new class of membranes. Karnik and O'Hern [370] observed the actual holes in the graphene membrane, looking at the material through a high powered electron microscope and found that the pores ranged in size from about 1 to 12 nm just wide enough to let some small molecules through the membrane.

Graphene can be prepared by simply heating and cooling down an SiC crystal [371]. In general, single layer or bilayer graphene forms on the Si face of crystal, whereas few-layer graphene grows on the C surface [372]. The results are highly dependent on the parameters used, like temperature, heating rate, or pressure. The other method is chemical vapor deposition, a well-known process in which a substrate is exposed to gaseous compounds. These compounds decompose on the surface in order to grow a thin film, where the by-products evaporate. Graphene can be grown by exposing a Ni film to a gas mixture of H_2, CH_4, and Ar at about 1,000 °C. The methane decomposes on the surface, so that the hydrogen evaporates. The carbon diffuses into the nickel. After cooling down in an Ar atmosphere, a graphene layer grows on the surface. The average number of layers depends on the Ni thickness and can be controlled in this way.

Porous graphene membranes have been suggested for the separation of hydrogen from methane, the separation of helium from other noble gases and methane, the selective passage of ions, the characterization of DNA, the filtration of water, and the separation of nitrogen from hydrogen. It was also reported that graphene pores are capable of separating fermionic He from bosonic He [373]. Surface adsorption effects have a significant influence on the gas permeability, especially at low temperatures [389]. Hauser and Schwerdtfeger [374] reported that nanoporous graphene membranes could be made CH_4 selective for gas purification by adjusting pore sizes in the graphene membrane.

Membranes act as selective barriers and play an important role in processes such as cellular compartmentalization and industrial-scale chemical and gas purification. The ideal membrane should be as thin as possible to maximize flux, mechanically robust to prevent fracture, and have well-defined pore sizes to increase selectivity. Graphene is an excellent starting point for developing size-selective membranes because of its atomic thickness, high mechanical strength, relative inertness and impermeability to all standard gases. However, pores that can exclude larger molecules but allow smaller molecules to pass through would have to be introduced into the material [375]. Koeing et al. [375] reported that ultraviolet-induced oxidative etching can create pores in micrometer-sized graphene membranes, and the resulting membranes can be used as molecular sieves. A pressurized blister test and mechanical resonance were used to measure the transport of a range of gases (H_2, CO_2, Ar, N_2, CH_4, and SF_6) through the pores. The experimentally measured leak rate, separation factors and Raman spectrum agree well with models based on effusion through a small number of ångstrom-sized pores. Du et al. [373] showed, by molecular dynamics simulations, that the stronger adsorption of molecular nitrogen to graphene leads to a permeation ratio $N_2/H_2 > 1$ for graphene pores with sizes above 4.48 Å.

As the permeability of a membrane is inversely proportional to its thickness [376] the permeability of a graphene-based membrane can be enhanced tremendously because of its one-atom thickness. Unfortunately, the pristine graphene is impermeable to gases as small as helium. This is due to the electron density of its aromatic rings, which is enough to repel atoms and molecules trying to pass through the hollows; it is therefore necessary to destroy its aromatic structures for gas permeability. Fischbein and Drndic [377] sculpted closely spaced nanopores within suspended graphene using a focused electron beam of a transmission electron microscope. In addition, porous 2D sheets have been created by assembling molecular building blocks. Improvements in these techniques may be helpful for creating ordered subnanometer-sized pores within graphene which may then be used as a 2D molecular-sieve membrane [378]. Jiang et al. [379] used a porous graphene with nitrogen functionalization to separate H_2 and CH_4.

Du et al. [373] presented porous graphene with various pore shapes for separation of hydrogen and nitrogen. The pore size and the functionalization of the pore introduced to graphene strongly affect the diffusion properties and the characteristics of the membrane. Blankenburg et al. [380] demonstrated that the porous graphene, which they fabricated, exhibits an extremely high selectivity in favor of H_2 and He among other atmospheric gases.

Qin et al. [378] proposed a new line defect-containing graphene as a gas filter for different gas species (He, H_2, O_2, N_2, CO, CO_2, and CH_4). They designed a new line defect consisting of a sequence of octagons and all-hydrogen passivated pores in graphene as a gas separation membrane using first-principles calculations. The all-hydrogen passivated pore produced a formidable barrier of 1.5 eV for CH_4 but an easily surmountable barrier of 0.12 eV for H_2. Hence it exhibited extremely high separation capability in favor of H_2 among all studied species with selectivity on the order of 10^{22} for H_2/CH_4. It was suggested by Qin et al. that such a line defect-containing a graphene-based membrane could play a great role on numerous clean energy applications.

Schrier [381] demonstrated that graphene could be permeable to gases and could be made selectively permeable by introduction of pores. Schrier proposed an economical means of separating He from the other noble gases and alkanes present in natural gases by using tailored graphene. Schrier and McLain [382] also demonstrated that isotope separation can be done by using graphene membranes. Jiang et al. [379] also investigated permeability and selectivity of graphene sheets with designed subnanometers pores using first principles density functional theory calculations. It was found that high selectivity on the order of 10^8 for H_2/CH_4 with a high H_2 permeance would be possible for a nitrogen-functionalized pore. Extremely high selectivity on the order of 10^{23} for H_2/CH_4 for an all-hydrogen passivated pore, whose small width (at 2.5 Å) presents a formidable barrier (1.6 eV) for CH_4, would be easily surmountable for H_2. These results suggest that these pores are far superior to traditional polymer and silica membranes, where bulk solubility and diffusivity dominate the transport of gas molecules through the material.

Jungthawan et al. [383] studied the structural deformation of porous graphene (PG) under tensile stress and the diffusion properties of H_2, O_2, and CO_2 through PG under different strain conditions using the first-principles density functional theory. It was observed that the application of a tensile stress can significantly increase the diffusion rate of H_2, O_2, and CO_2 in PG by 7, 13, and 20 orders of magnitude, respectively. Thus, the diffusion rate of gases through PG can be controlled by applying tensile stress. This technique will lead to wide range of energy and environmental applications.

Lee and Aluru [384] observed that by introducing a water slab between a gas mixture and the graphene membrane, the gas mixture can be separated based on the water-solubility of the gas molecule. Lee and Aluru separated CO_2 from CO_2/O_2, CO_2/N_2, and CO_2/CH_4 mixtures by using this technique. The separation ratio followed the water solubility of gas molecules in the mixture. The separation of gas mixtures can be controlled and the selectivity ratio can be enhanced with the water slab. With a thicker water slab, higher selectivity can be obtained. Graphene may become an ideal material for next-generation membranes due to its atomistic thickness, remarkable mechanical strength, and potential for size selective transport through nanometer-scale holes in its lattice. In the near future, graphene membranes could take the place of polymeric membranes for gas separation.

3.6 Gas Separation Membrane Structures

Transport of small gas molecules through polymers occurs by diffusion through transient free-volume elements or through cavities by random thermally stimulated motion of the flexible chains. Cavity sizes and shapes in rigid microporous inorganic materials such as zeolite [385] and carbon molecular sieve materials [386] are uniform, but not in the amorphous polymers. The cavity radius (r) of the most selective polymer such as polyimides, polysulfones and polycarbonates, as measured by positron annihilation life time spectroscopy (PALS), is 0.3 nm or less with a broad distribution of cavity sizes, and gas permeability is rather low [387]. The most permeable polymer, poly(1-trimethylsilyl-1-propyne) (PTMSP) has a cavity size distribution centered at around $r=0.3$ nm and $r=0.6$–0.7 nm [124]. Thus, among known polymers, free volume element size and distribution play a key role in determining permeability and separation characteristics. However, the broad size range of free volume elements in such materials precludes the preparation of polymers having both high permeability and high selectivity.

Park et al. [388] demonstrated that free-volume structures in dense vitreous polymers that enable outstanding molecular and ionic transport and separation performance surpass the limits of conventional polymers. The unusual microstructure in these materials can be systematically tailored by thermally driven segment rearrangement.

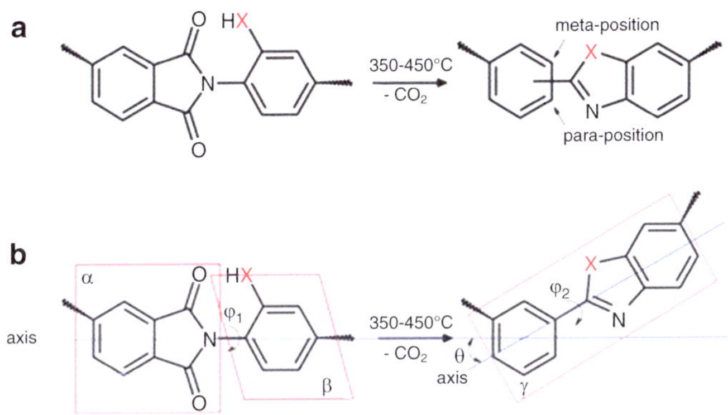

Fig. 3.69 Two major factors contributing to structural change during thermal chain rearrangement of polyimides containing ortho-positioned functional groups (X is O or S). (**a**) Change of chain conformation–polymer chains consisting of meta- and/or para-linked chain conformations can be created via rearrangement. (**b**) Spatial relocation due to chain rearrangement in confinement, which may lead to the generation of free-volume elements [α plane, phthalic imide ring; β plane, XH-containing phenylene ring; γ plane, newly created phenylene-heterocyclic ring (if X is O, benzoxazole-phenylene ring; if X is S, benzothiazole-phenylene ring); φ_1 and φ_2, dihedral angle; θ, tilting angle after transformation]

Free volume topologies can be tailored by controlling the degree of rearrangement, flexibility of the original chain, and judicious inclusion of small templating molecules. Thus, this rational tailoring of free-volume element architecture provides a route for preparing high performance polymers for molecular-scale separation. For their demonstration, Park et al. prepared completely aromatic, insoluble, infusible polymers from highly soluble precursors of aromatic polyimides containing ortho-positioned functional groups (e.g., –OH and –SH) by irreversible molecular rearrangement at about 350–450 °C (Fig. 3.69).

If managed properly, these changes in chain conformation and topology create well-connected, narrow size distribution of free volume elements (i.e., cavities) appropriate for molecular separation.

3.6.1 *Homogeneous Dense Membranes or Symmetric Membranes*

Symmetric membranes merely consist of a uniform structure (usually the cross-section of the membrane). Homogeneous and microporous are the two typical examples of symmetric membranes; in particular, a homogeneous membrane is referring to as a dense membrane, which has tremendous scientific value and are intensively used at the laboratory scale for the fundamental study of intrinsic membrane properties.

Non-porous or dense membranes provide high selectivity or separation of gases from their mixtures, but the rates of transport of gases are usually low. An important property of non-porous dense membranes is that even permeates of similar sizes may be separated if their solubility in the membrane differs significantly. Dense membranes can be prepared by melt extrusion, where a melt is envisioned as a solution in which the polymer is both a solute and solvent. In the solution-casting method, dense membranes are cast from polymer solution prepared by dissolution of a polymer in a solvent vehicle to form a sol. This is followed by complete evaporation of the solvent after casting.

3.6.2 *Asymmetric Membranes*

Asymmetric membranes are used primarily for pressure-driven membrane processes. In contrast to symmetric membranes, their structure consists of a very thin active skin layer on a highly porous substrate. The main purpose of this support layer is to provide the membrane with adequate mechanical strength and eliminate substantial structure resistance of gas transport through the polymer matrix. Phase inversion is mainly used to prepare asymmetric membranes.

In general, asymmetric membranes can be grouped into the following four basic structures [389]:

1. An integrally asymmetric membrane with a porous skin layer.
2. An integrally asymmetric membrane with a dense skin layer.
3. A thin-film porous composite membrane.
4. A thin-film dense composite membrane.

The membranes included in groups 1 and 2 are made of one material, whereas the membranes included in groups 3 and 4 are made of at least two different materials and consist of a thin top layer over a porous support or a backing material, which provides mechanical strength to the whole membrane while the membrane performance is controlled mainly by the top thin layer.

3.6.2.1 Integrally Skinned Bilayer Membranes

An integrally skinned asymmetric membrane consists of a relatively dense skin supported by a microporous sublayer as shown in Fig. 3.70a, b (bilayer). It is generally accepted that the bilayer membrane is formed by gelation and liquid–liquid demixing. When another layer, termed the nodular layer, is present between the skin and the sublayer, the membrane is called a trilayer membrane. A nodule denotes a fine spherical particle consisting of macromolecular aggregates. The skin layer itself consists of a closely packed array of nodules as shown in Fig. 3.70c. Most of integrally skinned asymmetric membranes have been prepared by an immersion precipitation process, which involves immersion of the polymer solution into a nonsolvent gelation bath.

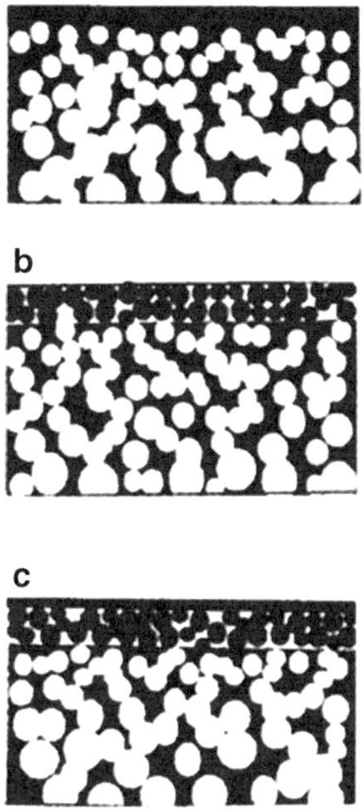

Fig. 3.70 Schematic drawings of various membrane morphology. (**a**) Bilayer; (**b**) bilayer; and (**c**) trilayer

Both the thermodynamics and kinetics—but predominantly the kinetics—of the immersion process play important roles in determining the membrane morphology [390].

Loeb and Sourirajan (1960) invented the first integrally skinned membrane for desalination by phase inversion of cellulose acetate sols. In the integrally skinned membrane, the skin and the porous substrate are composed of the same material. Differences in density between the two layers are the result of interfacial forces and the fact that solvent loss occurs rapidly from the air-polymeric solution and polymeric solution-coagulation media interfaces from the solution interior [391].

Kesting [392] demonstrated that there are four superimposed tires of structure in integrally skinned phase inversion membrane:

1. Macromolecules: A macromolecule is a very large molecule commonly created by polymerization of smaller subunits. The individual constituent molecules of polymeric macromolecules are called monomers.
2. Nodules: these are the macromolecular aggregates consisting of several tens of individual macromolecules and approximately 200 Å in diameter.
3. Nodule aggregates: 400–1,000 Å in diameter spherical clumps of nodules.
4. Supernodular aggregates: aggregates of nodule aggregates which constitute the walls of 0.1–2 μm in diameter open cells in the membrane structure.

The skins of integrally skinned membranes consist of a single layer, one nodule aggregate thick of coalesced and compacted nodule aggregates. Pores in the various separation regimes (gas separation (GS), RO, UF, and MF) can be seen as more or less static two-dimensional (or fractal) spaces that are bounded by progressively larger structural subelements and elements.

Two types of pore are found in the skins of GS membranes and these may account for the origin of dual mode sorption and the permeation of gases. The smaller pores, the Henry's mode sites, are the average interchain displacements between parallel chain segments within the nodule. The large pores, the Langmuir sorption sites, are the average displacements in the low density domain where nodules impinge on one another.

3.6.2.2 Integrally Skinned Trilayer Membranes

Figure 3.68c shows the structure of the integrally skinned trilayer membranes including

1. Skin
2. Nodular layer
3. Sublayer

To improve the permeation through a membrane is to reduce the thickness of the dense skin layer. To achieve this goal, integrally skinned asymmetric membranes [107, 390, 393] and multilayer composite membranes are most widely used [394]. As already mentioned, integrally skinned asymmetric membranes are made from one material and consist of a thin, essentially defect-free skin layer superimposed on a porous substrate; therefore, the gas permselectivity is determined by the intrinsic properties of the membrane material used. Multilayer composite membranes possess one or more selective layers on top of a microporous support layer, in which the selective layer and the support layer are made from different materials.

Figure 3.71 illustrates the schematic representation of an asymmetric bilayer and graded density skin asymmetric trilayer. The transition layer of the graded density skin asymmetric membrane corresponds to the nodular layer of the integrally skinned trilayer membrane.

Integrally skinned membranes can be formed by contacting the polymer solution with a non-solvent and forming the membrane in a one-step process. On contact with the non-solvent, mass transfer takes place between the non-solvent from the coagulation bath and the solvent in the nascent membrane resulting in micro-phase separation within the membrane. Depending on the pathway of phase separation, a dense layer, also called the skin layer, is believed to form on the surface of the membrane. The skin formation is hypothesized to occur when solvent outflow from the membrane exceeds the non-solvent inflow resulting in delayed demixing. This process increases the concentration of the polymer at the membrane-coagulant interface and forms the skin. An evaporative step can be included prior to the phase separation step to enhance skin formation by the evaporation of the volatile solvent from the nascent membrane, followed by a rapid phase separation of the underlying region to form a highly porous support [395]. Baker [396] predicted in 2002 that multilayer composite membranes will gradually displace simple Loeb-Sourirajan

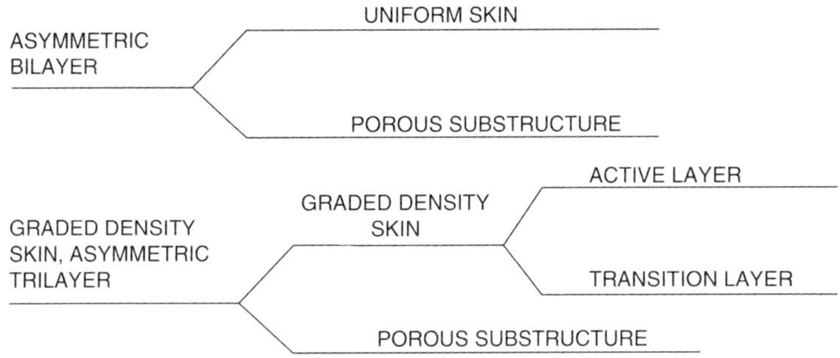

Fig. 3.71 Schematic representation of an asymmetric bilayer and graded density skin asymmetric trilayer membrane

membranes, which may be true for reverse osmosis (RO) and nanofiltration (NF) membranes but integrally skinned membranes are still dominant in manufacturing of other separation membranes. Even TFC RO and NF membranes use integrally skinned asymmetric membrane for the support layer.

Few patents are reported on integrally skinned membranes (trilayer)—mainly hollow fibers for gas separation [395, 397]. Fritzsche et al. [391] spun polysulfone hollow fiber membranes from propionic acid (*N*-methylpyrrolidone complex) and from a formylpiperidine/formamide mixture. The structure of these membranes was investigated as a function of progressive surface removal with oxygen plasma. A pure gas permeation rated was obtained on these samples. Li et al. [106] also studied integrally skinned hollow fibers with a defect free top layer for gas separation; however, not much research has been done so far on these membranes for gas separation.

3.6.2.3 Thin Film Composite Membranes (TFC)

Thin film, composite membranes consist of a thin polymer barrier formed on more porous support layers (almost always a different polymer from the surface layer). The skin layer determines the separation characteristics of the membrane; the porous backing serves only as a support for the selective layer and has no effect on membrane transport properties. In TFC membranes, it is possible to optimize the performance of the different materials independently (two steps). Composite membranes are less sensitive to the presence of humidity in the gas streams, avoiding the separate hydration step in their processing.

Thin-film composite membranes (TFC or TFM) are manufactured principally for use in water purification or water desalination systems. They also have use in chemical applications such as batteries and fuel cells. Essentially, TFC material is a molecular sieve constructed in the form of a film from two or more layered materials.

Thin film composite membranes are used:

- In water purification.
- As a chemical reaction buffer (batteries and fuel cells).
- In industrial gas separations.

Kusakabe et al. [398] coated the outer surface of a α-alumina tube with a γ-alumina film via the sol–gel process, then the tube was further coated with polycarbosilane (PC), which was cured at 473 K and pyrolyzed at 623–823 K. The composite membrane had 1–1.4 μm in thickness and had no pinholes larger than several nm. The permeation of H_2 and the separation factor of H_2 to N_2 were 5.5×10^{-7} mol m^{-2} s^{-1} Pa^{-1} and 7.2, respectively at 673 K.

Ren et al. [399] used poly(amide-6-b-ethylene oxide) (PEBA1657) copolymer to prepare a polyetherimide (PEI)/PEBA1657 (PEI as a support) composite membrane and ultra thin multilayer PEI/polydimethylsilicone (PDMS)/PEBA1657/PDMS composite membranes by dip-coating method for sour gas capturing. The gas permeation and transport characteristics of sour gases (CO_2, H_2S and SO_2) were investigated and analyzed. With the increase of transmembrane pressure difference, the permeation of H_2S, CO_2, and SO_2 increased due to a pressure-induced plasticization effect. The temperature dependency of H_2S permeance for PEI/PEBA1657 composite membranes changed from positive to negative when the transmembrane pressure difference increased from 3 to 7 atm. In the multilayer PEI/PDMS/PEBA1657/PDMS composite membrane, the transport resistance for CO_2 and H_2S was mainly from the PEBA1657 selective layer, and thus, the membranes had high permeances for CO_2 and H_2S and high selectivities for CO_2/N_2 and H_2S/N_2.

Jiang et al. [400] synthesized aqueous polyurethane dispersions (PUDs) with poly(dimethylsiloxane) (PDMS), or mixed PDMS/poly(ethylene glycol) (PDMS/PEG) as the soft segment and made thin film PUD-PVDF composite membranes for gas separation. PDMS/PEG-based PU was typically solubility-selective for condensable hydrocarbons, and nitrogen permeance was marginally enhanced in hydrocarbon–nitrogen mixtures. The copolymer membranes with both urethane and PEG segments could effectively tolerate the swelling caused by the condensable gases. The selectivities of propylene and propane to nitrogen were substantially improved, i.e., in a mixture of propylene (28 %) and nitrogen (72 %), the selectivity of propylene to nitrogen reached 29.2 with a propylene permeance of 34.4 GPU.

Gupta et al. [401] studied the gas separation on self-supported polyaniline films, and polyaniline nanomembranes with a selective layer thickness as thin as 300 nm supported on a porous polyvinylidene difluoride (PVDF). The selectivities, $\alpha_{A/B}$ (ideal) for the gas pairs H_2/N_2, H_2/O_2, H_2/CO_2, CO_2/O_2, CO_2/N_2, and O_2/N_2 were 348, 69.5, 8.6, 8.1, 40.4, and 7.1, respectively, achieved for the self-supported undoped polyaniline films. The authors claimed that these values are considerably higher than reported by other researchers. The gas transport rates for various gases for the dense polyaniline nano-films supported on polyvinylidene difluoride were the order of 10^5 times higher than those reported for self-supported polyaniline membranes. The higher transport rates for various gases through the membranes may have been due to a higher free volume due to increased crystallinity of the polyaniline.

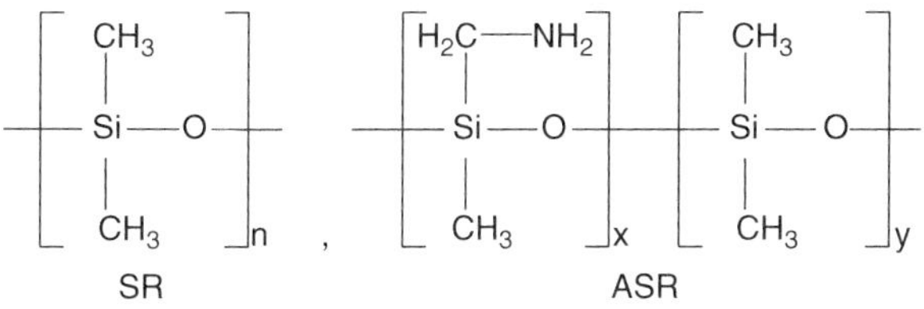

Fig. 3.72 Repeat unit structure of SR and ASR

Silicone rubber/poly(4-vinylpyridine)/polyetherimide (SR/P4VP/PEI) multilayer composite hollow fiber membranes were used for gas permeation performance (H_2, CO_2, O_2, N_2, CH_4) [394]. The PEI hollow fibers prepared from a PEI/PEG/NMP (23/0/77) spinning dope, after coating with 0.2 wt% P4VP and 3 wt% SR solutions, have gas permeances of $H_2=41$, $CO_2=7.4$, and $O_2=2.0$ GPU (1 GPU = 1×10^{-6} cm^3 (STP) cm^{-2} s^{-1} $cmHg^{-1}$) with selectivities of $H_2/N_2=117$, $CO_2/CH_4=62$, and $O_2/N_2=5.8$. Achalpurkar et al. [12] studied the gas permeation properties of dense and thin film composite membranes based on amine substituted silicon rubber (ASR) and unsubstituted silicon rubber (SR). Figure 3.72 shows the repeat unit structure of SR and ASR.

The ASR membrane exhibited higher CO_2 (15 %) as well as CH_4 (12 %) permeability as compared to the SR dense membrane, while the permeability for other gases (He, H_2, N_2, and O_2) were decreased up to 15 %. The permeance of TFC membranes based on different UF supports decreased in the order of decreasing porosity and increasing solution concentration (coating solution). The authors concluded that a careful variation of porosity (support material) and solution (coating solution) could lead to optimum combination of permeance and selectivity.

Monsanto-type silicon rubber-coated membranes are different from TFC membranes. High fluxes alone are not sufficient to make membrane gas separations competitive. Selectivity is also important. Membranes made from rubbery polymers, which usually are highly permeable, are not very selective. On the other hand glassy polymer membranes exhibit the opposite permeation characteristics. In order to take advantage of the high selectivities that glassy polymers afford, these materials have been spun into asymmetric hollow-fiber membranes, but such fibers normally contain surface pores or defects, which make the membranes nonselective, low-resistance channels for gas leakage. Because the permeabilities of glassy polymers are inherently low, a small number of surface pores in the skin of a glassy asymmetric fiber can significantly reduce its selectivity [402]. The Prism® gas separator developed by Monsanto [390, 403, 404], solved this problem by coating the skin of a glassy, asymmetric hollow fiber with an elastomer that serves to plug pores, thereby decreasing the effective permeability of these pores by four to five orders of magnitude. Although transport through pores that are coated and filled in this manner is still relatively nonselective, the volume of gas crossing the pores is greatly reduced by the coating process. Thus, the inherent selectivity of the glassy polymeric substrate can be approached. Unlike the TFC membranes in which the membrane properties are controlled by the top selective layer, the properties of

Monsanto's membrane are determined by the supporting membrane. In Monsanto's membrane, the silicon rubber layer does not function as a selective barrier but rather plugs the defects of the supporting membrane.

Reid et al. [405] studied the gas transport properties of surface treated poly (3-(2-acetoxyethyl)thiophene) (P3AcET). Hydrolysis of the ester group of P3AcET yields poly(3-(2-hydroxyethyl)thiophene) (P3HET) which is a highly permselective conducting polymer. Permeability coefficients were determined for as-cast P3AcET membranes ($P_{CO_2} = 1.42$, $P_{O_2} = 0.24$, $P_{N_2} = 0.05$, $P_{CH_4} = 0.08$ Barrers) and selectivity values were calculated ($O_2/N_2 = 5.1$, $CO_2/CH_4 = 18.5$). Formation of a thin selective P3HET surface layer resulted in an overall decrease in permeance accompanied by a dramatic increase in selectivity for both the base-treated ($O_2/N_2 = 12.9$, $CO_2/CH_4 = 20.0$) and acid-treated ($O_2/N_2 = 11.7$, $CO_2/CH_4 = 45.0$) composites.

3.7 Liquid Membranes for Gas Separation

A Liquid Membrane (LM) is just as it sounds—it is made of liquid. Because of the nature of a liquid, liquid membranes circumvent problems more conventional solid membranes encounter, but they have their own problems. One of the benefits of LMs is their high selectivity, and by using carriers, specific molecular recognition can be achieved. Despite high efficiency the lack of stability hampers LM's industrial applications.

Liquid membranes can be fashioned in two physical forms—immobilized on a solid support or as an emulsion, as shown in Fig. 3.70. Liquid membranes can perform in two modes: (1) with chemical carrier and (2) without chemical carrier. With a chemical carrier, diffusion of the permeant species is increased by diffusion of the reaction product. Without an active carrier, the liquid membrane relies on solubility differences and/or diffusion coefficient differences to separate components [406] (Fig. 3.73).

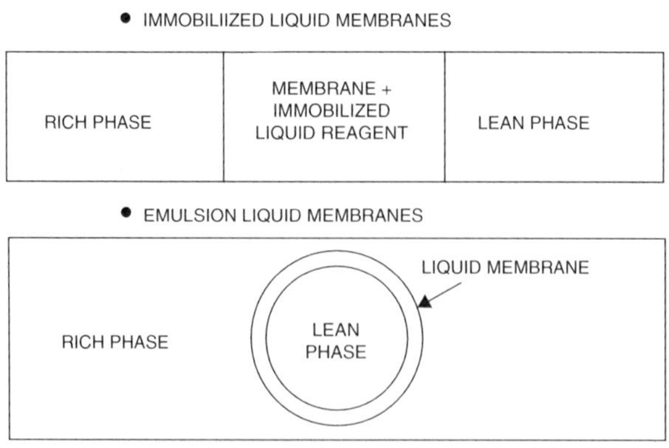

Fig. 3.73 The two forms of liquid membranes

There are two basic types of liquid membranes—emulsion liquid membrane (ELM) and immobilized liquid membrane (ILM), also called supported liquid membrane (SLM). It is necessary to stress that although supported liquid membrane and emulsion liquid membranes are conceptually similar—i.e., they employ liquid films as phase separators—the engineering aspects and applications of each are very different. Immobilized liquid membranes have been studied primarily for gas separations while emulsion liquid membranes have only been applied to liquid phase separations. ELMs remove the equilibrium limitation of solvent extractions by combining extraction and stripping in a single operation. ELMs have been successfully used to treat aqueous streams contaminated with heavy metals ions like Cu, Zn, Cd, Ni, Hg, Pb, and Cr.

3.7.1 Supported Liquid Membranes (SLM) or Immobilized Liquid Membranes (ILM)

Supported liquid membranes (SLMs) consist of an organic liquid immobilized in the pores of a support by means of capillary forces [407] and were reported for the first time in 1967 by Ward and Robb [408]. This first SLM consisted of an aqueous bicarbonate-carbonate solution fixed in a porous cellulose acetate film. Baker et al. [409] declared that supported liquid membranes (SLMs) are a possible solution to overcome low oxygen flux and selectivities.

In the early years, supported liquid membranes were called facilitated transport membranes. There are some disadvantages with supported liquid membranes: They degrade easily and have overly large membrane thicknesses, which influence the flux and selectivity of the membranes in a negative way. However, to eliminate this disadvantage, some developments have occurred. Lee et al. [410] developed SLM in which nanosized liquid domains were dispersed uniformly in the solid polymer matrix by using phase separation techniques to stabilize the supported liquid membrane. Ionic liquids were chosen as the liquid phase in the supported liquid membrane. In this system, the solvent was gradually evaporated under controlled conditions to induce the thermodynamically unstable state of the casting film. As a result, the ionic liquid domains were formed within the casting film and became bigger with decreasing solvent concentration in the cast film; the domain size can be controlled through determining the phase separation conditions in terms of the rate and time of solvent evaporation, temperature and quenching conditions. Several authors have reported a variety of active liquids that facilitate the transport of several gases. Cellulose acetate membrane used by Ward and Robb [408] for SLM was capable of separating CO_2 and O_2 with a selectivity of 4,100 using an aqueous bicarbonate-carbonate solution.

Chen et al. [411] developed a liquid membrane using hemoglobin as the carrier in a flat microporous sheet and obtained a maximum O_2 permeance of 1.4×10^{-9} mol m^{-2} s^{-1} Pa^{-1} at 720 Pa with a maximum O_2/N_2 selectivity of 18. Castro-Domínguez et al. [412] used perfluorotributylamine (PFTBA) supported on porous alumina for the separation of O_2/N_2 and H_2/N_2 at 40 °C and 1 atm. The membrane had an average O_2/N_2 separation factor of about 60 with an O_2 permeance of 8×10^{-10} mol m^2 s^{-1} Pa^{-1}, an average H_2/N_2 separation factor of 100, and a H_2 permeance of 1×10^{-9} mol m^2 s^{-1} Pa^{-1}. The permeance of the perfluorocarbon membrane was correlated with the gas molecular size as $H_2>O_2>N_2$, suggesting that the liquid forms pockets for accommodating these gases. Although the PFTBA SLM presented an excellent separation performance, the poor stability of the membrane is a serious issue that requires significant improvements.

Deetz [413] studied the limitations of ILMs and recommended the following:

1. The stability of ILMs can be significantly improved by selecting a support with very small pore diameters.
2. Ultrathin liquid membranes can be fabricated down to 1 μm in thickness by forming in the skin layer of an porous asymmetric polymer membranes by the methods in which the liquid is selectivity deposited in the skin rather than the backing support. The wide variety of pore sizes and membrane configurations available in asymmetric membranes allows for good flexibility in the design of an ultrathin liquid membrane system.

Deetz also describes the advantages of utilizing liquids as the membrane substrate rather than solids as follows:

1. *High selectivity*. The large differences in the gas/vapor solubilities of various liquid phases allow for the development of high selective membranes. It is possible to fabricate stable ILMs composed of homogeneous liquids with selectivity ratios greater than 100,000 to 1.
2. *High flux*. Because of their high gas diffusion coefficients (1,000× greater than in solids) and solubilities, liquid membranes are inherently more permeable. Homogeneous liquid membranes with permeabilities approaching those of microporous membranes (100,000 Barrer) are possible. The fabrication of ILMs in ultrathin form enhances the already high flux such that, in some cases, the boundary layer of gas passing over the membrane acts as a greater barrier than the membrane itself. In this case the design of a system in which the boundary layer thickness is minimized becomes the paramount concern.
3. *No pinhole problems*. When solids are cast very thin, pinhole problems frequently occur. The occurrence of pinholes results in the convective transfer of gases across the membrane and, thus, a reduction in selectivity.
4. *Short development time*. Due to the extensive database available on liquid systems, the performance of an ILM can be predicted or easily determined. No new materials need to be developed. Because of these factors the time required to develop a highly selective membrane can be short.

The most common mass transfer measurement made in liquid membrane research is flux. Several techniques have been reported for flux measurement. Ward [414] measured the pressure drop as a function of time across a liquid membrane and calculated a mass flux from the measurement. If the low pressure side of the membrane is swept by steam or an inert gas, the concentration of the permeant species can be determined by a standard method such as chromatography. The mass or molar flux could be calculated by a material balance. To measure the flux directly, Donaldson and Quinn described a method based on radioisotope tracer technique [415]. Both sides of a permeation cell containing the liquid membrane was charged with an equal partial pressure of carbon dioxide. A small quantity $^{14}CO_2$ was introduced into the lower half of the cell and the rate of accumulation of $^{14}CO_2$ was measured by Geiger–Müller tube. This method does not create pH gradients across the membrane.

Since the major mechanism of transport in liquid membranes is diffusion for both of the transported species and the carrier complex, a knowledge of diffusion coefficients is essential for accurate description of the system performance. Many theoretical works for binary diffusion have been reported in the literature. Most theoretical developments for binary diffusion coefficients are based on modifications to the Stokes equation and are applicable to certain solvents and/or components [416, 417]. Reid et al. made a good description of predictive equations and their applicability and accuracy. In general, accuracy is decreased as the solutions become more non-ideal and/or the viscosity increases [418].

The supported ionic liquid membranes (SILMs), in which porous supports are filled with an ionic liquid (IL), have received significant attention during recent years. These membranes have been shown to be an attractive way for highly selective transport of organic compounds involved in the synthesis of pharmaceutical and fine chemicals. They can also be used for gas separation [419]. Seeberger et al. [420] showed the possibility of separating continuously gaseous compounds like CO_2, H_2S, THT (tetrahydrothiophene), and SO_2 from N_2 or CH_4 with supported ionic liquid membranes. A polymer film as support was coated by ionic liquids and tested for continuous separation of CO_2 and sulfur compounds from different gas mixtures. The influence of support properties, ionic liquid and gas flow on the achievable degree of separation, i.e., permeabilities and selectivities, was studied. The results indicated that competitive selectivities and permeabilities can be achieved, as compared to industrial processes based on polymeric membranes.

Cserjési and Vass [421] introduced four types of novel ionic liquids named as VACEM type ionic liquids (Table 3.27) and impregnated them in a porous hydrophobic PVDF membrane. VACEM type ionic liquids, which were built up of a common hexafluorophosphate anion and different cations, were tailored in order to dissolve CO_2. The permeability of H_2, N_2, and CO_2 was investigated under various gas phase pressures (2.2, 1.8, 1.4 bar) and temperatures (30, 40, 50 °C). It was observed that permeability of CO_2 was much higher than the permeability of the other two gases through the membranes prepared with VACEM type ionic liquids.

Gan et al. [422] studied the permeability of H_2, O_2, N_2, CO, and CO_2 through SILMs supported on nanofiltration membranes, applying four types of ionic liquids—(C_4-mim [NTf_2], C_{10}-mim[NTf_2], N_{8881}[NTf_2], C_8P_y[NTf_2])—with a common anion but different cations supported on nanofiltration membranes. The molecular structure of C_4-mim [NTf_2], C_{10}-mim[NTf_2], N_{8881}[NTf_2], and C_8P_y[NTf_2] are given in Fig. 3.74.

Table 3.27 Names and structural formula of the VACEM type ionic liquid

Name		Cation	Anion
VACEM 42	1-(2-Etoxy-ethyl)-3-methylimidazolium hexafluorophosphate	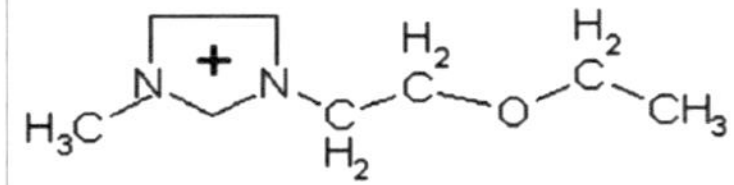	PF_6^-
VACEM 44	1-{2-[2-(Etoxy)-etoxy]-ethyl}-3-methylimidazolium hexafluorophosphate	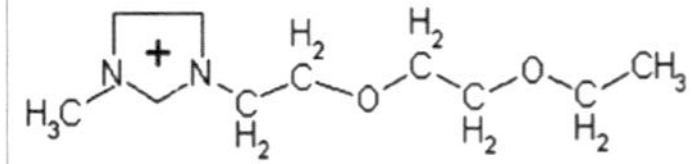	PF_6^-
VACEM 47	1-[2-{2-[2-(2-Ethoxy)-ethoxy]-ethoxy}-ethyl]-3-methylimidazolium hexafluorophosphate	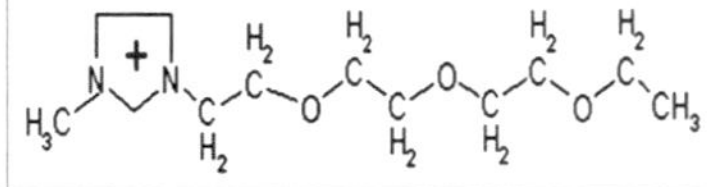	PF_6^-
VACEM 58	1-(2-Methoxy-ethyl)-3-methylimidazolium hexafluorophosphate	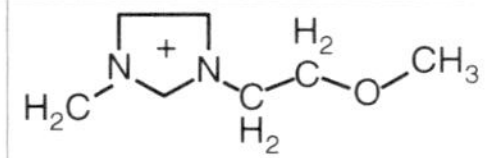	PF_6^-

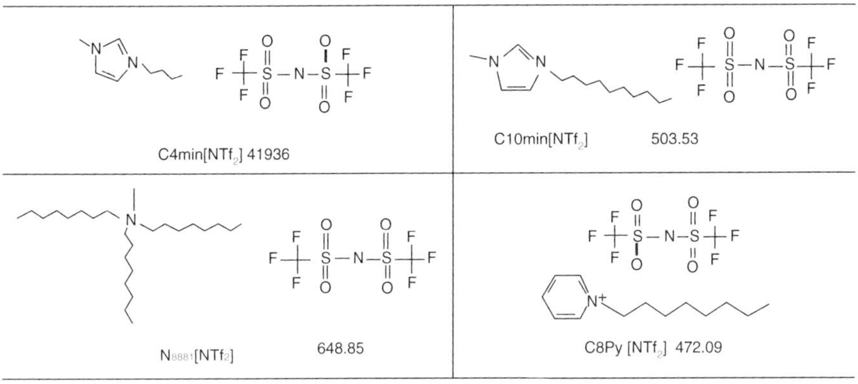

Fig. 3.74 The molecular structure of C_4-mim[NTf_2], C-mim[NTf_2], N_{8881}[NTf_2], and C_8Py[NTf_2]

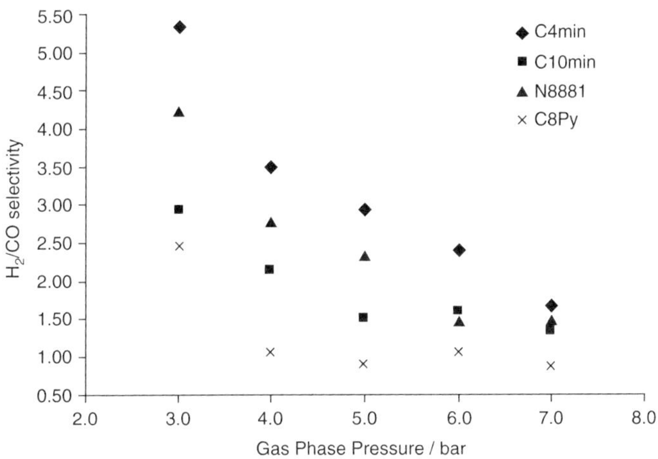

Fig. 3.75 Calculated H_2/CO selectivity based on single gas permeation measurements $T-20$ °C

Figure 3.75 shows the calculated H_2/CO selectivity based on single gas permeation measurements at 20 °C.

N_{8881}[NTf_2] had the best H_2/CO selectivity but offered a permeation rate far less than C_8Py[NTf_2], which had the best permeation performance but the worst selectivity. For all four ionic liquids, analysis of H_2/CO selectivity in single as well as binary gas feed systems established a trend of better H_2/CO selectivity associated with lower permeability at lower gas phase pressures, in contrast to lower selectivity associated with greater permeability at high pressures.

Neves et al. [423, 424] studied H_2, N_2, and CO_2 permeability and CO_2/H_2, CO_2/N_2, and H_2/N_2 ideal selectivity for (bmim)(PF_6), (omim)(PF_6), (hmim)(PF_6), and (bmim)(BF_4) IL based SILMs. All supported ionic liquid membranes (SILMs) had the highest permeability for CO_2 and the permeability was affected by the alkyl chain length of the IL cation and by the anion. It was concluded that SILMs could be used in gas separation due to their adequate permeability and high selective values.

Cserjési et al. [425] prepared supported liquid membranes (SLMs) with 12 different types of ionic liquids: two well-known ILS—([bmim][BF_4]) and ([emim][CF_3SO_3])—and ten different types of commercially available novel ILS (Ammoeng™ 100, Ecoeng™111P, Cyphos 102, Cyphos 103, Cyphos 104, Cyphos 106, Cyphos 166, Cyphos 163, Cyphos 169, and [Set_3][NTf_2]). The chemical structure, water content, purity, and the sources of the used ILS are given in Table 3.28. The supporting phase was hydrophobic porous PVDF flat sheet membrane.

Tables 3.29 and 3.30 contain the N_2, H_2, CH_4, and CO_2 permeabilities and the selectivities of the investigated SILMs.

All SILMs have the highest permeability values for CO_2 and the lowest for N_2 and the membrane permeabilities vary in the range of $37.5–210 \times 10^{18}$ $m^2\ s^{-1}\ Pa^{-1}$ for N_2, $90–840 \times 10^{18}$ $m^2\ s^{-1}\ Pa^{-1}$ for H_2, $45–847.5 \times 10^{18}$ $m^2\ s^{-1}\ Pa^{-1}$ for CH_4, and 705–$5{,}602.5 \times 10^{18}$ $m^2\ s^{-1}\ Pa^{-1}$ for CO_2. Ammoeng™ 100 has the lowest and [Set_3][NTf_2] the highest permeability for the four gases studied.

From Tables 3.29 and 3.30, it is clear that all of the SILMs are highly selective for CO_2 over the other three gases, selective for H_2 over N_2 and CH_4, and selective for CH_4 over N_2. However, there is a significant variance in the ideal selectivity results. There is a 500 % selectivity difference in CO_2/N_2, 260 % in CO_2/H_2, 406 % in CO_2/CH_4, 388 % in H_2/N_2, 187 % in H_2/CH_4, and 427% in CH_4/N_2. On comparing these preliminary selectivity results to the upper-bound values for the selectivity vs. permeability of polymer membranes for some common gas pairs (i.e., O_2/N_2, H_2/N_2, CO_2/CH_4) given by Robeson [426], most of these SILMs have better ideal-selection properties than the commonly used, industrial polymer membranes.

Hanioka et al. [427] demonstrated a support liquid membrane (SLM) based on task-specific ionic liquid to achieve the selective and facilitated CO_2 transport through the membrane. For this purpose three ionic liquids were synthesized:

1. 1-Butyl-3-methylimidazolium bis(trifluoromethylsulfonyl)imide [C_4mim][Tf_2N].
2. *N*-Aminopropyl-3-methylimidazolium bis(trifluoromethylsulfonyl)imide [C_3NH_2min][Tf_2N], and
3. *N*-Aminopropyl-3-methylimidazolium trifluoromethanesulfone [C_3NH_2min][CF_3SO_3].

Table 3.31 shows the molecular structures and abbreviations of the ionic liquids.

The porous hydrophilic polytetrafluoroethylene (PTFE) membrane was used as the support of the SLM.

Supported ionic liquid membranes (SILMs) have been used in gas separation of various gases, including CO_2/N_2, CO_2/CH_4, CO_2/H_2, CO/H_2, CO_2/He, and SO_2/CH_4. However, so far SILMs have not been used for industrial applications due to the membrane liquid loss under the high cross-membrane pressure difference (δp). It has been generally considered that the membrane liquid loss of SLMs is attributed to the high δp over the capillary force that the membrane can sustain. The maximum δp that a SLIM can resist is related to the maximum pore size of the membrane and the pore structure, the interfacial tension of membrane liquids, and the contact angle.

Table 3.28 The ionic liquids used in SILMs [425]

Short name	Name	Chemical structure	Water content (w/w, %)	Purity	Source
[bmim][BF_4]	1-Butyl-3-methylimidazolium tetrafluoroborate	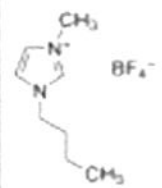	0.13	>97 %	Solvent Innovation GMbh, Cologne, Germany
Ammoeng™ 100		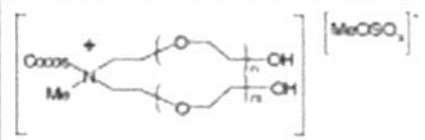	0.21	>95 %	
Ecoeng™ 111P	1,3-Dimethylimidazolium dimethylphosphate	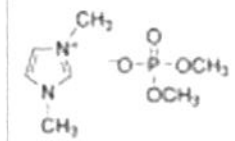	0.78	>98 %	
Cyphos 102	Trihexyltetradecylphosphonium bromide	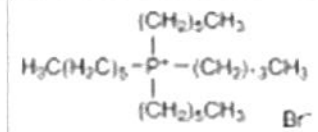	0.04	>85 %	
Cyphos 103	Trihexyltetradecylphosphonium decanoate	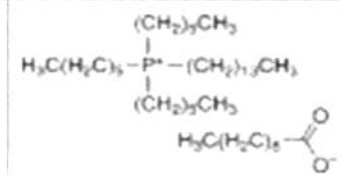	4.68	>95 %	IoLiTec GMbH & Co., KG, Germany
Cyphos 104	Trihexyltetradecylphosphonium bis(2,4,4- trimethylpentyl)phosphinate	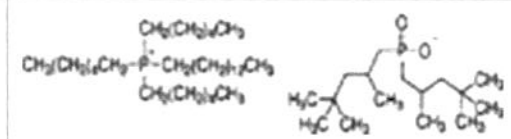	0.07	>95 %	
Cyphos106	Trisobutylmethylphosphonium tosylate	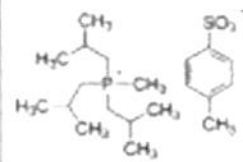	1.22	>95 %	
Cyphos 166	Tetraoctylphosphonium bromide	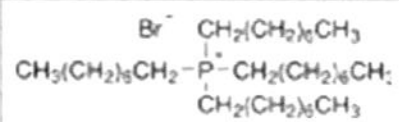	0.01	>95 %	

(continued)

Table 3.28 (continued)

Short name	Name	Chemical structure	Water content (w/w, %)	Purity	Source
Cyphos 163	Tetrabutylphosphonium bromide	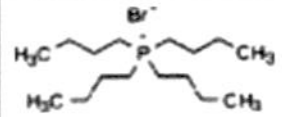	0.01	>95 %	
Cyphos 169	Ethyltributylphosphonium diethyl phosphate	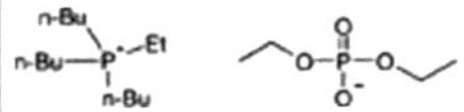	0.01	>97 %	
[emim][CF_3SO_3]	1-Ethyl-3-methylimidazolium triflate	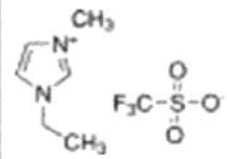	0.83	>99 %	
[Set_3][NTf_2]	Triethylsulfonium bis(trifluoromethylsulfonyl) imide	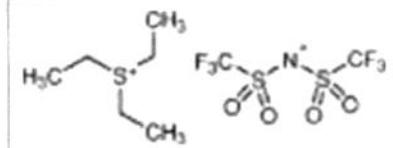	0.32	>99 %	

Table 3.29 Permeability results of the SILMs [425]

	Permeability (barrer)			
SILMs	N_2	H_2	CH_4	CO_2
[bmim][BF_4]	5.04	32.2	20.3	93.9
Ammeoeng™ 100	1.79	11.9	5.76	93.9
Ecoeng™	11.6	19.9	15.6	127
Cyphos 102	15.3	92.6	76.5	637
Cyphos 103	11.3	86.6	65.1	487
Cyphos 104	20.3	124	113	642
[emim][CF_3SO_3]	14.3	37.2	21.1	486
[Set_3][NTf_3]	28.4	112	81.2	747

Table 3.30 Selectivities results of the SILMs

	Selectivity					
SILMs	CO_2/N_2	CO_2/H_2	CO_2/CH_4	H_2/N_2	H_2/CH_4	CH_4/N_2
[bmim][BF_4]	52.3	8.18	12.9	6.40	1.58	4.04
Ammoeng™ 100	52.6	7.93	16.5	6.64	2.06	3.22
Ecoeng™	10.9	6.38	8.12	1.71	1.27	1.35
Cyphos 102	41.5	6.87	8.32	6.03	1.21	4.99
Cyphos 103	43.1	5.62	7.49	7.67	1.33	5.76
Cyphos 104	31.6	5.17	5.66	6.11	1.10	5.58
[emim][CF_3SO_3]	34.0	13.1	23.1	2.61	1.77	1.47
[Set_3][NTf_3]	36.2	6.67	9.290	3.93	1.38	2.85

Thus, several researchers have studied the stability of SILMs for gas separation. Zhao et al. [428] reported that the loss of the SILMs can be attributed to two reasons—membrane liquid loss from membrane compression and from the large pores. The latter can lead to SILM degradation, but the former does not lead to degradation owing to the decrease of membrane pore diameter under compression. Both the thickness and pore diameter of the SILM can be compressed due to δp, including the decrease in porosity. Luisa et al. [429] obtained the permeabilities of air, CO_2 and 10 vol% SO_2–air by using different SILMs. The permeability of air was one order of magnitude lower than CO_2 permeability and was also lower than the permeability of the mixture of air and 10 vol% SO_2.

Theoretically, SLM technology is one of the most efficient membrane-based methods of separation. It does not use pressure or voltage but is based on the difference of chemical energy as a driving force of the process; for example, the use of a simple H^+ concentration difference. Coupled co- or counter-ion transport allows for an active transport of the targeted species from dilute solutions into more concentrated solutions and for collecting toxic or precious species in a small volume of the acceptor solution.

Table 3.31 Ionic liquid tested in the present study

Abbreviation	Molecular structure	Water content (%)	Viscosity (mPas)
[C_3NH_2min][CF_3SO_3]	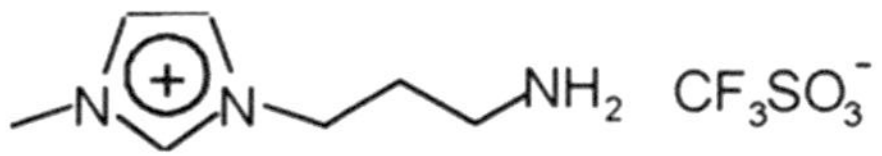	1.4	3,760
[C_3NH_2min][Tf_2N]	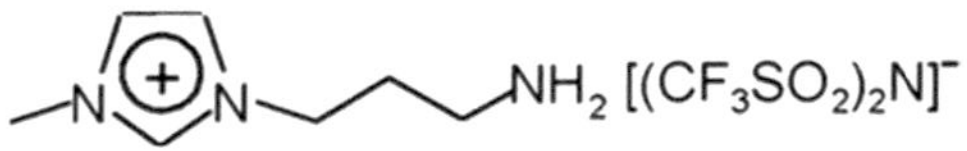	5.7	2,180
[C_4mim][Tf_2N]	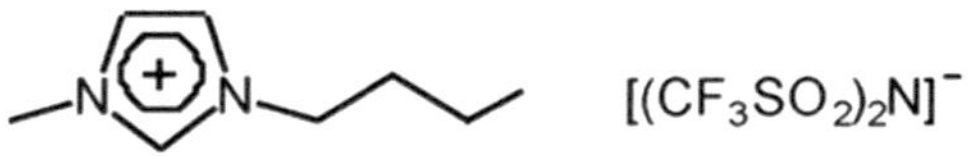	1.8	70

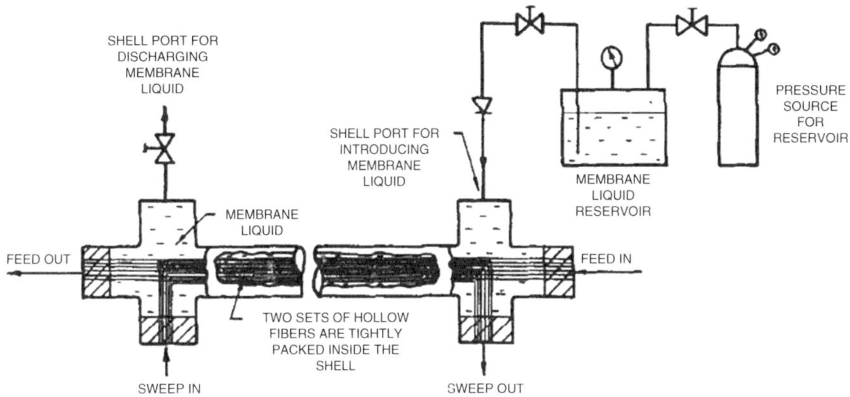

Fig. 3.76 HFCLM permeator with membrane liquid reservoir and pressure source

The primary drawback of supported liquid membrane is that over time the liquid carrier evaporates or is pushed out of membrane pores, resulting in a nonselective transport. Thick membranes could improve stability and allow a reasonable life time but with a sacrifice in gas flux. Room temperature ionic liquids with negligible vapor pressure may overcome the problem of volatility. Majumdar et al. [430] developed a new membrane separation technique for gas mixtures. In this technique feed and sweep gases flow through the lumen of two different sets of hydrophobic microporous hollow fibers while a liquid on the shell side acts as the membrane. This membrane was named the contained liquid membrane (CLM). The details are shown in Fig. 3.76.

The aqueous liquid membrane, generally maintained at a pressure higher than the feed gas and permeate gas pressures, prevents physical mixing of the feed gas or the sweep (permeate) gas. The aqueous liquid was introduced to the permeator shell side from a membrane liquid reservoir under pressure. There are two sets of hollow fibers, tightly packed together in the permeator shell, which provides a very high ratio of membrane surface area to volume. Experimental studies have been made with different CO/N, feed mixtures and a pure helium sweep stream, with special emphasis on model landfill gas purification. The experimental data showed good agreement with the theoretical predictions [430].

References

1. Sing KSW, Everett DH, Haul RAW, Moscou L, Pierotti RA, Rouquérol J, Siemieniewska T (1985) Reporting physisorption data for gas/solid systems with special reference to the determination of surface area and porosity. Pure Appl Chem 57:603–619
2. Rouquerol JJ, Avnir D, Fairbridge CW, Everett DH, Haynes JM, Pernicone N, Ramsay JDF, Sing KSW, Unger KK (1994) Recommendations for the characterization of porous solids (Technical Report). Pure Appl Chem 66:1739–1758

3. Robeson LM (1991) Correlation of separation factor versus permeability for polymeric membranes. J Membr Sci 62:165–185
4. Aoki T (1999) Macromolecular design of permselective membranes. Prog Polym Sci 24: 951–993
5. Schmeling N, Konietzny R, Sieffert D, Rölling P, Staudt C (2010) Functionalized copolyimide membranes for gaseous and liquid mixtures. Beiltein J Org Chem 6:789–800
6. Budd PM, McKeown NB (2010) High permeable polymers for gas separation membranes. Polym Chem 1:63–68
7. Powell CE, Qiao GG (2006) Polymeric CO_2/N_2 gas separation membranes for the capture of carbon dioxide from power plant flue gases. J Membr Sci 279:1–49
8. Gantzel PK, Merten U (1970) Gas separations with high flux cellulose acetate membranes. Ind Eng Chem Process Des Dev 9:331–332
9. Nunes SP, Pinemann K-V (2001) Membrane materials and membrane preparation. In: Nunes SP, Pinemann K-V (eds) Membrane technology in chemical industry. Wiley-VCH, Weinheim, FRG, pp 1–67
10. Bernardo P, Drioli E, Golemme G (2009) Membrane gas separation: a review/state of the art. Ind Eng Chem Res 48:4638–4663
11. Reddy BSR, Senthilkumar U (2003) Prospects of siloxane membrane technology for gas separation—a review. J Sci Ind Res 62:666–677
12. Achalpurkar MP, Kharul UK, Lohokare HR, Karadkar PB (2007) Gas permeation in amine functionalized silicon rubber membranes. Sep Purif Technol 57:304–313
13. Stern SA, Shah VM, Hardy BJ (1987) Structure-permeability relationships in silicone polymers. J Polym Sci 25:1263–1298
14. Furuzono T, Seki K, Kishida A, Ohshige T-A, Waki K, Maruyama I, Akashi M (1996) Novel functional polymers: poly(dimethylsiloxane)-polyamide multiblock copolymer. III. Synthesis and surface properties of disiloxane-aromatic polyamide multiblock copolymer. J Appl Polym Sci 59:1059–1065
15. Loeb S, Sourirajan S (1960) Sea water demineralization by means of a semipermeable membrane. Sea Water Research Report 60-60. UCLA, Department of Engineering
16. Dortmundt D, Doshi K (1999) Recent developments in CO_2 removal membrane technology. UOP LLC, Des Plaines, IL
17. White L (2010) Evolution of natural gas treatment with membrane systems. In: Yampolskii Y, Freeman B (eds) Membrane gas separation. John Wiley & Sons, Ltd, Chichester, UK
18. Puleo AC, Paul DR, Kelley SS (1989) The effect of degree of acetylation on gas sorption and transport behaviour in cellulose acetate. J Membr Sci 47:301–332
19. Scholes CA, Stevens GW, Kentish SE (2012) Membrane gas separation applications in natural gas processing. Fuel 96:15–28
20. Minhas BS, Matsuura T, Sourirajan S (1987) Formation of asymmetric cellulose acetate membranes for the separation of carbon dioxide–methane gas mixtures. Ind Eng Chem Res 26: 2344–2348
21. Houde AY, Krishnakumar B, Charati SG, Stern SA (1996) Permeability of dense (homogeneous) cellulose acetate membranes to methane, carbon dioxide, and their mixtures at elevated pressures. J Appl Polym Sci 62:2181–2192
22. Donohue MD, Minhas BS, Lee SY (1989) Permeation behaviour of carbon dioxide–methane mixtures in cellulose acetate membranes. J Membr Sci 42:197–214
23. Houde AY, Stern SA (1997) Solubility and diffusivity of light gases in ethyl cellulose at elevated pressures: effects of ethoxy content. J Membr Sci 127:171–183
24. Park J, Steward MG (1968) Film-forming cellulose compounds. British Patent 1120373A, 17 June 1968
25. Cooley TE, Coady AB (1978) Removal of H_2S and/or CO_2 from a light hydrocarbon stream by use of gas permeable membrane. US Patent 4130403A, 19 Dec 1978
26. Sharma AK (1985) Fluorinated cellulose acetate polymers. US Patent 4549012A, 22 Oct 1985
27. Rahman SA, Ismail AF, Abdul-Rahman WAW (2001) Formation of cellulose acetate membrane for gas separation from binary dope system: effect of shear rate. Paper presented at

Regional Symposium on Membrane Science and Technology, Puteri Pan Pacific Hotel, Johor Bharu, Malaysia, 21–25 Apr 2004
28. Vorotyntsev IV, Drozdov PN, Karyakin NV (2006) Ammonia permeability of a cellulose acetate membrane. Inorg Mater 42:231–235
29. Tanioka A, Ishikawa K, Kakuta A, Kuramoto M, Ohno M (1984) Mixed gas separation by fine porous freeze-dried cellulose acetate membrane. J Appl Polym Sci 29:583–594
30. Kim W, Lee JS, Bucknall DG, Koros WJ, Nair S (2013) Nanoporous layered silicate AMH-3/cellulose acetate nanocomposite membranes for gas separations. J Membr Sci 441:129–136
31. Scholes CA, Kentish SE, Stevens GW (2008) Separation through polymeric membrane systems for flue gas applications. Recent patents on chemical engineering 1:52–66
32. Ward WJ, Browall WR, Salemme RM (1976) Ultrathin silicon/polycarbonate membranes for gas separation processes. J Membr Sci 1:99–108
33. Acharya NK, Yadav PK, Vijay YK (2004) Study of temperature dependent gas permeability of polycarbonate membrane. Ind J Pure Appl Phys 43:179–181
34. Vijay YK, Acharya NK, Wate S, Avasthi DK (2004) Characterization of track etched membrane by gas separation. Int J Hydrogen Energy 29:515–519
35. Fu YJ, Chen JT, Chen CC, Liao KS, Hu CC, Lee KR, Lai JY (2013) Characterization of morphology and gas separation performance of dry-cast polycarbonate membranes. Polym Eng Sci 53:1623–1630
36. Hacarlioglu P, Toppare L, Yilmaz L (2003) Polycarbonate-polypyrrole mixed matrix gas separation membranes. J Membr Sci 225:51–62
37. Sen D, Kalipecilar H, Yilmaz L (2006) Development of zeolite filled polycarbonate mixed matrix gas separation membranes. Desalination 200:222–224
38. López-González M, Saiz E, Guzmán J, Riande E (2001) Experimental and simulation studies on the transport of gaseous diatomic molecules in polycarbonate membranes. J Chem Phys 115:6728–6736
39. Yampol'skii YP, Bespalova NB, Finkel'shtein ES, Bondar VI, Papov AV (1994) Synthesis, gas permeability, and gas sorption properties of fluorine-containing norbornene polymers. Macromolecules 27:2872–2878
40. Tetsuka H, Hagiwara M, Kaita S (2011) Addition-type poly(norbornene)s with siloxane substituents: synthesis, properties and nanoporous membrane. Polym J 43:97–100
41. Tetsuka H, Isobe K, Hagiwara M (2009) Synthesis and properties of addition-type poly(norbone)s with siloxane substituents. Polym J 41:643–649
42. Dorkenoo KD, Pfrommm PH, Rezac ME (1998) Gas transport properties of a series of high T_g polynorbornenes with aliphatic pendant groups. Polym Sci B Polym Phys 36:797–803
43. Deniz S (2006) Effect of nonsolvent type on the surface morphology and preparation of microporous membranes from blends of poly(phenylene oxide) and poly(p-phenylene oxide sulfone) or polysulfone. Desalination 200:52–54
44. Paul DR, Yampoliskii Y (eds) (1994) Polymer gas separation membranes. CRC Press, Boca Raton, FL
45. Khulbe KC, Matsuura T, Lamarche G, Kim HJ (1997) The morphology characterization and performance of dense PPO membranes for gas separation. J Membr Sci 135:211–223
46. Hamad F, Matsuura T (2005) Performance of gas separation membranes made from brominated high molecular weight poly(2,4-dimethyl-1,6-phenylene oxide). J Membr Sci 253:183–189
47. Khulbe KC, Chowdhury G, Kruczek B, Vujosevic R, Matsuura T, Lamarche G (1997) Characterization of the PPO dense membrane prepared at different temperatures by ESR, atomic force microscope and gas permeation. J Membr Sci 126:115–122
48. Hamad F, Khulbe KC, Matsuura T (2002) Characterization of gas separation membranes prepared from brominated poly(phenylene oxide) by infrared spectroscopy. Desalination 148:369–375
49. Yu B, Cong H, Zhao X (2012) Hybrid brominated sulfonated poly(2,6-diphenyl-1,4-phenylene oxide) and SiO_2 nanocomposite membranes for CO_2/N_2 separation. Prog Nat Sci Mater Int 22:661–667

50. Langsam M (1996) Polyimides for gas separation. In: Ghosh MK, Mittal KL (eds) Polyimides: fundamental and application. Mercel Dekker, New York
51. Yoon JC, Park HB (2011) Gas separation properties of triptycene-based polyimide membranes. In: Escobar IC, Bruggen BV (eds) Modern applications in membrane science and technology. ACS symposium series, vol 1078. Oxford, Washington, DC; American Chemical Society, New York, pp 107–128
52. Kim KJ, Park SH, So WW, Ahn DJ, Moon SJ (2003) CO_2 separation performances of composite membranes of 6FDA-based polyimides with a polar group. J Membr Sci 211:41–49
53. Li DF, Chung TS, Wang R, Liu Y (2002) Fabrication of fluoropolyimide/polyethersulfone (PES) dual-layer asymmetric hollow fiber membranes for gas separation. J Membr Sci 198:211–223
54. Tin PS, Chung TS, Liu Y, Wang R, Liu SL, Pramoda KP (2003) Effects of cross-linking modification of Matrimid membranes. J Membr Sci 225:77–90
55. Kapantaidakis GC, Koops GH, Wessling M, Kaldis SP, Sakellaropoulos GP (2003) CO_2 plasticization of polyethersulfone/polyimide gas-separation membranes. AIChE J 49:1702–1711
56. Bos A, Punt I, Wessling M, Strathmann H (1998) Suppression of CO_2-plasticization by semi interpenetrating polymer network formation. J Polym Sci B Polym Phys 36:1547–1556
57. Faiz R, Li K (2012) Polymeric membranes for light olefin/paraffin separation. Desalination 287:82–97
58. Dong G, Li H, Chen V (2010) Factors affect defect-free Matrimid® hollow fiber gas separation performance in natural gas purification. J Membr Sci 353:17–27
59. Peng N, Chung TS (2008) The effects of spinneret dimension and hollow fiber dimension on gas separation performance of ultra-thin defect-free Torlon® hollow fiber membrane. J Membr Sci 310:455–465
60. Chung TS, Shao L, Tin PS (2006) Surface modification of polyimide membranes by diamines for H_2 and CO_2 separation. Macromol Rapid Commun 27:998–1003
61. Liu Y, Wang R, Chung TS (2001) Chemical cross-linking modification of polyimide films for gas separation. J Membr Sci 189:231–239
62. Wang L, Cao Y, Zhou M, Qiu X, Yuan Q (2009) Synthesis, characterization, and gas permeation properties of 6FDA-2,6-DAT/mPDA copolyimides. Front Chem Chin 4:215–221
63. Barsema JN, Kapantaidakis GC, van der Vegt NFA, Koops GH, Wessling M (2003) Preparation and characterization of highly selective dense and hollow fiber asymmetric membranes based on BTDA-TDI/MDI co-polyimide. J Membr Sci 216:195–205
64. Kneifel K, Peinemann KV (1992) Preparation of hollow fiber membranes from polyetherimide for gas separation. J Membr Sci 65:295–307
65. Wang D, Teo WK, Li K (2002) Permeation of H_2, N_2, CH_4, C_2H_6 and C_3H_8 through asymmetric poly(etherimide) hollow-fiber membrane. J Appl Polym Sci 86:698–702
66. Wang D, Li K, Teo WK (1998) Preparation and characterization of polyetherimide asymmetric hollow fiber membranes for gas separation. J Membr Sci 138:193–201
67. Wang D, Li K, Teo WK (2002) Preparation of asymmetric polyetherimide hollow fibre membrane with high gas selectivity. J Membr Sci 208:419–426
68. Pientka Z, Brožová L, Bleha M, Puri P (2003) Preparation and characterization of ultrathin polymeric films. J Membr Sci 214:157–161
69. Bruma M, Hamciuc E, Yampolskii Y, Alentiev A, Ronova IA, Rojkov EM (2004) Polyetherimides for gas separation membranes. Mol Cryst Liq Cryst 418:11–19
70. Squire EN (1988) Amorphous copolymers of perfluoro-2,2-dimethyl-1,3-dioxole. EP0645406 B1, 11 Apr 2001
71. Markel TC, Pinnau I, Prabhakar R, Freeman B (2006) Gas and transport properties of perfluoropolymers. In: Yampolskii Y, Pinnau I, Freeman B (eds) Materials science of membrane for gas and vapor separation. John Wiley & Sons, Chichester, UK
72. Pinnau I, Toy LG (1996) Gas and vapour transport properties of amorphous perfluorinated copolymer membranes based on 2,2-bistrifluoromethyl-4,5-difluoro-1,3-dioxole/tetrafluoroethylene. J Membr Sci 109:125–133
73. Alentiev AY, Yampolskii YP, Shantarovich VP, Nemser SM, Plate NA (1997) High transport parameters and free volume of perfluoroxole copolymers. J Membr Sci 126:123–132

74. Nemser SM, Roman IC (1991) Polymer of perfluoro-2,2-dimethyl-1,3-dioxole membranes. US Patent 5051114A, 24 Sept 1991
75. Masaru N, Isamu K, Kazuya O, Gen K, Masashi M, Shunichi S, Motoi K (1990) Novel fluorine-containing cyclic polymer. US Patent 4,897,457, 30 Jan 1990
76. Avery DL, Shanbhag PV (2002) Designed selectivity gas permeable membranes. US Patent 6406517B1, 18 June 2002
77. Tokarev A, Friess K, Machkova J, Sipek M, Yamapolskii Y (2006) Sorption and diffusion of organic vapors in amorphous Teflon AF2400. J Polym Sci B Polym Phys 44:832–844
78. Jansen JC, Macchione M, Drioli E (2005) High flux asymmetric gas separation membranes of modified poly(ether ether ketone) prepared by the dry phase inversion technique. J Membr Sci 255:167–180
79. Galland G, Lam TM (1993) Permeability and diffusion of gases in segmented polyurethanes: structure–properties relations. J Appl Polym Sci 50:1041–1058
80. Chen SH, Yu KC, Houng SL, Lai JY (2000) Gas transport properties of HTPB based polyurethane/cosalen membrane. J Membr Sci 173:99–106
81. Sadeghi M, Semsarzadeh MA, Barikani M, Chenar MP (2011) Gas separation properties of polyether-based polyurethane–silica nanocomposite membranes. J Membr Sci 376: 188–195
82. Talakesh MM, Morteza S, Chenar MP, Afsaneh K (2012) Gas separation properties of poly(ethylene glycol)/poly(tetramethylene glycol) based polyurethane membranes. J Membr Sci 415–416:469–477
83. Liang W, Martin CR (1991) Gas transport in electronically conductive polymers. Chem Mater 3:390–391
84. Martin CR, Liang W, Menon V, Parthasarathy R, Parthasarathy A (1993) Electronically conductive polymers as chemically-selective layers in membrane-based separations. Synth Met 55–57:3766–3773
85. Anderson MR, Mattes BR, Reiss H, Kaner RB (1991) Conjugated polymer films for gas separations. Science 252:1412–1415
86. Sairam M, Nataraj SK, Aminabhavi TM, Roy M, Madhusoodana CD (2006) Polyaniline membranes for separation and purification of gases, liquids, and electrolyte solutions. Sep Purif Rev 35:249–283
87. Blinova NV, Frantisek S (2012) Functionalized polyaniline-based composite membranes with vastly improved performance for separation of carbon dioxide from methane. J Membr Sci 423–424:514–521
88. Kuwabata S, Martin CR (1994) Investigation of the gas-transport properties of polyaniline. J Membr Sci 91:1–12
89. Lee YM, Ha SY, Lee KY, Suh DH, Hong SY (1999) Gas separation through conductive polymer membranes. 2. Polyaniline membranes with high oxygen selectivity. Ind Eng Chem Res 38:1917–1924
90. Illing G, Hellgardt K, Wakeman RJ, Jungbauer A (2001) Preparation and characterisation of polyaniline based membranes for gas separation. J Membr Sci 184:69–78
91. Hasbullah H, Kumbharkar S, Ismail AF, Li K (2011) Preparation of polyaniline asymmetric hollow fiber membranes and investigations towards gas separation performance. J Membr Sci 366:116–124
92. Julian H, Wenten G (2012) Polysulfone membranes for CO_2/CH_4 separation: state of the art. IOSR J Eng 2:484–495
93. Dai Y, Guiver MD, Robertson GP, Kang YS, Lee KJ, Jho JY (2004) Preparation and characterization of polysulfones containing both hexafluoroisopropylidine and trimethylsilyl groups as gas separation membrane materials. Macromolecules 37:1403–1410
94. Dai Y, Guiver MD, Robertson GP, Kang YS, Lee KJ (2003) Enhancement in the gas permeabilities of novel polysulfones with pendant 4-trimethylsilyl-α-hydroxylbenzyl substituents. Macromolecules 36:6807–6816
95. Mc Hattie JS, Koros WJ, Paul DR (1991) Gas transport properties of polysulphones: 1. Role of symmetry of methyl group placement on bisphenol rings. Polymer 32:840–850

96. Marchese J, Ochoa N, Pagliero C (1995) Preparation and gas separation performance of silicone-coated polysulfone membranes. J Chem Technol Biotechnol 63:329–336
97. Wang D, Teo WK, Li K (2002) Preparation and characterization of high-flux polysulfone hollow fiber gas separation membranes. J Membr Sci 204:247–256
98. Ahn J, Chung WJ, Pinnau I, Guiver MD (2008) Polysulfone/silica nanoparticle mixed-matrix membranes for gas. J Membr Sci 314:123–133
99. Weng TH, Tseng HH, Wey MY (2009) Preparation and characterization of multi-walled carbon nanotube/PBNPI nanocomposite membrane for H_2/CH_4 separation. Int J Hydrogen Energy 34:8707–8715
100. Mao Z, Jie X, Cao Y, Wang L, Li M, Yuan Q (2011) Preparation of dual-layer cellulose/polysulfone hollow fiber membrane and its performance for isopropanol dehydration and CO_2 separation. Sep Puri Technol 77:179–184
101. Arahman N, Arifin B, Mulyati S, Ohmukai Y, Matsuyama H (2012) Structure change of polyethersulfone hollow fiber membrane modified with pluronic F127, polyvinylpyrrolidone, and Tetronic 1307. Mater Sci Appl 3:72–77
102. Wang D, Li K, Teo WK (2000) Highly permeable polyethersulfone hollow fiber gas separation membranes prepared using water as non-solvent additive. J Membr Sci 176:147–158
103. Borneman Z, Vant's Hof JA, Smolders CA, Van vee HM (1986) Hollow fiber gas separation membranes: structure and properties. In: Proceedings of the fourth BOC Priestley conference. Royal Society of Chemistry, London, 16–18 Sept 1986
104. Vant's Hof JA (1988) Wet spinning of asymmetric hollow fiber membranes for gas separation. Ph.D. Thesis, Twente University, The Netherlands
105. Van't Hof JA, Reuvers AJ, Boon RM, Rolevink HHM, Smolders CA (1992) Preparation of asymmetric gas separation membranes with high selectivity by a dual-bath coagulation method. J Membr Sci 70:17–30
106. Li SG, Koops GH, Mulder MHV, Van den Boomgaard T, Smolders CA (1994) Wet spinning of integrally skinned hollow fiber membranes by a modified dual-bath coagulation method using a triple orifice spinneret. J Membr Sci 94:329–340
107. Kesting RE, Fritzsche AK, Murphy MK, Handermann AC, Cruse CA, Malon RF (1989) Process for forming asymmetric gas separation membranes having graded density skins. US Patent 4 871,494, 3 Oct 1989
108. Fritzsche AK, Cruse CA, Murphy MK, Kesting RE (1990) Polyethersulfone and polyphenylsulfone fiber trilayer membranes spun from Lewis acid:base complexes—structure determination by SEM, DSC and oxygen plasma ablation. J Membr Sci 54:29–50
109. Wang D (1995) Polyethersulfone hollow fiber gas separation membranes prepared from solvent systems containing nonsolvent additives. Ph.D. Thesis, National University of Singapore
110. Wang D, Li K, Teo WK (1996) Polyethersulfone hollow fiber gas separation membranes prepared from NMP/alcohol solvent systems. J Membr Sci 115:85–108
111. Kim DH, Ko YH, Kim TW, Park JS, Lee HK (2012) Separation of N_2/SF_6 binary mixtures using polyethersulfone (PESf) hollow fiber membrane. Korean J Chem Eng 29:1081–1085
112. Jiang L, Chung TS, Li DF, Cao C, Kulprathipanja S (2004) Fabrication of Matimid/polyethersulfone dual-layer hollow fiber membranes for gas separation. J Membr Sci 240: 91–103
113. Ismail AF, Norida R, Rahman WAW, Matsuura T, Hashemifard SA (2011) Preparation and characterization of hyperthin-skinned and high performances asymmetric polyethersulfone membrane for gas separation. Desalination 273:93–104
114. Pesiri DR, Jorgensen B, Dye RC (2003) Thermal optimization of polybenzimidazole meniscus membranes for the separation of hydrogen, methane, carbon dioxide. J Membr Sci 218: 11–18
115. Berchtold KA, Young JS, Dudeck KW (2006) High temperature separation membranes of hydrogen purification and carbon capture. LALP-06-043, Mar 2006
116. Choi S, Coronas J, Lai Z, Yust D, Onorato F, Tsapatsis M (2008) Fabrication and gas separation properties of polybenzimidazole (PBI)/nanoporous silicates hybrid membranes. J Membr Sci 316:145–152

117. Hosseini SS, Peng N, Chung TS (2010) Gas separation membranes developed through integration of polymer blending and dual-layer hollow fiber spinning process for hydrogen and natural gas hollow fiber spinning process for hydrogen and natural gas enrichments. J Membr Sci 349:156–166
118. Kumbharkar SC, Liu Y, Li K (2011) High performance polybenzimidazole based asymmetric hollow fiber membranes for H_2/CO_2 separation. J Membr Sci 375:231–240
119. Young JSY, Long GS, Espinoza BF (2006) Cross-linked polybenzimidazole membrane for gas separation. US Patent 20060021502A1, 2 Feb 2006
120. Kong J, Li K (2001) Preparation of PVDF hollow-fiber membranes via immersion precipitation. J Appl Polym Sci 81:1643–1653
121. Shen Y, Lua AC (2012) Preparation and characterization of mixed matrix membranes based on PVDF and three inorganic fillers (fumed nonporous silica, zeolite 4A and mesoporous MCM-41) for gas separation. Chem Eng J 192:201–210
122. Consolati G, Pegoraro M, Quasso F, Severini F (2001) Chlorinated PTMSP membranes: permeability, free volume and physical properties. Polymer 42:1265–1269
123. Masuda T, Isobe E, Higashimura T, Takada K (1983) Poly[1-(trimethylsilyl)-1-propyne]: a new high polymer synthesized with transition-metal catalysts and characterized by extremely high gas permeability. J Am Chem Soc 105:7473–7474
124. Nagai K, Masuda T, Nakagawa T, Freeman BD, Pinnau I (2001) Poly[1-(trimethylsilyl)-1propyne and related polymers: synthesis properties and functions. Prog Polym Sci 26: 721–798
125. Ichiraku Y, Stern SA, Nakagawa T (1987) An investigation of the high gas permeability of poly(1-trimethylsilyl-1-propyne). J Membr Sci 34:5–18
126. Merkel TC, He Z, Pinnau I, Freeman BD, Meakin P, Hill AJ (2003) Effect of nanoparticles on gas sorption and transport in poly(1-trimethylsilyl-1-propyne). Macromolecules 36: 6844–6855
127. Woo M, Choi J, Tsapatsis M (2008) Poly(1-trimethylsilyl-1-propyne)/MFI composite membranes for butane separation. Microporous Mesoporous Mater 110:330–338
128. Qiu J, Zheng JM, Peinemann KV (2006) Gas transport properties in a novel poly (trimethylsilylpropyne) composite membrane with nanosized organic filler trimethylsilylglucose. Macromolecules 39:4093–4100
129. Peter J, Peinemann KV (2009) Multilayer composite membrane for gas separation based on crosslinked PTMSP gutter layer and partially crosslinked Matrmid® 5218 selective layer. J Membr Sci 340:62–72
130. Vopiča O, De Angelis MG, Sarti GC (2014) Mixed gas sorption in glassy polymeric membranes: I. CO_2/CH_4 mixtures sorption in poly(1-trimetylsilyl-1-propyne) (PTMSP). J Membr Sci 449:97–108
131. Xiao S, Feng X, Huang RYM (2007) Trimesoyl chloride crosslinked membranes for CO_2/N_2 separation and pervaporation dehydration of isopropanol. J Membr Sci 306:36–46
132. Papanceaa A, Valente AJM, Patachia S, Lobo VMM (2009) Poly (vinyl alcohol) (PVA)-based polymer membranes. Nova, NY
133. Zou J, Ho WSW (2006) CO_2-selective polymeric membranes containing amines in cross-linked poly(vinyl alcohol). J Membr Sci 286:310–332
134. Matsuyama H, Terada A, Nakagawara T, Kitamura Y, Teramoto Y (1999) Facilitated transport of CO_2 through polyethylenimine/poly(vinyl alcohol) blend membrane. J Membr Sci 163:221–227
135. Park YI, Lee KH (2001) Preparation of water-swollen hydrogel membranes for gas separation. J Appl Polym Sci 80:1785–1791
136. Jenkins AD, Kratochvíl P, Stepto RFT, Suter UW (1996) Glossary of basic terms in polymer science. Pure Appl Chem 68(12):2287–2311
137. Paul DR, Newman S (eds) (1978) Polymer blends. Academic, London
138. Schmidt JJ, Gardella JA Jr, Salvati L Jr (1989) Surface studies of polymer blends. 2. An ESCA and IR study of poly(methyl methacrylate)/poly(vinyl chloride) homopolymer blends. Macromolecules 22:4489–4495

139. Coleman MM, Painter PC (1976) Fourier transforms infrared studies of polymeric materials. J Macromol Sci C 16:197–313
140. Yoshino M, Ito K, Okamoto KI (2000) Effects of hard-segment polymers on CO_2/N_2 gas-separation properties of poly(ethyleneoxide)-segmented copolymers. J Polym Sci B Polym Phys 38:1707–1715
141. Zimmerman CM, Koros WJ (1999) Polypyrrolones for membrane gas separation. I. Structural comparison of gas transport and sorption properties. J Polym Sci B Polym Phys 37: 1235–1249
142. Zimmerman CM, Koros WJ (1999) Polypyrrolones for membrane gas separation. II. Activation energies and heat of sorption. J Polym Sci B Polym Phys 37:1251–1265
143. Patil VE, van der Broeke LJP, Vercauteren FF, Keurentjes JTF (2006) Permeation of supercritical carbon dioxide through polymeric hollow fiber membranes. J Membr Sci 271:77–85
144. Wang M, Yang D, Wang Z, Wang J, Wang S (2010) Effects of pressure and temperature on fixed-site carrier membrane for CO_2 separation from natural gas. Front Chem Eng Chin 4:127–132
145. Semsarzadeh MA, Ghalei B (2012) Characterization and gas permeability of polyurethane and polyvinyl acetate blend membranes with polyethylene oxide-polypropylene oxide block copolymer. J Membr Sci 401–402:97–108
146. Car A, Stropnik C, Yave W, Peinemann KV (2008) PEG modified poly(amide-b-ethylene oxide) membranes for CO_2 separation. J Membr Sci 307:88–95
147. Yave W, Car A, Peinemann KV, Shaikh MQ, Rätzke K, Faupel F (2009) Gas permeability and free volume in poly(amide-b-ethylene oxide)/polyethylene glycol blend membranes. J Membr Sci 339:177–183
148. Yave W, Car A, Funari SS, Nunes SP, Peinemann KV (2010) CO_2-philic polymer membrane with extremely high separation performance. Macromolecules 43:326–333
149. Madaeni SS, Nooripour RM, Vatanpour V (2012) Preparation and characterization of polyimide and polyethersulfone blend membranes for gas separation. Asia-Pacific J Chem Eng 7:747–754
150. Kapantaidakis GC, Koops GH, Wessling M (2002) Preparation and characterization of gas separation hollow fiber membranes based on polyethersulfone-polyimide miscible blends. Desalination 145:353–357
151. Kapantaidakis GC, Koops GH (2002) High flux polyethersulfone-polyimide blend hollow fiber membrane for gas separation. J Membr Sci 204:153–171
152. Kapantaidakis GC, Koops GH, Wessling M (2002) Effect of spinning conditions on the structure and the gas permeation properties of high flux polyethersulfone-polyimide blend hollow fibers. Desalination 144:121–125
153. Koros WJ, Woods DG (2001) Elevated temperature application of polymer hollow-fiber membranes. J Membr Sci 181:157–166
154. Seo Y, Kim S, Hong SU (2006) Highly selective polymeric membranes for gas separation. Polymer 47:4501–4504
155. Okamoto K, Tanaka K, Muraoka M, Kita H, Maruyama Y (1992) Gas permeability and permselectivity of fluorinated polybenzoxazoles. J Polym Sci B 30:1215–1221
156. McKeown NB (1998) Phthalocyanine materials: synthesis, structure and function. CUP, Cambridge, UK
157. Ilinitch OM, Fenelonov VB, Lapkin AA, Okkel LG, Terskikh VV, Zamaraev KI (1999) Intrinsic microporosity and gas transport in polyphenylene oxide polymers. Microporous Mesoporous Mater 31:97–110
158. McKeown NB, Budd PM (2009) Polymers of intrinsic microporosity. In: Encyclopedia of polymer science and technology. John Wiley & Sons, NY
159. McKeown NB, Budd PM, Msayib KJ, Ghanem BS, Kingston HJ, Tattershall CE, Makhseed S, Reynolds KJ, Fritsch D (2005) Polymers of intrinsic microporosity (PIMs): bridging the void between microporous and polymeric materials. Chem Eur J 11:2610–2620
160. McKeown NB (2012) Review article: Polymers of intrinsic microporosity. ISRN Mater Sci. Article ID 513986, 16p

161. Makhseed S, McKeown NB, Msayib K, Bumajdad A (2005) Inducing solid-state isolation of the phthalocyanine macro-cycle by its incorporation within rigid, randomly shaped oligomers. J Mater Chem 5:1865–1870
162. Msayib K, Makhseed S, McKeown NB (2001) Synthetic strategies towards macrodiscotic materials. Can a new dimension be added to liquid crystal polymers? J Mater Chem 11:2784–2789
163. Du NY, Song J, Robertson GP, Pinnau I, Guiver M (2008) Linear high molecular weight ladder polymer via fast polycondensation of 5,5′,6,6′-tetrahydroxy- 3,3,3′,3′-tetramethylspirobisindane with 1,4-dicyanotetrafluoroben-zene. Macromol Rapid Commun 29:783–788
164. Du NY, Cin MMD, Pinnau I, Nicalek A, Robertson GP, Guiver MD (2011) Azide-based cross-linking of polymers of intrinsic microporosity (PIMs) for condensable gas separation. Macromol Rapid Commun 32:631–636
165. Du N, Park HB, Robertson GP, Dal-Cin MM, Visser T, Scoles L, Guiver MD (2011) Polymer nanosieve membranes for CO_2-capture applications. Nat Mater 10:372–375
166. Koros WJ, Mahajan R (2000) Pushing the limits on possibilities for large scale gas separation: which strategies? J Membr Sci 175:181–196
167. Roualdes S, Lee AVD, Berjoan R, Sanchez J, Durand J (1999) Gas separation properties of organosilicon plasma polymerized membranes. AIChE J 45:1566–1575
168. Won J, Kim MH, Kang YS, Park HC, Kim UY, Choi SC, Koh SK (2000) Surface modification of polyimide and polysulfone membranes by ion beam for gas separation. J Appl Polym Sci 75:1554–1560
169. Maya EM, Munoz DM, de la Campa JG, de Abajo J, Lozano AE (2006) Thermal effect on polyethyleneoxide-containing copolyimide membranes for CO_2/N_2 separation. Desalination 199:188–190
170. Li Y, Cao C, Chung TS, Pramoda KP (2004) Fabrication of dual-layer polyethersulfone (PES) hollow fiber membranes with an ultrathin dense-selective layer for gas separation. J Membr Sci 245:53–60
171. Castro-Domínguez B, Leelachaikul P, Takagaki A, Sugawara T, Kikuchi R, Oyama ST (2013) Perfluorocarbon-based supported liquid membranes for O_2/N_2 separation. Sep Purif Technol 116:19–24
172. McLeay EE, Jansen JC, Kapteijn F (2006) Zeolite based films, membranes and membrane reactors: progress and prospects. Microporous Mesoporous Mater 90:198–220
173. Barrer RM (1939) Activated diffusion in membranes. Trans Faraday Soc 35:644–656
174. Van Den Broeke LJP, Bakker WJW, Kapteijn F, Moulijn JA (1999) Binary permeation through a silicalite-1 membrane. AIChE J 45:976–985
175. Petersa TA, Fontalvoa J, Vorstmana MAG, Benesa NE, van Damb RA, Vroonb ZAEP, van Soest-Vercammenc ELJ, Keurentjesa JTF (2005) Hollow fibre microporous silica membranes for gas separation and pervaporation: synthesis, performance and stability. J Membr Sci 248:73–80
176. Uhlhorn RJR, Keizer K, Burggraaf AJ (1992) Gas transport and separation with ceramic membranes. Part II. Synthesis and separation properties of microporous membranes. J Membr Sci 66:271–287
177. de Lange RSA, Hekkink JHA, Keizer K, Burggraaf AJ (1995) Formation and characterization of supported microporous ceramic membranes prepared by sol–gel modification techniques. J Membr Sci 99:57–75
178. de Vos RM, Verweij H (1998) Improved performance of silica membranes for gas separation. J Membr Sci 143:37–51
179. Pohl PI, Heffelfinger GS (1999) Massively parallel molecular dynamics simulation of gas permeation across porous silica membrane. J Membr Sci 155:1–7
180. Zhang K, Sunarso J, Shao Z, Zhou W, Sun C, Wang S, Liu S (2011) Research progress and materials selection guidelines on mixed conducting perovskite-type ceramic membranes for oxygen production. RSC Adv 1:1661–1676

181. Kharton VV, Yaremchenko AA, Kovalevsky AV, Viskup AP, Naumovich EN, Kerko PF (1999) Perovskite-type oxides for high-temperature oxygen separation membranes. J Membr Sci 163:307–317
182. Ayral A, Julbe A, Roualdes S, Rouessac V, Durand J, Sala B (2006) Silica membranes-basic principles. Period Polytech Ser Chem Eng 50:67–79
183. Ramsay JDF (1999) Characterization of the pore structure of membranes. MRS Bull 24: 36–40
184. Topuz B, çiftçioglu M (2006) Permeation of pure gases through silica membranes with controlled pore size. Desalination 200:80–82
185. Shelekhin AB, Dixon AG, Ma YH (1992) Adsorption, permeation, and diffusion of gases in microporous glass membranes. J Membr Sci 75:233–244
186. Naskar MK, Kundu D, Chatterjee M (2009) Silicate-1 zeolite membranes on unmodified and modified surfaces of ceramic supports: a comparative study. Bull Mater Sci 32:537–541
187. Marković A, Stoltenberg D, Enke D, Schlünder E-U, Seidel-Morgenstern A (2009) Gas permeation through porous glass membranes. Part 1. Mesoporous glasses-Effect of pore diameter and surface properties. J Membr Sci 336:17–31
188. Structure-zeolites. www.ch.ic.ac.uk/vchemlib/course/zeolite/structure.html
189. Nomura M, Yamaguchi T, Nakao S (1997) Silicalite membranes modified by counterdiffusion CVD technique. Ind Eng Chem Res 36:4217–4223
190. Zeolite—Wikipedia, the free encyclopedia. International Zeolite Association, database of zeolite structures
191. Zeolite—Wikipedia, the free encyclopedia. Webmineral Zeolites, Dana Classification
192. Caro J, Noack M, Kolsch P, Schafer R (2000) Zeolite membranes—state of their development and perspective. Microporous Mesoporous Mater 38:3–24
193. Worathanakul P, Kongkachuichay P (2008) New SUZ-4 zeolite membrane from sol-gel technique. World Acad Sci Technol 2:11–21
194. Wong WC, Au LTY, Ariso CT, Yeung KL (2001) Effects of synthesis parameters on the zeolite membrane growth. J Membr Sci 191:143–146
195. Horri K, Tanaka K, Kita K, Okamoto K (1994) In: Proceedings of the 26th autumn meeting of Soc. Chem. Eng., Japan, p 99
196. Bernal MP, Xometritakis G, Tsapatsis M (2001) Tubular MFI zeolite membranes made by secondary (seeded) growth. Catal Today 67:101–107
197. Caro J, Noack M (2008) Zeolite membranes—recent developments and progress. Microporous Mesoporous Mater 115:215–233
198. Tompsett GA, Conner WC, Yngvesson KS (2006) Microwave synthesis of nanoporous materials. Chem Phys Chem 7:296–319
199. Li Y, Yang W (2008) Microwave synthesis of zeolite membranes: a review. J Membr Sci 316:3–17
200. Cundy CS (1998) Microwave techniques in the synthesis and modification of zeolite catalyst. A review. Collect Czech Chem Commun 63:1699–1723
201. Li Y, Chen H, Liu J, Yang W (2006) Microwave synthesis of LTA zeolite membranes without seeding. J Membr Sci 277:230–239
202. Choi J, Jeong HK, Snyder MA, Stoeger JA, Masel RI, Tespatsis M (2009) Grain boundary defect elimination in a zeolite membrane by rapid thermal processing. Science 325:590–593
203. Varoon K, Zhang X, Elyassi B, Brewer DD, Gette M, Kumar S, Lee A, Maheshwari S, Mittal A, Sung CY, Cococcioni M, Francis LF, McCormick AV, Mkhoyan A, Tsapatsis M (2011) Dispersible exfoliated zeolite nanosheets and their application as a selective membrane. Science 334:72–75
204. Liu L, Cheng M, Ma D, Hu G, Pan X, Bao X (2006) Synthesis, characterization, and catalytic properties of MWW zeolite with variable Si/Al ratios. Microporous Mesoporous Mater 94: 304–312
205. Huang A, Wang N, Caro J (2012) Stepwise synthesis of sandwich-structured composite zeolite membranes with enhanced separation selectivity. Chem Commun 48:3542–3544

206. Cheng ZL, Liu Z, Wan HL (2005) Microwave-heating synthesis and gas separation performance of NaA zeolite membrane. Chin J Chem 23:28–31
207. Yuwen L, Zhu M, Su H, You X, Deng C, Lv X (2011) Effects of synthesis parameters on hydrothermal synthesis of NaA zeolite. Adv Mater Res 148:1444–1448
208. Xu X, Yang W, Liu J, Lin L, Stroh N, Brunner H (2000) Synthesis and gas permeation properties of an NaA zeolite membrane. Chem Commun :603–604
209. Aoki K, Kusakabe K, Morooka S (1998) Gas permeation properties of A-type zeolite membrane formed on porous substrate by hydrothermal synthesis. J Membr Sci 141:197–205
210. Chen X, Yang W, Liu J, Xu X, Huang A, Lin L (2002) Synthesis of NaA zeolite membrane with high performance. J Mater Sci Lett 21:1023–1025
211. Dey KP, Kundu D, Chatterjee M, Naskar MK (2013) Preparation of NaA zeolite membranes using poly(ethyleneimine) as buffer layer, and study of their permeation behavior. J Am Chem Soc 96:68–72
212. Gies H (1986) Studies on clatherasis. VI: Crystal structure of decadodecasil 3R the missing link between zeolites and clathrasils. Z Kristallogr 175:93–104
213. Nakayama K, Suzuki K, Yoshida M, Tomita T (2006) Method of preparing DDR type zeolite membrane, DDR type zeolite membrane, and composite DDR type zeolite membrane, and method for preparation thereof. US Patent 7014680B2, 21 Mar 2006
214. van den Bergh J, Zhu W, Kapteijn F, Moulijn JA, Yajima K, Nakayama K, Tomita T, Yoshida S (2007) Natural gas purification wth a DDR zeolite membrane: permeation modeling with Maxwell–Stefan equations. Stud Surf Sci Catal 170:1021–1027
215. van den Bergh J, Zhu W, Gascon J, Moulinj JA, Kapteijn F (2008) Separation and permeation characteristics of a DDR zeolite membrane. J Membr Sci 316:35–45
216. Tomita T, Nakayama K, Sakai H (2004) Gas separation characteristics of DDR type zeolite membrane. Microporous Mesoporous Mater 68:71–75
217. Kanezashi M, O'Brien-Abraham J, Lin YS (2008) Gas permeation through DDR-type zeolite membranes at high temperatures. AICHE J 54:1478–1486
218. Xiao J, Wei J (1992) Diffusion mechanism of hydrocarbons in zeolites. 1. Theory. Chem Eng Sci 47:1123–1141
219. Himeno S, Tomita T, Suzuki K, Nakayama K, Yajima K, Yoshida S (2007) Synthesis and permeation of a DDR-type zeolite membrane for separation of CO_2/CH_4 gaseous mixtures. Ind Eng Chem Res 46:6989–6997
220. Himeno S, Takeya K, Fujita S (2010) Development of biogas separation process using DDR-type zeolite membrane. Kagaku Kogaku Ronbun 36:545–551
221. Hong M, Li S, Falconer JL, Noble RD (2008) Hydrogen purification using a SAPO-34 membrane. J Membr Sci 307:277–283
222. Li S, Falconer JL, Noble RD (2006) Improved SAPO-34 membranes for CO_2/CH_4 separations. Adv Mater 18:2601–2603
223. Li S, Martinek JG, Falconer JL, Noble RD, Gardner TQ (2005) High-pressure CO_2/CH_4 separation using SAPO-34 membranes. Ind Eng Chem Res 44:3220–3228
224. Poshusta JC, Tuan VA, Pape EA, Noble RD, Falconer JL (2000) Separation of light gas mixtures using SAPO-34 membranes. AIChE J 46:779–789
225. Zhou R, Ping EW, Funke HH, Falconer JL, Noble RD (2013) Improving SAPO-34 membrane synthesis. J Membr Sci 444:384–393
226. Li S, Carreon MA, Zhang Y, Funke HH (2010) Scale-up of SAPO-34 membranes for CO_2/CH_4 separation. J Membr Sci 352:7–13
227. Ping WE, Zhou R, Funke HH, Falconer JL, Noble RD (2012) Seeded-gel synthesis of SAPO-34 single channel and monolith membranes, for CO_2/CH_4 separations. J Membr Sci 415–416:770–775
228. Poshusta JC, Noble RD, Falconer JL (2001) Characterization of SAPO-34 membranes by water adsorption. J Membr Sci 186:25–40
229. Carreon ML, Li S, Carreon MA (2012) AlPO-18 membranes for CO_2/CH_4 separation. Chem Commun 48:2310–2312

230. Wikipedia, the free encyclopedia. ZSM-5. http://en.wikipedia.org/wiki/ZSM-5. Accessed
231. Poshusta JC, Noble RD, Falconer JL (1999) Temperature and pressure effects on CO_2 and CH_4 permeation through MFI zeolite membranes. J Membr Sci 160:115–123
232. Takata Y, Tsuru T, Yoshioka T, Asaeda M (2002) Gas permeation properties of MFI zeolite membranes prepared by the secondary growth of colloidal silicate and application to the methylation of toluene. Microporous Mesoporous Mater 54:257–268
233. Kwon WT, Kim SR, Kim EB, Bae SY, Kim Y (2011) H_2/CO_2 gas separation characteristics of zeolite membrane at high temperature. Adv Mater Res 26–28:267–270
234. Welk ME, Nenoff TM (2004) H_2 separation through zeolite thin film membranes. Prep Pap Am Chem Soc Div Fuel Chem 40:889–890
235. Richter H, Voigt I, Fischer G, Puhlfürβ P (2003) Preparation of zeolite membranes on the inner surface of ceramic tubes and capillaries. Sep Purif Technol 32:133–138
236. Aoki K, Tuan VA, Falconer JL, Noble RD (2000) Gas permeation properties of ion-exchanged ZSM-5 zeolite membranes. Microporous Mesoporous Mater 39:485–492
237. Tuan VA, Noble RD, Falconer JL (2000) Boron-substituted ZSM-5 membranes: preparation and separation performance. AIChE J 46:1201–1208
238. Wang H, Lin YS (2012) Synthesis and modification of ZSM-5/silicate bilayer membrane with improved hydrogen separation performance. J Membr Sci 396:128–137
239. Cheng Y, Li JS, Wang LJ, Sun XY, Liu XD (2006) Synthesis and characterization of Ce-ZSM-5 zeolite membranes. Sep Purif Technol 51:210–218
240. Hasegawa Y, Tanaka T, Watanabe K, Jeong BH, Kusakabe K, Morooka S (2002) Separation of CO_2/CH_4 and CO_2/N_2 systems using ion-exchaged FAU-type zeolite membranes with different Si/Al ratios. Korean J Chem Eng 19:309–313
241. Gu X, Dong J, Nenoff TM (2005) Synthesis of defect-free FAU-type zeolite membranes and separation for dry and moist CO_2/N_2 mixtures. Ind Eng Chem Res 44:937–944
242. Kumar P, Sung CY, Muraza O, Cococcioni M, Hashimi SA (2011) H_2S adsorption by Ag and Cu ion exchanged faujasite. Microporous Mesoporous Mater 156:127–133
243. Julbe A, Motuzas J, Cazeville F, Volle G, Guizard C (2003) Synthesis of sodalite/α-Al_2O_3 composite membranes by microwave heating. Sep Purif Technol 32:139–149
244. Xu X, Bao Y, Song C, Yang W, Liu J, Lin L (2004) Microwave-assisted hydrothermal synthesis of hydroxyl-sodalite zeolite membrane. Microporous Mesoporous Mater 75:173–181
245. Cui Y, Kita H, Okamoto KI (2004) Preparation and gas separation performance of zeolite T membrane. J Mater Chem 14:924–952
246. Chen X, Wang J, Yin D, Yang J, Lu J, Zhang Y, Chen Z (2013) High performance zeolite T membrane for dehydration of organics by a new varying temperature hot-dip coating method. AIChE J 59:936–947
247. Barrer RM, Villger H (1969) The crystal structure of the synthetic zeolite L. Z Kristallogr Bd 128:352–370
248. Tsapatsis M, Lovallo M, Okubo T, Davis ME, Sadakata M (1995) Characterization of zeolite L nanoclusters. Chem Mater 7:1734–1741
249. Yin X, Wang X, Chu N, Yang J, Lu J, Zhang Y, Yin D (2010) Zeolite L/carbon nanocomposite membrane on the porous alumina tubes and their gas separation properties. J Membr Sci 348:181–189
250. Corma A, Rey F, Rius J, Sabater MJ, Valencia S (2004) Supramolecular self-assembled molecules as organic directing agent for synthesis of zeolites. Nature 431:287–290
251. Casado-Coterillo C, Sato J, Jimare MT, Valencia S, Corma A (2012) Preparation and characterization of ITQ-29/polysulfone mixed-matrix-membranes for gas separation: effect of zeolite composition crystal size. Chem Eng Sci 73:116–122
252. Tiscornia I, Valencia S, Corma A, Téllez C, Coronas J, Santamaria J (2008) Preparation of ITQ-29 (Al-free zeolite A) membranes. Microporous Mesoporous Mater 110:303–309
253. Huang A, Caro J (2010) Steam-stable hydrophobic ITQ-29 molecular sieve membrane with H_2 selectivity prepared by secondary growth using Krytofix 222 as SDA. Chem Commun 46:7748–7750

254. Moscoso JG, Lewis GJ, Miller MA, Jan DY, Patton RL, Rohde LM (2003) UZM-5, UZM-5P and UZM-6: crystalline aluminosilicate zeolite and processes using the same. US Patent 6613302 B1, 2 Sept 2003
255. Blackwell CS, Broach RW, Gatter MG, Holmgren JS, Jan DY, Lewis GJ, Mezza BJ, Mezza TM, Miller MA, Moscoso JG, Patton RL, Rohde LM, Schoonover MW, Sinkler W, Wilson BA, Wilson ST (2003) Open-framework materials synthesized in the TMA+TEA+Mixed-Template system: the new low Si/Al ratio zeolites UZM-4 and UZM5. Angew Chem Int Ed 42:1737–1740
256. Liu C, Moscoso JG, Wilson ST (2012) Microporous UZM-5 inorganic zeolite membranes for gas, vapor, and liquid separations. US Patent 20120240763 A1, 27 Sept 2012
257. Maghsoodloorad H, Mirfendereski SY, Mohammadi T, Pak A (2011) Effects of gel parameters on the synthesis and characteristics of W-type zeolite nanoparticles. Clay Clay Miner 59:328–335
258. Mohammdi T, Maghsoodloorad H (2012) Synthesis and characterization of ceramic membranes (W-Type) zeolite membrane. Int J Appl Ceram Technol 1:1–11
259. Li T, Pan Y, Peinemann KV, Lai Z (2013) Carbon dioxide selective mixed matrix composite membrane containing ZIF-7 nano-fillers. J Membr Sci 425–426:235–242
260. Bae TH, Lee JS, Qiu W, Koros WJ, Jones CW, Nair S (2010) A high performance gas-separation containing submicrometer-sized metal-organic framework crystals. Angew Chem Int Ed 49:9863–9866
261. Bux H, Liang F, Li Y, Cravillon J, Wiebcke M, Caro J (2009) Zeolitic imidazolate membrane with molecular sieving properties by microwave-assisted solvothermal synthesis. J Am Chem Soc 131:1600–1601
262. Li YS, Liang FY, Bux H, Feldhoff A, Yand WS, Caro J (2010) Molecular sieve membrane: supported metal-organic framework with high hydrogen selectivity. Angew Chem Int Ed 49:548–551
263. Liu Y, Hu E, Khan EA, Lai Z (2010) Synthesis of ZIF-69 membranes and separation of CO_2/CO mixture. J Membr Sci 153:36–40
264. Venna SR, Carreon MA (2010) Highly permeable zeolite imidazolate framework-8 membranes for CO_2/CH_4 separation. J Am Chem Soc 132:76–78
265. Sandström L, Sjöberg E, Hedlund J (2011) Very high flux MFI membrane for CO_2 separation. J Membr Sci 380:232–240
266. Nair S, Lai Z, Nikolakis V, Xomeritakis G, Bonilla G, Tsapatsis M (2001) Separation of close-boiling hydrocarbon mixtures by MFI and FAU membranes made by secondary growth. Microporous Mesoporous Mater 48:219–228
267. Bétard A, Bux HG, Henke S, Zacher D, Caro J, Fischer RA (2012) Fabrication of a CO_2-selective membrane by step-wise liquid-phase deposition of an alkylether functionalized pillared-layer metal-organic framework [Cu_2L_2P]*n* on a macroporous support. Microporous Mesoporous Mater 150:76–82
268. Bennett TD, Goodwin AL, Dove MT, Keen DA, Tucker MG, Barney ER, Soper AK, Bithell EG, Tan JC, Cheetham AK (2010) Structure and properties of an amorphous metal-organic framework. Phys Rev Lett 104:115503–115506
269. Ranjan R, Tsapatsis M (2009) Microporous metal organic framework membrane on porous support using the seeded growth method. Chem Mater 21:4920–4924
270. Bohrman JA, Carreon MA (2012) Synthesis of CO_2/CH_4 separation performance of Bio-MOF-1 membranes. Chem Commun 48:5130–5132
271. An J, Rosi NL (2010) Tuning MOF CO_2 adsorption properties via cation exchange. J Am Chem Soc 132:5578–5579
272. An J, Shade CM, Chengelis-Czegan DA, Petoud S, Rosi NL (2011) Zinc-adeninate metal–organic framework for aqueous encapsulation and sensitization of near-infrared and visible emitting lanthanide cations. J Am Chem Soc 133:1220–1223
273. Xomeritakis G, Naik S, Braunbarth CM, Cornelius CJ, Pardey R, Brinker CJ (2003) Organic-templated silica membranes. I. Gas and vapor transport properties. J Membr Sci 215:225–233

274. Li Y, Chung TS (2008) Exploratory development of dual-layer carbon-zeolite nanocomposite hollow fiber membrane with high performance for oxygen enrichment and natural gas separation. Microporous Mesoporous Mater 113:315–324
275. Alfaro S, Valenzuela A (2006) Zeolite membrane prepared by the dry gel method for gas separation. Adv Mater Technol Mater Proc J 8:63–66
276. Kuznicki SM (1990) Preparation of small-pored crystalline titanium molecular sieve zeolites. US Patent 4938939 A, 3 July 1990
277. Stoeger JA, Veziri CM, Palomino M, Corma A, Kanellopoulos NK, Tsapatsis MN, Karanikolos G (2012) On stability and performance of highly c-oriented $AlPO_4$-5 and CoAPO-5 membranes. Microporous Mesoporous Mater 147:286–294
278. Nagase T, Kiyozumi Y, Hasegava Y, Inoue T, Ikeda T (2007) Dehydration of concentrated acetic acid solutions by pervaporation using novel MER zeolite membranes. Chem Lett 36:594–595
279. Hasegawa Y, Nagase T, Kiyozumi Y, Mizukami F (2010) Preparation, characterization, and dehydration performance of MER-type zeolite membranes. Sep Purif Technol 73:25–31
280. Kim SJ, Yang S, Reddy GK, Smirniotis P, Dong J (2013) Zeolite membrane reactor for high-temperature water-gas shift reaction: effects of membrane properties and operating conditions. Energy Fuel 27:4471–4480
281. Yang S, Lin X, Lewis W, Suyetin M, Bichoutskaia E, Parker JE, Tang CC, Allan DR, Rizkallah PJ, Hubberstey P, Champness NR, Thomas KM, Blake AJ, Schrőder M (2012) A partially interpenetrated metal-organic framework for selective hysteretic sorption of carbon dioxide. Nat Mater 11:710–716
282. Radnedge S (2012) New holey material soaks up CO_2/News/gasworld. 18 June 2012
283. http://www.gasworld.com/news/regions/west-europe/new-holey-mat
284. Shah M, McCarthy MC, Sachdeva S, Lee AK, Jeong HK (2012) Current status of metal-organic framework membranes for gas separations: promises and challenges. Ind Eng Chem Res 51:2179–2199
285. Rowsell JLC, Yaghi OM (2004) Metal-organic frameworks: a new class of porous materials. Microporous Mesoporous Mater 73:3–14
286. Yoo Y, Lai Z, Jeong HK (2009) Fabrication of MOF-5 membranes using microwave-induced rapid seeding and solvothermal secondary growth. Microporous Mesoporous Mater 123: 100–106
287. Liu Y, Ng Z, Khan AE, Jeong HK, Ching CB, Lai Z (2009) Synthesis of continuous MOF-5 membranes on porous α-alumina substrates. Microporous Mesoporous Mater 118:296–301
288. Klinowski J, Paz FAA, Silva P, Rocha J (2011) Microwave-assisted synthesis of metal-organic frameworks. Dalton Trans 40:321–330
289. Hu Y, Dong X, Nan J, Jin W, Ren X, Xu N, Lee YM (2011) Metal-organic framework membranes fabricated *via* reactive seeding. Chem Commun 47:737–739
290. Schoedel A, Scherb C, Bein T (2010) Oriented nanoscale films of metal-organic frameworks by temperature gel-layer synthesis. Angew Chem Int Ed 49:7225–7228
291. Lu H, Zhu S (2013) Interfacial synthesis of free standing metal-organic framework-membranes. Eur J Inorg Chem 2013:1294–1300
292. Ben T, Lu C, Pei C, Xu S, Qiu S (2012) Polymer-supported and free-standing metal-organic framework membrane. Chem Eur 18:10250–10253
293. Ai X, Hu X (2003) Study on organic-inorganic hybrid membranes. Huxue Jinzhan 16:83–89
294. Chung TS, Jiang LY, Kulprathipanja S (2007) Mixed matrix membranes (MMMs) comprising organic polymers with dispersed inorganic fillers for gas separation. Prog Polym Sci 32:483–507
295. Bouma RHB, Checchetti A, Chidichimo G, Drioli E (1997) Permeation through a heterogeneous membrane: the effect of the dispersed phase. J Membr Sci 128:141–149
296. Funk CV, Lloyd DRE (2008) Zeolite-filled microporous mixed matrix (ZeoTIPS) membranes: prediction of gas separation performance. J Membr Sci 313:224–231

297. Paul DR, Kemps DR (1973) The diffusion time lag in polymer membranes containing adsorptive fillers. J Polym Sci Polym Phys 41:79–93
298. Kulprathipanja S, Neuzil RW, Li NN (1988) Separation of fluids by means of mixed matrix membranes. US Patent 4740219, 26 Apr 1988
299. Kulprathipanja S, Neuzil RW, Li NN (1992) Separation of gases by means of mixed matrix membranes. US Patent 5127925, 7 July 1992
300. Hussain M, König A (2012) Mixed-matrix membranes for gas separation, polydimethylsiloxane filled with zeolite. Chem Eng Technol 35:561–569
301. Ismail AF, Kusworo TD, Mustafa A (2008) Enhanced gas permeation of polyethersulfone mixed matrix hollow fiber membranes using novel Dynasylan Ameo silane agent. J Membr Sci 319:306–312
302. Widjojo N, Chung TS, Kulprathipanja S (2008) The fabrication of hollow fiber membranes with double-layer mixed-matrix materials for gas separation. J Membr Sci 325:326–335
303. Chaidou CI, Pantoleontos G, Koutsonikolas DE, Kaldis SP, Sakellaropoulos GP (2012) Gas separation properties of polyimide-zeolite mixed matrix membrane. Sep Purif Technol 47: 950–962
304. Boroglu MS, Gurkaynak MA (2011) Fabrication and characterization of silica modified polyimide-zeolite mixed matrix membranes for gas separation properties. Polym Bull 66: 463–478
305. Karkhanechi H, Kazemian H, Nazockdast H, Mozdianfard MR, Bidoki SM (2012) Fabrication of homogeneous polymer-zeolite nanocomposites as mixed-matrix membranes for gas separation. Chem Eng Technol 35:885–888
306. Adams R, Carson C, Ward J, Tannenbaum R, Koros W (2010) Metal organic framework mixed matrix membranes for gas separation. Microporous Mesoporous Mater 131:13–20
307. Nik OG, Chen XY, Kaliaguine S (2012) Functionalized metal organic framework-polyimide mixed matrix membranes for CO_2/CH_4 separation. J Membr Sci 413–414:48–61
308. Tanh Jeazet HB, Staudt C, Janiak C (2012) Metal–organic frameworks in mixed-matrix membranes for gas separation. Dalton Trans 41:14003–14027
309. Hu J, Cai H, Ren H, Wei Y, Xu Z, Liu H, Hu Y (2010) Mixed-matrix membrane hollow fibers of $Cu_2(BTC)_2$ MOF and polyimide for gas separation and adsorption. Ind Eng Chem Res 49:12605–12612
310. Basu S, Cano-Odena A, Vankelecom IFJ (2011) MOF-containing mixed-matrix membranes for CO_2/CH_4 and CO_2/N_2 binary gas mixture separations. Sep Purif Technol 81:31–40
311. Li Y, Chung TS, Huang Z, Kulprathipanja S (2008) Dual-layer polyethersulfone (PES)/ BTDA-TDI/MDI co-polyimide (P84) hollow fiber membrane with submicron PES-zeolite beta mixed matrix dense-selective layer for gas separation. J Membr Sci 277:28–37
312. Jiang LY, Chung TS, Kulprathipanja S (2006) Fabrication of mixed matrix hollow fibers with intimate polymer-zeolite interface for gas separation. AIChE J 52:2898–2908
313. Cong H, Radosz M, Towler BF, Shen Y (2007) Polymer-inorganic membrane for gas separation. Sep Purif Technol 55:281–291
314. Liu H (1997) Synthesis of TiO_2 nanopowder enwrapped by organic membrane with microwave induced plasma method. Huaxue Tongbao 10:44–46
315. Iwata M, Adahi T, Tomidokoro M, Ohta M, Kobayashi T (2003) Hybrid sol-gel membranes of polyacrylonitrile-tetraethoxysilane composites for gas perm selectivity. J Appl Polym Sci 88:1752–1759
316. Ahmad J, Deshmukh K, Hägg MB (2013) Influence of TiO_2 on the chemical, mechanical, and gas separation properties of polyvinyl alcohol-titanium dioxide (PVA-TiO_2) nanocomposite membrane. Int J Polym Anal Charact 18:287–296
317. Perez EV, Balkus KJ, Ferraris JP, Musselman IH (2009) Mixed-matrix membranes containing MOF-5 for gas separation. J Membr Sci 328:165–173
318. Morooka S, Kusakabe K (1999) Ceramics: getting into the 2000s, Part D. In: Advances in science and technology (Faenza, Italy), vol 16, pp 389–400

319. Gopolan S (2002) Using ceramic mixed ionic and electronic conductors for gas separation. JOM 54:26–29
320. Kulprathipanja A, Alptekin GO, Falconer JL, Way JD (2005) Pd and Pd-Cu membranes: inhibition of H_2 Permeation by H_2S. J Membr Sci 254:49–62
321. Kluiters SCA (2004) Status review on membrane system for hydrogen preparation. Intermediate Report EU Project MIGREYD NNES-2001-670, ECNC-04-102
322. Fuertes AB (2000) Adsorption-selective carbon membranes for gas separation. J Membr Sci 177:9–16
323. Jones CW, Koros WJ (1994) Carbon molecular sieve gas separation membranes. Part I. Preparation and characterization based on polyimide precursors. Carbon 32:1419–1425
324. Ismail AF, David LIB (2001) A review of the latest development of carbon membranes for gas separation. J Membr Sci 193:1–18
325. Sauf SM, Ismail AF (2004) Fabrication of carbon membranes for gas separation—a review. Carbon 42:241–259
326. Saufi SM, Ismail AF (2002) Development and characterization of polyacrylonitrile (PAN) based carbon hollow fiber membrane. Songklanakarin J Sci Technol 24:843–854
327. Song C, Wang T, Qiu Y, Qiu J, Cheng H (2009) Effect of carbonization atmosphere on the structure changes of PAN carbon membranes. J Porous Mater 16:197–203
328. Grainger D, Hägg MB (2008) The recovery of carbon molecular sieve membranes of hydrogen transmitted in natural gas networks. Int J Hydrogen Energy 33:2379–2388
329. Favas EP, Kapantaidakis GC, Nolan JW, Mitropoulos AC, Kanellopoulos NK (2007) Preparation, characterization and gas permeation properties of carbon hollow fiber membranes based on Matrimid® 5218 precursor. J Mater Process Technol 186:102–110
330. Salleh WNW, Ismail AF (2011) Fabrication and characterization of PEI/PVP-based carbon hollow fiber membranes for CO_2/CH_4 and CO_2/N_2 separation. AIChE J 58:3167–3175
331. Favvas EP, Kouvelos EP, Romanos GE, Pillatos GL, Mitropoulos AC, Kanellopoulos NK (2008) Characterization of highly selective microporous carbon hollow fiber membranes prepared from commercial co-polyimide. J Porous Mater 15:625–613
332. Rao PS, Wey MY, Tseng HH, Kumjar IA, Weng TH (2008) A comparison of carbon/nanotube molecular sieve membranes with polymer blend carbon molecular sieve membranes for the gas permeation application. Microporous Mesoporous Mater 113:499–510
333. Kim YK, Park HB, Lee YM (2004) Carbon molecular sieve membranes derived from thermally labile polymer containing blend polymers and their gas separation properties. J Membr Sci 343:9–17
334. Kim YK, Park HB, Lee YM (2005) Gas separation properties of carbon molecular sieve membranes derived from polyimide/polyvinylpyrrolidone blends: effect of the molecular weight of polyvinylpyrrolidone. J Membr Sci 251:159–167
335. Lee HJ, Suda H, Haraya K, Moon SH (2007) Gas permeation properties of carbon molecular sieving membranes derived from the polymer blend of polyphenylene oxide (PPO)/polyvinylpyrrolidone (PVP). J Membr Sci 296:139–146
336. Zhang B, Wang T, Wu Y, Liu Q, Liu S, Qiu J (2008) Preparation and gas permeation of composite carbon membranes from poly(phthalazinone ether sulfone ketone). Sep Purif Technol 60:259–263
337. Yoshimune M, Fujiwara I, Suda H, Haraya K (2005) Novel carbon molecular sieve membranes derived from poly (phenylene oxide) and its derivatives for gas separation. Chem Lett 34:958–959
338. Hatori H, Yamada Y, Shiraishi M (1992) Preparation of macroporous carbon films from polyimide by phase inversion method. Carbon 30:303–304
339. Hatori H, Shiraishi M, Nakata H, Yoshitomi S (1992) Carbon molecular sieve films from polyimide. Carbon 30:305–306
340. Rao MB, Sirkar S (1993) Nanoporous carbon membranes for separation of gas mixtures by selective surface flow. J Membr Sci 85:253–254
341. Rao MB, Sirkar S (1996) Performance and pore characterization of nanoporous carbon membrane for gas separation. J Membr Sci 110:109–118

342. Rao MB, Sirkar S (1993) Nanoporous carbon membrane for gas separation. Gas Sep Purif 7:279–284
343. Centeno TA, Vilas JL, Fuertes AB (2004) Effect of phenolic resin pyrolysis conditions on carbon membrane performance for gas separation. J Membr Sci 228:45–54
344. Acharya M, Foley HC (1991) Spray-coating of nanoporous carbon membranes for air separation. J Membr Sci 161:1–5
345. Centeno TA, Fuertes AB (2000) Carbon molecular sieve gas separation membranes based on poly(vinylidene chloride-co-vinyl chloride). Carbon 38:1067–1073
346. Park HB, Lee YM (2003) Pyrolytic carbon-silica membrane: a promising membrane for improved gas separation. J Membr Sci 213:263–272
347. Hosseini SS, Chung TS (2009) Carbon membranes from blends of PBI and polyimides for N_2/CH_4 and CO_2/CH_4 separation and hydrogen purification. J Membr Sci 328:174–185
348. Zhang B, Shen G, Wu Y, Wang T, Qiu J, Xu T, Fu C (2009) Preparation and characterization of carbon membranes derived from poly(phthalazinone ether) for gas separation. Ind Eng Chem Res 48:2886–2890
349. Iijima S (1991) Helical microtubules of graphitic carbon. Nature 354:56–58
350. Iijima S, Ichihashi T (1993) Single-shell carbon nanotubes of 1-nm diameter. Nature 663:603–605
351. Cho SJ, Shrestha SP, Lee SB, Boo JH (2014) Electrical characteristics of carbon nanotubes by plasma and microwave surface treatments. Bull Korean Chem Soc 35(3):905–907
352. Wang X, Li Q, Xie J, Jin Z, Wang J, Li Y, Jiang K, Fan S (2009) Fabrication of ultralong and electrically uniform single-walled carbon nanotubes on clean substrates. Nano Lett 9: 3137–3141
353. Zhao YL, Stoddart JF (2009) Noncovalent functionalization of single-walled carbon nanotubes. Acc Chem Res 42:1161–1171
354. Bernhole J, Brenner D, Buongiorno Nardelli M, Meunier V, Roland C (2002) Mechanical and electrical properties of nanotubes. Annu Rev Mater Sci 32:347–375
355. Sholl DS, Johnson JK (2006) Making high-flux membranes with carbon nanotubes. Science 312:1003–1004
356. Noy A (2013) Kinetic model of gas transport in carbon nanotube channels. J Phys Chem C 117:7656–7660
357. Majumder M, Ajayan PM (2010) Carbon nanotube membranes: a new frontier in membrane science. In: Drioli E, Giorno L (eds) Comprehensive membrane science and engineering. Elsevier Science, Amsterdam
358. Bruggen BV (2012) The separation power of nanotubes in membranes: a review. ISRN Nanotechnol 2012:1–17
359. Arora G, Sandler SI (2006) Air separation by single wall carbon nanotubes: mass transport and kinetic selectivity. J Chem Phys 124:084702
360. Chen H, Sholl DS (2006) Prediction of selectivity and flux for CH_4/H_2 separations using single walled carbon nanotubes as membranes. J Membr Sci 269:152–162
361. Ackerman DM, Skoulidas AI, Sholl DS, Johnson JK (2003) Diffusivities of Ar and Ne in carbon nanotubes. Mol Simulat 29:677–684
362. Skoulidas AI, Ackerman DM, Johnson JK, Sholl DS (2002) Rapid transport of gases in carbon nanotubes. Phys Rev Lett 89:185901/1–4
363. Kim S, Pechar TW, Marand E (2006) Poly(imide siloxane) and carbon nanotube mixed matrix membranes for gas separation. Desalination 192:330–339
364. Tseng HH, Kumar IA, Weng TH, Lu CY, Wey MY (2009) Preparation and characterization of carbon molecular sieve membranes for gas separation—the effect of incorporated multi-wall carbon nanotubes. Desalination 240:40–45
365. Kusworo TD, Ismail AF, Widiasa IN, Johari S, Sunarso S (2010) CO_2 removal from biogas using carbon nanotubes mixed matrix membranes. Int J Sci Eng 1:1–6
366. Geim AK, Novoselov KS (2007) The rise of grapheme. Nat Mater 6:183–191
367. Meyer JC, Geim AK, Katsnelson MI, Novoselov KS, Booth TJ, Roth S (2007) The structure of suspended graphene sheets. Nature 446:60–63

368. Noble Foundation announcement (2010) On line at: http://www.nobelprize.org/nobel_prizes/physics/laureates/2010/press.html
369. Bunch JS, Verbridge SS, Alden JS, van der Zande AM, Parpia JM, Craighead HG, McEuen PL (2008) Impermeable atomic membranes from graphene sheets. Nano Lett 8:2458–2462
370. Jennifer C (2012) MIT news, tiny pores in graphene could give rise to membranes: new membranes may filter ware or separate biological samples. MIT News Office, 23 Oct 2012. http://newsoffice.mit.edu/2012/graphene=pores-and-membrans-1023
371. Forbeaux I, Themlin J-M, Debever J-M (1998) Heteroepitaxial graphite on 6H-SiC(0001): interface formation through conduction-band electronic structure. Phys Rev B 58: 16396–16406
372. Cambaz ZG, Yushin G, Osswald S, Mochalin V, Gogotsi Y (2008) Noncatalytic synthesis of carbon nanotubes, graphene and graphite on SiC. Carbon 48:841–849
373. Du H, Li J, Zhang J, Su G, Li X, Zhao Y (2011) Separation of hydrogen and nitrogen gases with porous graphene membrane. J Phys Chem C 115:23261–23266
374. Hauser AW, Schwerdtfeger P (2012) Methane-selective nanoporous graphene membranes for gas purification. Phys Chem Chem Phys 14:13292–13298
375. Koeing SP, Wang L, Pellegrino J, Bunch S (2012) Selective molecular sieving through porous graphene. Nat Nanotechnol 7:728–732
376. Oyama ST, Lee D, Hacarlioglu P, Saraf RF (2004) Theory of hydrogen permeability in non-porous silica membranes. J Membr Sci 244:45–53
377. Ying J, Peng C, Lei F, Huiqian L, Huan Y, Cong R, Lei S, Changzhi G, Hai-Hu W (2008) Critical fields and anisotropy of $NdFeAsO_{0.82}F_{0.18}$ single crystals. Appl Phys Lett 93:032503
378. Qin X, Meng Q, Feng Y, Gao Y (2013) Graphene with line defect as a membrane for gas separation: design via a first-principles modeling. Surf Sci 607:153–158
379. Jiang DE, Cooper VR, Dai S (2009) Porous graphene as the ultimate membrane for gas separation. Nano Lett 9:4019–4024
380. Blankenburg S, Bieri M, Fasel R, Mullen K, Pignedoli CA, Passerone D (2010) Graphene as an atmospheric nanofilter. Small 6:2266–2271
381. Schrier J (2010) Helium separation using porous graphene membranes. J Phys Chem Lett 1:2284–2287
382. Schrier J, McClain J (2012) Thermally-driven isotope separation across nanoporous graphene. Chem Phys Lett 521:118–124
383. Jungthawan S, Reunchan P, Limpijumnog S (2013) Theoretical study of strained porous structures and their gas separation properties. Carbon 54:359–364
384. Lee J, Aluru NR (2013) Water-solubility-driven separation of gases using graphene membrane. J Membr Sci 428:546–553
385. Lai Z, Bonilla G, Diaz I, Nery JG, Sujaoti K, Amat MA, Kokkoli E, Terasaki O, Thompson RW, Tsapatsis M, Vlachos DG (2003) Microstructural optimization of a zeolite membrane for organic vapor separation. Science 300:456–460
386. Park HB, Lee YM (2005) Fabrications and characterization of nanoporous carbon/silica membrane. Adv Mater 17:477–483
387. Yampolskii Y, Pinnau I, Freeman B (eds) (2006) Materials science of membrane for gas and vapor separation. John Wiley & Sons, Chichester, UK
388. Park HB, Jung CH, Lee YM, Hill AJ, Pas SJ, Mudie ST, Wagner EV, Freeman BD, Cookson DJ (2007) Polymers with cavities tuned for fast selective transport of small molecules and ions. Science 318:254–258
389. Khayet M, Matsuura T (2011) Membrane distillation, principles and applications. Elsevier, Amsterdam
390. Henis JMS, Tripodi MK (1980) Multicomponent membranes for gas separation. US Patent 4230463A, 28 Oct 1980
391. Fritzsche AK, Cruse CA, Kesting RE, Murphy MK (1990) Hollow fiber membranes spun from lewis acid:base complexes. I. Structure determination by oxygen plasma ablation. J Appl Polym Sci 40:19–40

392. Kesting RE (1990) The four tires of structure in integrally skinned phase inversion membranes and their relevance to the various separation regimes. J Appl Polym Sci 41: 2739–2752
393. Chung TS, Teoh SK, Hu X (1997) Formation of ultrathin high-performance hollow fiber membranes. J Membr Sci 133:161–175
394. Shieh JJ, Chung TS, Wang R, Srinivasan MP, Paul DR (2001) Gas separation performance of poly(4-vinylpyridine)/polyetherimide composite hollow fibers. J Membr Sci 182:111–123
395. Husain S (2013) Methods of preparing a crosslinked fiber membrane. US Patent 20130239805 A1, 19 Sept 2013
396. Baker WR (2002) Future directions of membrane gas separation technology. Ind Eng Chem Res 41:1393–1411
397. Kesting RE, Cruse CA, Fritzsche AK, Malon RF, Murphy MK, Handermann AC (1992) Asymmetric gas separation membranes having graded density skins. EP0257012 B1, 7 Oct 1992
398. Kusakabe K, Li ZY, Maeda H, Morooka S (1995) Preparation of supported composite membrane by pyrolysis of polycarbosilane for gas separation at high temperature. J Membr Sci 103:175–180
399. Ren X, Ren J, Deng M (2012) Poly(amide-6-b-ethylene oxide) membranes for sour gas separation. Sep Purif Technol 89:1–8
400. Jiang X, Ding J, Kumar A (2008) Polyurethane-poly(vinylidene fluoride) (PU-PVDF) thin film composite membranes for gas separation. J Membr Sci 323:371–378
401. Gupta Y, Hellgardt K, Wakeman RJ (2006) Enhanced permeability of polyaniline based nano membranes for gas separation. J Membr Sci 282:60–70
402. Lopez JL, Matson SL, Marchese J, Quinn JA (1986) Diffusion through composite membranes: a two dimensional analysis. J Membr Sci 27:301–325
403. Henis JMS, Tripodi MK (1980) A novel approach to gas separations using composite hollow fiber membranes. Sep Sci Technol 15:1059–1068
404. Henis JMS, Tripodi MK (1981) Composite hollow fiber membranes for gas separation: the resistance model approach. J Membr Sci 8:233–246
405. Reid BD, Ebron VHM, Musselman IH, Ferraris JP, Balkus JKJ (2002) Enhanced gas selectivity in thin film composite membranes of poly(3-(2-acetoxyethyl)thiophene). J Membr Sci 195:181–192
406. Way JD, Noble RD, Flynn TM, Sloan ED (1982) Liquid membrane transport: a survey. J Membr Sci 12:239–259
407. Kocherginsky NM, Yang Q, Seelam L (2007) Recent advances in supported liquid membrane technology. Sep Purif Technol 53:171–177
408. Ward WJ, Robb WL (1967) Carbon dioxide-oxygen separation: facilitated transport of carbon dioxide across a liquid film. Science 156:1481–1484
409. Baker RW, Roman I, Lonsdale HK (1987) Liquid membranes for the production of oxygen enriched air. J Membr Sci 31:15–29
410. Lee SH, Kim BS, Lee EW, Park YI, Lee JM (2006) The removal of acid gases from crude natural gas by using novel supported liquid membranes. Desalination 200:21–22
411. Chen XS, Nishide H, Tsuchida E (1996) Analysis of facilitated oxygen transport through in a liquid membrane of hemoglobin. Bull Chem Soc Jpn 69:255–259
412. Castro-Domínguez B, Leelachaikul P, Takagaki A, Sugawara T, Kikuchi R, Oyama ST (2013) Perfluorocarbon-based supported liquid membrane for O_2/N_2. Sep Purif Technol 118:19–24
413. Deetz DW (1987) Stabilized ultrathin liquid membranes for gas separation. In: Noble RD, Way JD (eds) Liquid membranes, ACS symposium series. American Chemical Society, Washington, DC
414. Ward WJ (1970) Analytical and experimental studies of facilitated transport. AIChE J 16:405–410
415. Donaldson TL, Quinn JA (1975) Carbon dioxide transport through enzymatically active synthetic membranes. Chem Eng Sci 30:103–115

416. Ghai RK, Ertl H, Dullien FAL (1973) Liquid diffusion of nonelectrolytes, Part I. AIChE J 19:881–900
417. Ghai RK, Ertl H, Dullien FAL (1974) Liquid diffusion of nonelectrolytes, Part II. AIChE J 20:1–20
418. Reid RC, Prausnitz JM, Sherwood TK (1977) The properties of gases and liquids. McGraw-Hill, New York
419. Cserjési P, Nemestóthy N, Vass A, Csanádi Z, Bélafi-Bako K (2009) Study on gas separation by supported liquid membranes applying novel ionic liquids. Desalination 246:370–374
420. Seeberger A, Kern C, Uerdingen M, Jess A (2007) Gas separation by supported ionic liquid membranes. In: DGMK-conference, opportunities and challenges at the interface between petrochemistry and refinery, 10–12 Oct 2007, Hamburg, Germany
421. Cserjési P, Nemestóthy N, Vass A, Csanádi Z, Bélafi-Bakó K (2009) Study on gas separation by supported liquid membranes applying novel ionic liquids. Desalination 245:743–747
422. Gan Q, Rooney D, Xue M, Thompson G, Zou Y (2006) An experimental study of gas transport and separation properties of ionic liquids supported on nanofiltration membranes. J Membr Sci 280:948–956
423. Neves L, Nemestóthy N, Alves VD, Cserjési P, Bélafi-Bakó K, Coelhoso I (2009) Separation of bio-hydrogen by supported ionic liquid membranes. Desalination 240:311–315
424. Neves L, Dabek W, Coelhoso IM, Crespo JG (2006) Design of new selective Nafion membranes using room temperature ionic liquids. Desalination 199:525–526
425. Cserjési P, Nemestóthy N, Bélafi-Bakó K (2010) Gas separation properties of supported liquid membranes prepared with unconventional ionic liquids. J Membr Sci 349:6–11
426. Robeson LM (2008) The upper bound revisited. J Membr Sci 320:390–400
427. Hanioka S, Maruyama T, Sotani T, Teramoto M, Matsuyama H, Nakashima K, Hanaki M, Kuboto F, Goto M (2008) CO_2 separation facilitated by task-specific ionic liquids using a supported liquid membrane. J Membr Sci 314:1–4
428. Zhao W, He G, Nie F, Zhang L, Feng H, Liu H (2012) Membrane liquid loss mechanism of supported ionic liquid membrane for gas separation. J Membr Sci 411–412:73–80
429. Luis P, Neves LA, Afonso CAM, Coelhoso IM, Crespo JG, Garea A, Irabiena A (2009) Facilitated transport of CO_2 and SO_2 through supported ionic liquid membranes (SILMs). Desalination 245:485–493
430. Majumdar S, Guha AK, Sirkar KK (1998) A new liquid membrane technique for gas separation. AIChE J 34:1135–1245

Chapter 4
Membrane Fabrication/Manufacturing Techniques

The escalating research in membrane fabrication for gas separation applications signifies that membrane technology is currently growing and becoming the major focus for industrial gas separation processes. Material selection and method of preparation are the most important parts in fabricating a membrane. Different preparation methods result in various isotropic and anisotropic membranes, which are related to different membrane processes. The commercial value of a membrane is determined by its transport properties—permeability and selectivity. The method of membrane fabrication can have considerable influence on its effectiveness and there are a range of techniques available to create membranes, such as melt-pressing, solution casting, phase inversion, sputtering, extruding, and interfacial polymerization. Membranes can be fabricated either in a hollow fiber or spiral wound format.

Synthetic membranes are fabricated in two main geometries:

1. Flat sheet: utilized in the construction of flat sheet, disc, spirally wound, plate, and frame modules.
2. Cylindrical: utilized in tubular and capillary, or hollow fiber modules.

Membranes are prepared from both polymeric and inorganic materials. Inorganic materials have higher chemical and thermal stability in comparison with polymeric membranes, but polymeric materials are mostly organic compounds. A number of different techniques are used to prepare synthetic membranes.

4.1 Polymeric Membranes

4.1.1 Phase Inversion Membranes

The invention of the first anisotropic cellulose acetate membrane via phase inversion by Loeb and Sourirajan in the 1960s [1] opened a new avenue for the preparation of different kinds of membranes. Different techniques such as solution casting,

A.F. Ismail et al., *Gas Separation Membranes*,
DOI 10.1007/978-3-319-01095-3_4

interfacial polymerization and plasma polymerization were invented to fabricate selective, permeable anisotropic membranes, and all of them are still in practice; however, the phase inversion technique is still the most popular and important preparation method, especially for commercial membrane production. In the process of phase inversion, the polymer solutions may be precipitated by different approaches, including cooling, immersion in a nonsolvent coagulant bath, evaporation, and vapor adsorption.

Phase inversion is the most frequently used method for creating the current range of commercially available polymer membranes. One well-known technique to prepare asymmetric polymeric membranes is to use phase inversion induced by immersion precipitation. In this process, a polymer solution is cast onto a suitable support by using a casting knife, and then immersed into a coagulation bath. This process results in an asymmetric membrane with a dense top layer and a porous sublayer. The formation of both layers is controlled by numerous variables in the polymer dope solution, such as composition, coagulant temperature, and organic/inorganic additives. During this so-called demixing stage, the polymer solidifies into a solid matrix. This method can be used to prepare both porous and non-porous membranes. To achieve a desired membrane morphology and performance, the phase inversion process must be carefully controlled. The original Loeb–Sourirajan membranes were made by casting 22–25 % cellulose acetate solution on a glass plate. After solvent evaporation for a pre determined time, the cast film was then immersed in a water bath to precipitate and form the membrane. This method has been used for the preparation of flat sheet membranes in the laboratory, but for commercial production, a casting machine is usually used [2]. The nascent film is cast onto a moving nonwoven fabric web by the casting blade, exposed it to a gaseous environment for solvent evaporation prior to entering the coagulant bath, and then the film is finally formed by quenching in the coagulation bath. In the phase inversion process, the formation of asymmetric membranes is controlled by both the thermodynamics and the kinetics involved in the phase inversion.

Figure 4.1 shows a typical diagram of the fabrication of flat sheet membranes. Fabricating a flat sheet membrane is a complicated process, as it involves the dope preparation, the rheology of the casting solution air gap, and immersion precipitation in the coagulant bath. The process has three main concerns—rheology of the dope at the casting window, the formation of nascent flat sheet membranes in the air gap/different environment, and solidification or vitrification of flat sheet membranes in the coagulation bath [2].

Many mass transfer models have been proposed to predict membrane formation during phase inversion [3]. The pioneers who proposed mass transfer models to explain membrane formation at the interface between the polymer solution and coagulation bath are Cohen et al. [4], the McHugh group [5, 6], and Reuvers et al. [7, 8].

Kim et al. [9] discussed a model of the spinodal decomposition mechanism based on the mass transfer that occurs during asymmetric membrane formation. Alternatively, Termonia [10] simulated polymer coagulation processing using the

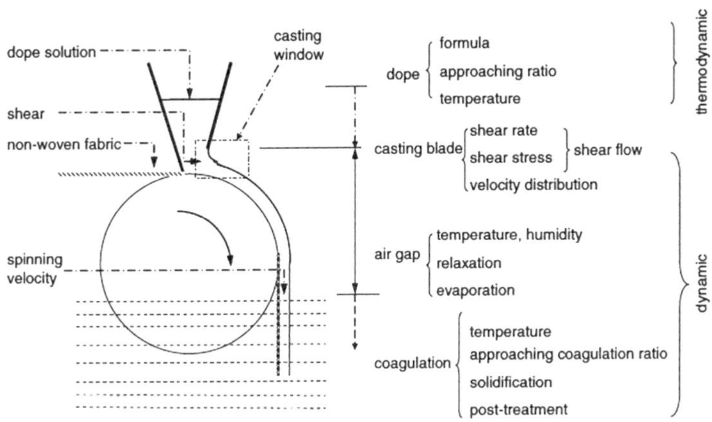

Fig. 4.1 Schematic diagram of flat sheet membrane fabrication [2]

Monte Carlo diffusion model. Wang et al. [11] attempted to investigate the effect of polymer chain length and solvent size on the kinetics of membrane formation during phase inversion using dynamic simulation. However, according to Peng, it is uncertain how the selective skin layer is formed. It could be any of the following routes [3]:

1. Nucleation and growth of a small polymer-rich phase followed by coarsening of the nuclei.
2. Spinodal decomposition to form nodules at the interface.
3. Spinodal decomposition followed by capillary force driven densification in the air-gap region of the dry-wet process.
4. Gelation induced via vitrification (glass transition).
5. Crystallization (for crystallizable polymers) interrupted liquid–liquid demixing.

4.1.2 *Precipitation by Solvent Evaporation*

For fabrication of asymmetric membranes, a solvent evaporation technique is used. In this technique a casting solution containing a polymer dissolved in a mixture of good volatile solvent and less volatile non-solvent (mostly water or alcohol) is prepared. When a film of this solution is cast and evaporation is allowed to take place, the good volatile solvent leaves first. Thus, the film becomes enriched in the non-solvent and finally precipitates. The casting-solution can be precipitated using imbibitions of water, either as a vapor form, or a humid atmosphere, or by immersion in a water bath.

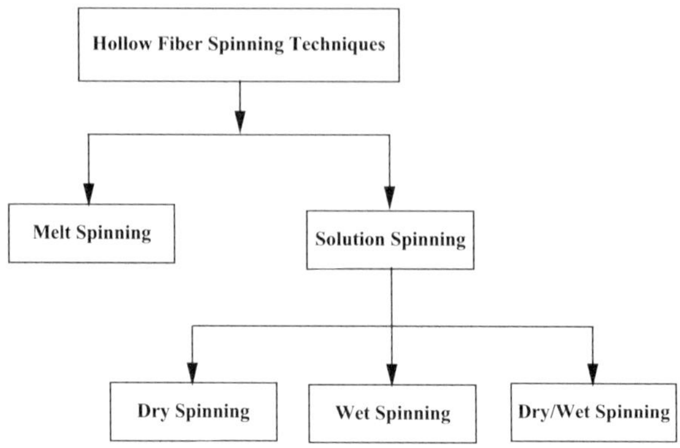

Fig. 4.2 Major techniques used in the fabrication of hollow fiber membranes for gas separation applications

4.1.3 *Preparation of Hollow Fiber Membranes*

The major techniques used in the fabrication of hollow fiber membranes for gas separation applications are summarized in Fig. 4.2.

Hollow fiber membranes can be fabricated in the form of dense and asymmetric structures. These structures differ only in the method used to solidify the gel filament. A dense structure is usually fabricated by melt spinning while solution spinning (phase inversion) yields asymmetric membranes.

Hollow fiber spinning is a tricky physical process and generally involves following four steps:

1. Solution formulation
2. Extrusion
3. Coagulation
4. Treatment of coagulation fiber

For polymeric hollow fiber membranes, spinning parameters are crucial factors that must be controlled during the preparation of membranes. These parameters include the amount and type of polymers, solvents, and additives mixed into the spinning dope solution, the dope and bore fluid rate, the kind of bore fluid, the fiber take-up velocity, the air-gap distance (unless wet spinning is used), the coagulant bath temperature and the kind of coagulant bath.

Melt spinning: In melt spinning, a molten polymer is extruded through a spinneret and the solid fiber is formed by cooling the gel filament in a medium such as air, gas, or liquid. Melt spinning processes have limited applications in gas separation. It is the most economic method of hollow fiber membrane production, since no wastewater or harmful by-products are involved. Hollow fibers of thermoplastic polymers can be made by this process. Polyolefines have been found most suitable for making hollow fiber membranes. One polymer of this group, poly(4-methyl-1-pentene) (PMP), shows particularly good gas permeation properties [12].

Solution spinning: Solution spinning processes are based on the phase inversion technique. Hollow fiber membranes produced from solution spinning can be asymmetric, dense or microporous depending on the conditions employed. This process is based on the phase inversion technique and is generally classified into three techniques namely, dry spinning, wet spinning, and dry/wet spinning:

1. *Dry spinning technique*: In this technique, the polymer solution is dissolved in an appropriate solvent and extruded through a spinneret. The dope solution is usually above ambient temperature and pressure. Solidification of the fibers is due to evaporation of solvent just after the filament emerges from the spinneret. Critical factors in determining the fiber properties are rheologically induced molecular orientation in the solidifying skin layer and phase inversion conditions. At present there is no reported commercial membrane gas separation system produced using dry spinning.
2. *Wet spinning technique*: In wet spinning, fibers are formed by extruding a dope solution through a spinneret into a coagulation bath. Skin formation and fiber solidification depend on the nature of the phase inversion. Any type of membrane morphology can be obtained with this technique since many parameters involved can be varied. Recent studies have reported that defect-free skin layers in asymmetric hollow fiber membranes were produced using this method via a modified dual bath coagulation method [13]. Selectivities above the recognized intrinsic value of the polymer were achieved by this method.
3. *Dry/wet spinning technique*: Dry/wet spinning is a combination of certain favorable features of the dry and wet spinning methods. The extruded fiber passes through an air gap of a certain length prior to immersion into the coagulation medium. The fiber is then collected on a wind-up drum and further washed before being subjected to any post treatment process. In the dry/wet spinning method, the air gap region is responsible for the formation of an ultrathin defect-free skin layer [14].

Recently, super selective and ultrathin skin layer hollow fiber membranes have been developed for carbon dioxide/methane separation using this dry/wet forced convective evaporation method [15].

4.1.3.1 Methods for Spinning

The structure and the permeation properties of the hollow fibers are highly dependent on the spinning conditions. Polymeric hollow membranes are prepared by extruding a polymer solution through an annular spinneret and bore fluid flows in the annular center. Figure 4.3 shows an example of hollow fiber spinning equipment.

The spinning of hollow fiber membranes is a very complicated process. Similar to the flat sheet membrane casting (Fig. 4.1), the fabrication of hollow fiber involves many spinning parameters as shown in Fig. 4.4 [2].

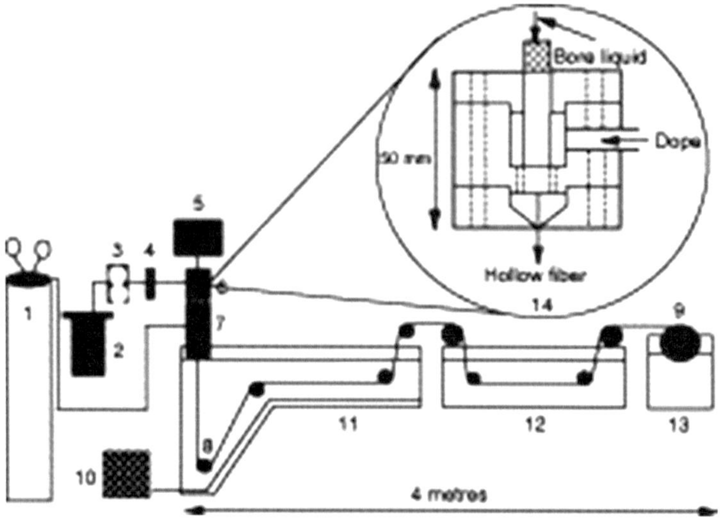

Fig. 4.3 Schematic diagram of hollow fiber spinning system. (*1*) Nitrogen cylinder; (*2*) dope reservoir; (*3*) gear pump; (*4*) online filter, 7 mm; (*5*) syringe pump; (*6*) spinneret; (*7*) forced convective tube; (*8*) roller; (*9*) wind-up drum; (*10*) refrigeration/heating unit; (*11*) coagulation bath; (*12*) washing/treatment bath; (*13*) wind-up bath; and (*14*) schematic spinneret [16]

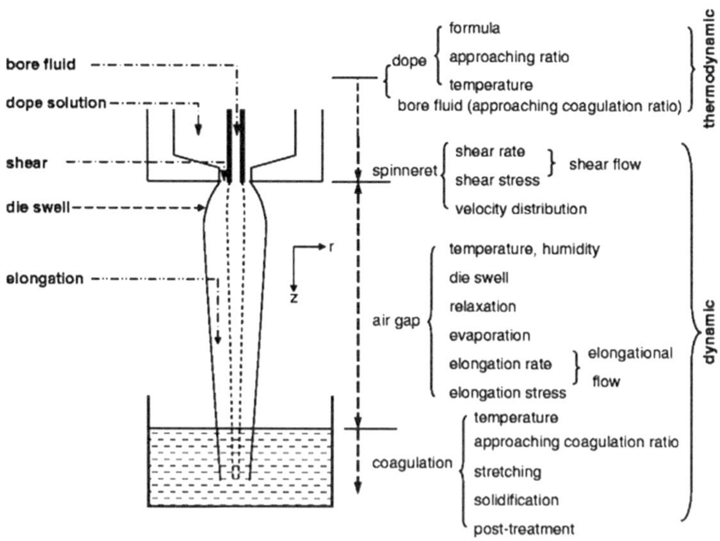

Fig. 4.4 Spinning parameters for the formation of hollow fiber

The fabrication of hollow fiber membranes involves the thermodynamics of the polymer solution and the phase inversion process, the rheologies of the polymer solution inside the spinneret and at the air gap, and other spinning conditions.

The dry-jet spinning process mainly concerns three steps:

1. Rheology of the dope solution in the spinneret.
2. Formation of nascent hollow fiber membranes in the air gap.
3. Solidification or vitrification of hollow fiber membranes in the coagulation bath.

Two different flow patterns of shear flow and elongation flow exhibit inside the spinneret and in the air gap, respectively [17]. Chung et al. [18] made one of the pioneering works in fabricating high performance hollow fiber membranes from a polymer/solvent binary system by producing polyethersulfone hollow fibers with an ultra-thin dense layer of 474 Å. The proposed key parameters were (1) controlling the bore fluid chemistry and flow rate, and (2) using a dope exhibiting significant chain entanglement. However, Chung et al.'s hollow fiber membranes had many defects, and application of silicone rubber coating was necessary in order to regenerate their membranes' selectivity [3].

Wallace et al. [19] discussed the relationship between polymer properties (solvent and non-solvent interactions, viscosity, etc.) and the techniques of fiber spinning (dope formulation, spinning process, solvent exchange/drying). Cao et al. [20] reported the influence of elongation and shear rates induced by the geometry of spinnerets on gas performance of PES hollow fiber membranes. It was concluded that the elongation rate has more of an influence on permselectivity than permeance and that the shear rate has more of an influence on permeance than permselectivity. A hypothetical mechanism was discussed to explain the effects of elongation and shear rates on the conformation changes of the polymer chain. Figure 4.5 shows such a hypothetical mechanism on conformation changes to the polymer chains induced by elongation and shear rate.

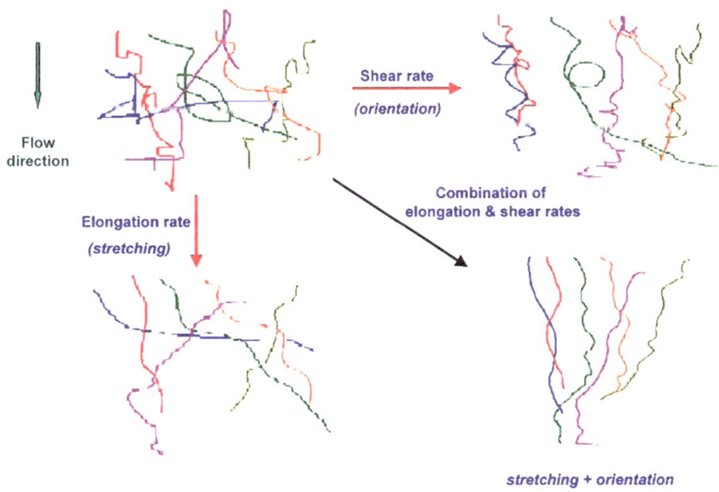

Fig. 4.5 A hypothetic mechanism illustrating conformation changes to polymer chains induced by elongation and shear rate [20]

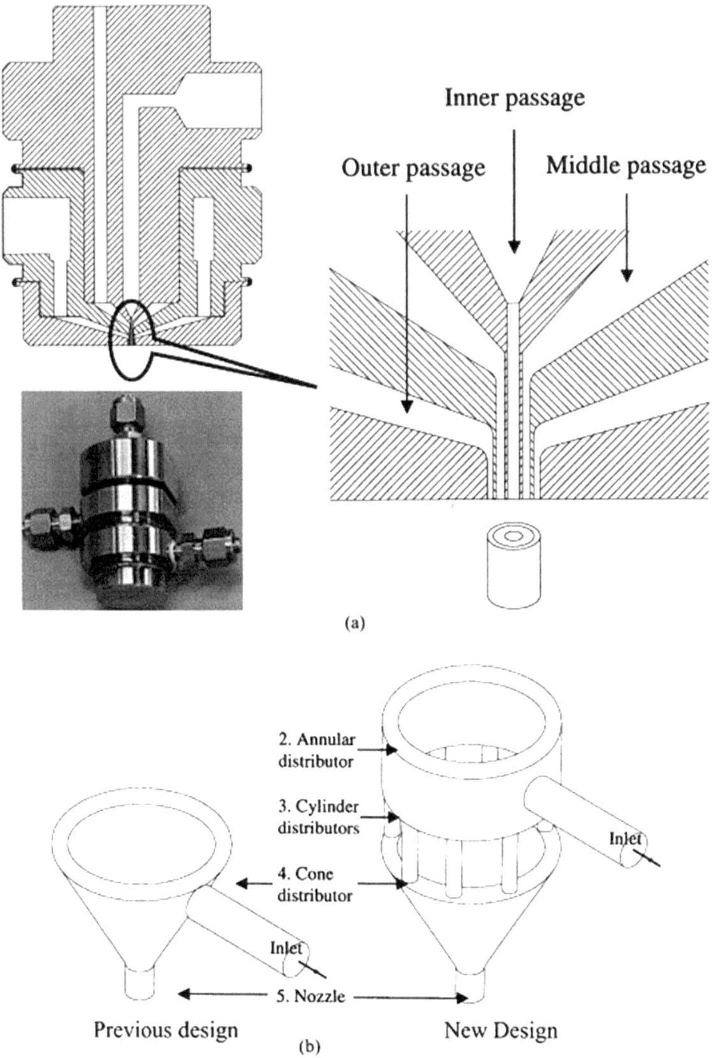

Fig. 4.6 Structure of (**a**) dual-layer spinneret, (**b**) outer passage of dual layer spinneret

Li et al. [21] successfully fabricated dual-layer PES hollow fiber membranes with an ultrathin dense-selective layer of 407 Å by using co-extrusion and dry-wet-spinning phase inversion techniques with the aid of heat-treatment. Figure 4.6 shows their dual-layer asymmetric hollow fiber spinning process.

To run the dual-layer hollow fiber spinneret, two metering pumps (Zenith® B-Series from Parker Hannifin Corporation Zenith Pumps Division) were employed to deliver inner and outer dopes. For the inner dope, the module number of metering pump was BPB5596 with a flow rate of 0.160 ml/rev; for the outer dope, the module

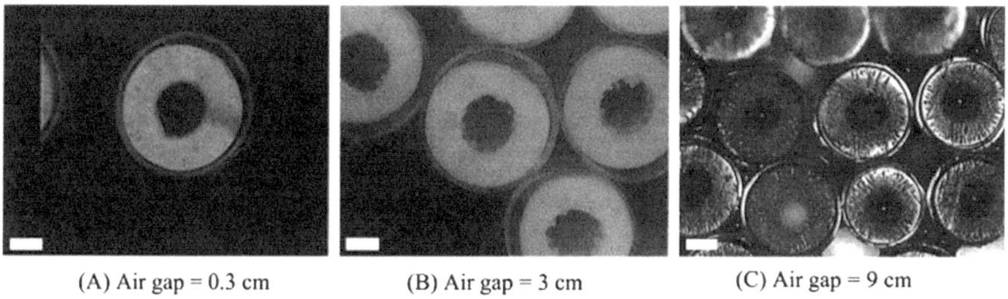

Parameters	Values
Outer-layer dope composition:	26% 6FDA durene/mPDA (50:50) in the mixture of NMP / THF = 5 / 3
Outer-layer dope flow rate:	0.2 ml/min
Inner-layer dope Composition:	28% PES in the mixture of NMP / water = 10 / 1
Inner-layer dope flow rate:	0.6 ml/min
Bore fluid composition:	95 wt% NMP in water
Bore fluid flow rate:	0.2 ml/min
Spinneret temperature:	50 °C
Air gap length:	0.3 - 9 cm
Take up rate:	300 - 500 cm/min

Fig. 4.7 Delamination—effect of air-gap

number was BPB5566 with a flow rate of 0.066 ml/rev. The bore fluid was delivered using a syringe pump (ISCO100DX from ISCO Inc.). Three pieces of 15 μm online filter (from Swagelok Marketing Co.) was installed in the system to prevent the spinneret from blocking. As an auxiliary device, a solvent cleaning system was embedded into the spinning system so that the spinneret could be cleaned up immediately after spinning.

In the spinning process for the formation of the dual-layer asymmetric hollow fiber, delamination was observed between the dual layers. The delamination resulted in weaker mechanical properties and also created a dense layer structure which caused additional substructure resistance. The main cause for delamination may have been due to the difference in shrinkage rates between outer and inner layers when the dual layer asymmetric hollow fiber membrane was formed by the phase inversion mechanism. Several approaches were explored to overcome the problem.

Figure 4.7 shows the cross section image of dual-layer asymmetric hollow fiber membranes as a function of air gap. The air gap changed from 0.3 to 9 cm. No improvement on delamination was observed until the air gap was significantly increased (i.e. 9 cm). This may be due to the fact that the gravity-induced elongation speeds up the spinning process, thinning down the fiber dimension and wall thickness. This elongation causes the tightening of the gap between the dual layers. The side effect of this gravity-induced elongation is the introduction of unstable spinning. Figure 4.7c illustrates the uneven distribution of fiber size [21].

Pereira et al. [22] studied the simultaneous spinning of two polymer solutions and the conditions that promote the adhesion of the two layers were determined. The stability period of the region between the two polymer solutions seems to be the major factor controlling interpenetration of the solutions and, consequently, the adhesion. Two variables are important: (a) the polymer solutions and bore liquid compositions, since the miscibility gap of each solution and the inflow of solvent to the interfacial region contributes to creating a longer period before precipitation, and (b) the distance between the spinneret and the external coagulation bath contributes to the interpenetration of the polymer solutions, since it may allow a longer time for mass exchange in the interfacial region.

4.1.3.2 Thermally Induced Phase Separation (TIPs)

Microporous membranes are generally prepared by the TIPs process which is based on the solvent quality decreasing when the temperature decreases [23]. On removing the thermal energy by cooling or quenching a polymer–diluent solution, phase separation occurs. After the phase separation, the diluent is removed, typically by solvent extraction, and the extractant is evaporated to yield a microporous structure. Typically, the TIPS process has been used to produce isotropic structures; that is, the pore size does not vary with direction in the membrane. A few studies have reported on the formation of anisotropic and asymmetric membranes by the TIPs process. The formation of anisotropic or asymmetric structures can be accomplished by imposing a temperature gradient across the membrane during the cooling process. High cooling rates bring about smaller pores.

The formation of the hollow fiber membrane via the TIPs process has not been studied very much. Kim et al. [24] prepared a polypropylene hollow fiber membrane from a polypropylene/soybean oil mixture. The membrane was formed by the TIPs process and subsequent cold-stretching. The hollow fiber became more oriented by increasing the melt-draw ratio (defined as the ratio of take-up speed) to the extrusion rate of the polymer solution. The cold-stretching of the hollow fiber membranes remarkably increased the membrane porosity.

4.1.4 Other Techniques

4.1.4.1 Coating

Coating methods are especially important for dense polymeric membranes and inorganic composite membranes. To get mechanical strength for reduced membrane thickness, composite membranes are used (supported by a porous sublayer). Coating procedures in use include dip coating, plasma polymerization, interfacial polymerization, and in situ polymerization [25]. Composite membranes are typically prepared by post treatment of a porous support, for instance dip-coating with a dilute polymer solution and subsequent solvent evaporation [26] by phase interfacial polymerization [27]. In the case of hollow fibers, the composite membrane can be prepared by direct spinning with a triple orifice spinneret [13].

Dip Coating

Dip-coating is a convenient membrane preparation technology and has been widely applied to fabricate ceramic membranes with macroporous to microporous levels upon porous supports. During the dip-coating process, a wet layer of ceramic particles is deposited upon a porous support by coating the dry support surface with a particle-dispersed suspension, or sol, followed by a controlled sintering process. However, the repeated coating-sintering procedure has many drawbacks, such as the need for additional preparation steps, the reduction of the membrane permeation properties and the extra energy consumption of the sintering process [28].

Zhu et al. suggested a modified dip-coating method to prepare pinhole-free ceramic membranes. In their method, tangential flow of suspension was used against the support dipping in order to assist the capillary-filtration effect. This modified dip-coating method for preparation of pinhole-free membranes was very effective and was able to resist the presence of pinhole defects in the supports and avoid new pinhole occurrences. The modified dip-coating method had several advantages compared with the repeated coating-sintering procedure, such as not producing additional undesired membrane thicknesses, not reducing the membrane permeation properties, and requiring only a single coating-sintering procedure.

4.1.4.2 Interfacial Polymerization

Interfacial polymerization (polycondensation) (IP) is polymerization that takes place at the interface between two immiscible phases upon contact. In the IP method, reactive monomers are dissolved in two immiscible phases and the polymerization of the reactive monomers occurs on the surface of the porous support membrane. The advantage of IP is that the reaction is self-inhibiting through passage of a limited supply of reactants through the already formed layer resulting is an extremely thin film with thickness in the 50 nm range [29].

Preparation of a thin film composite (TFC) membrane occurs during the IP reaction between two monomers. Figure 4.8 is schematic diagrams of TFC membrane preparation using the IP technique [30]. To prepare a very thin PA active layer on top of a supporting membrane, first the substrate typically will be immersed into an aqueous solution consisting of an amine monomer, prior to immersion in a second organic solution containing an acryl chloride monomer. The membrane is then subjected to heat treatment to densify the polymerization properties of the PA layer and/or enhance adhesion of the PA thin layer to the surface of the support membrane. Due to the significant advantages of the IP technique in optimizing the properties of the skin layer and microporous substrate layer, a wide variety of membranes have been successfully developed by many companies, allowing for the application of membranes for various industrial separation processes [30].

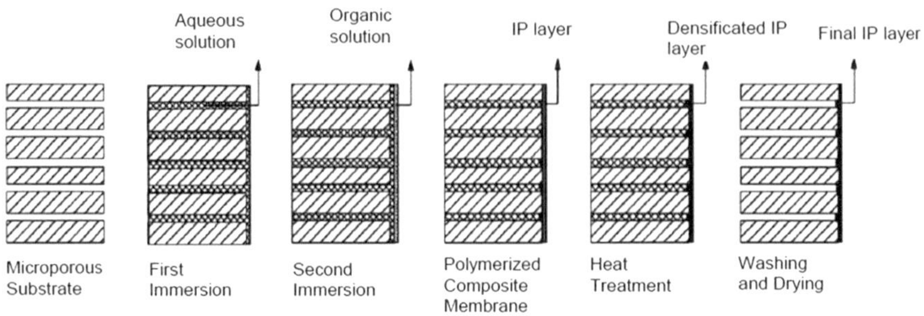

Fig. 4.8 Schematic diagrams of the preparation process of TFC membranes by conventional IP technique

4.1.4.3 Plasma Polymerization

When a vacuum is maintained inside a tubular reactor and a high frequency electric field is applied outside, a glow discharge is generated inside the reactor. Plasma that consists of various ions, radicals, electrons, and molecules is formed in the glow discharge. When a porous substrate membrane is placed in the plasma, the surface of the membrane is subjected to various changes corresponding to the properties of the plasma. The substrate surface can be etched and/or chemically active sites can be introduced to the surface, and, upon contact with organic compounds, an irregular polymerization can occur at the substrate surface. This is called plasma polymerization.

Plasma polymerization (or glow discharge polymerization) uses plasma sources to generate a gas discharge that provides energy to activate or fragment gaseous or liquid monomer, often containing a vinyl group, in order to initiate polymerization. Polymers formed from this technique are generally highly branched and highly cross-linked, and adhere to solid surfaces well. The biggest advantage to this process is that polymers can be directly attached to a desired surface while the chains are growing, which reduces steps necessary for other coating processes such as grafting. This is very useful for pinhole-free coatings (of 100 pm to 1 μm thicknesses) on solvent insoluble polymers [31].

The formation of plasma for polymerization depends on many of the following conditions. First, an electron energy of 1–10 eV is required, with electron densities of 10^9–10^{12} per cubic centimeter, in order to form the desired plasma state. The formation of a low-temperature plasma is important; the electron temperatures are not equal to the gas temperatures and have a ratio of T_e/T_g of 10–100, so that this process can occur at near ambient temperatures, which is advantageous because polymers degrade at high temperatures. If a high-temperature plasma were used the polymers would degrade after formation or would never be formed [32].

It has been claimed that flawless thin polymeric coatings could be formed on metals, although for very thin films (<10 nm), but this assertion has recently been shown to be an over simplification [33, 34]. By selecting the monomer type and the energy density per monomer—known as the Yasuda parameter—the chemical composition and structure of the resulting thin film can be varied with a wide range. These films are usually inert, adhesive, and have low dielectric constants [31]. Some common monomers polymerized by this method include styrene, ethylene, methacrylate and pyridine, to name a few. The 1970s brought about many advances in plasma polymerization, including the polymerization of different types of monomers. The mechanisms of deposition, however, have been largely ignored until more recently. Most attention had been devoted to plasma polymerization in the fields of coatings, but since controlling polymer structure is difficult, plasma coating has had limited applications.

Plasma polymerization offers a number of advantages over other polymerization methods. The most significant advantage of plasma polymerization is its ability to produce polymer films of organic compounds that do not polymerize under normal chemical polymerization conditions [35]. Nearly all monomers, even saturated hydrocarbons and organic compounds without a polymerizable structure such as a double bond, can be polymerized with this technique [36]. While coating a substrate with conventional polymers requires a number of steps, plasma polymerization accomplishes all these in essentially a single step [31].

Plasma polymerization leads to a cleaner and "greener" synthesis and coating process for membrane preparation; no solvent is needed during the membrane preparation, no cleaning is needed, and no initiator is needed. The resultant polymer coatings also have a number of advantages over typical coatings. These advantages include being nearly pinhole free, high density, and the ability to vary the thickness of the coating [37]. However, because of the complexity of the process, it is not easy to achieve good control over the chemical composition of the surface of the membrane.

4.1.4.4 Graft Polymerization

The surface of a porous substrate membrane is irradiated with γ-rays, which causes the generation of radicals on the membrane surface. Then, the membrane is immersed in a monomer solution. The graft polymerization of the monomers is initiated at the membrane surface. By choosing a very hydrophilic monomer, the hydrophilicity of the surface is increased considerably.

4.1.4.5 Particle Leaching

Particle (particulate, salt, porogen) leaching is applied in combination with various different techniques such as solvent casting, compression molding or foaming. In particle leaching, particles (e.g., salt, sugar, or specially prepared spheres) are incorporated in a polymer sample. After processing the polymer sample in the final form, the particles are dissolved and washed out creating (additional) porosity in the scaffold [38]. This method ensures that membranes with highly controlled porosity and pore sizes are produced [39]. However, this technique may not be applicable to all materials (e.g., soluble protein scaffolds); the washing out post-process is also time-consuming, and there is a risk of residues remaining from the method of processing (i.e., organic solvents) [38]. This method is preferred for polymers that are not soluble in common organic solvents [38, 39]. Porous membranes produced via this method include polyethylene membranes using tapioca starch as the leachable component [30] and 2,3-dialdehydecellulose membranes with sodium chloride (NaCl) used as the leachable component [40].

4.1.4.6 Track Etching

In track etching, a sheet of polymeric film moves underneath a radiation source and is irradiated by high-energy particles. The spots that are subjected to bombardment of the particles are degraded and chemically altered during this process. Then, the film undergoes an etching process in an alkaline or hydrogen peroxide bath (depending on the material), where the polymer is etched along the path of high energy particles.

4.1.5 Polyelectrolyte Multilayer Membranes

Photoelectrolytes are polymers whose repeating units bear an electrolyte group. Polycations and polyanions are polyelectrolytes. These groups dissociate in water and make the polymer charged. Thus, polyelectrolytes resemble both salts (electrolytes) and polymers (high molecular weight compounds). Sometimes polyelectrolytes are also called polysalts. Like salts, their solution is electrically conductive. Electrostatic layer-by-layer assembly of oppositely charged polyelectrolytes has been demonstrated. The present literature shows that a significantly large number of depositions of anionic/cationic polyelectrolytes are often needed in order to achieve high membrane selectivity, especially when the substrate membrane is microporous [41]. Decher and Hong [42] described a novel method for preparation of polyelectrolyte membrane with controlled thickness in the 10–100 nm range.

Tieke et al. [43] suggest that alternating electrostatic adsorption of cationic and anionic polyelectrolytes on porous supports is a versatile method to prepare composite membranes with ultrathin, pore-free separation layers. By careful choice of the polyelectrolytes and the supporting membranes, and by optimization of the processing and operating parameters, composite membranes with excellent separation capability for mixtures of polar liquids and ions of different charge density can be tailored. The authors also show that the transport of small molecules is possible.

Soon after the method of layer by layer assembly was reported, ultrathin polyelectrolyte multilayers were studied as permselective membranes. In a typical process of membrane preparation, a negatively charged porous or nonporous supporting membrane is dipped into a dilute aqueous solution of a positively charged polyelectrolyte so that the polymer is adsorbed at the substrate as a molecularly thin film and the surface charge is reverted. After washing, the coated substance is dipped into the aqueous solution of a negatively charged polyelectrolyte so that this polymer is adsorbed on top of the previous one and the surface charge is reversed again. By repeating the adsorption steps several times, a polyelectrolyte multilayer is obtained, whose thickness is adjustable between a few nanometers and about half a micrometer by varying the number of dipping cycles.

4.2 Inorganic Membranes

Metallic membranes (pure or alloys) are widely used for the separation of hydrogen. Metallic membrane material can be classified as follows:

1. Pure (single element)
2. Crystalline
3. Amorphous

There are basically two types of inorganic membranes: (1) dense (non-porous), and (2) porous membranes. Examples of commercial porous inorganic membranes are ceramic membranes (such as alumina, silica, titania, and zeolite) and glass porous metal (such as stainless steel and silver). Dense inorganic membranes are very specific in their separation behaviors; for example, Pd-metal-based membranes are hydrogen specific and metal oxide membranes are oxygen specific.

4.2.1 Preparation of Inorganic Membranes

Several methods can be used to prepare porous membranes from inorganic materials, and the choice of methods depends upon the desired membrane material and the pore size. Since many membranes are multilayered composite membranes, different methods may also be used to prepare the support material and separating layer. Preparation and fabrication techniques are as follows [44]:

1. Extrusion
2. Powder suspension
3. Molten salt inclusion
4. Phase separation and leaching
5. Nuclear track etching
6. Dynamic deposition
7. Anodic oxidation
8. Pyrolysis
9. Particle dispersion/slip casting
10. Thin film deposition

For preparing commercial membranes, the primary methods are phase separation and leaching, anodic oxidation, particle dispersion, particle dispersion/slip casting and pyrolysis.

New and novel alloys are being fabricated by sputtering, thermal evaporation, arc-melting, die cast techniques, and electro-deposition. However, the most common methods for preparing novel alloys of variable structures and diverse compositions are melt-spinning and arc-melting [45]. Alloys based on Ni, Ti, Zr and Cu have all been developed as bulk metallic glasses (BMGs) and BMG matrix composites. Nanocrystalline alloy membranes are significantly attractive due to their high resilience to degradation [45, 46]. Alloying is primarily employed to improve a pure metal's physical characteristics (e.g., strength, durability, degradation resistance) while maintaining a single-phase bee structure that is required for high hydrogen permeation.

4.2.1.1 Chemical Vapor Deposition

Chemical vapor deposition (CVD) is a chemical process used to produce high-purity, high-performance solid materials. The chemical vapor deposition (CVD) technique is used to deposit thin oxide layers on porous substrates. In this process, the desired films can be prepared at high temperatures obviating the drying and calcinations required in film formation by other methods (precipitation by solvent evaporation). The removal of solvent or condensation products accompanying drying and calcinations often causes shrinking and crack formation. Thus, films produced by CVD are generally denser and more uniform than those produced by the liquid-phase technique. In CVD, a metal organic component is vaporized in the carrier gas, from one side of the membrane (porous substrate), and the other reactants enter from the other side of the membrane. Reactant diffuses into the pores and reacts there. The product is deposited on the pores.

Chemical vapor deposition (CVD) methods to prepare a membrane on porous substrate are classified into two types, based on the supplying configuration of the precursors. In the first type, the precursors are provided from one side of the substrate, while the other side of the substrate is usually vacuumed to obtain a pin-hole free membrane. The second method is counter diffusion CVD where two kinds of reactants are supplied from the opposite sides of the substrate. Pore sizes and effective membrane thickness can be controlled by changing reactants and reaction conditions [45]. One of the first gas-phase methods to be developed was generation of a silica-modified membrane by a high temperature atmospheric CVD on Vycor glass [47]. Gavalas and co-workers [48, 49] deposited SiO_2 films within the walls of a porous Vycor tube by SiH_4 oxidation in an opposing reactant geometry. In this method SiH_4 was passed inside the Vycor tube (pore size 4 nm) while O_2 was passed outside and reacted within a narrow front inside the tube wall to form a thin SiO_2 film. Once the pores were plugged, the reactants could not reach each other and the reaction stopped. Other CVD methods use simple thermal decomposition or oxidation of the precursor with oxygen or ozone [50].

Metals can be deposited on a substrate by means of physical vapor deposition (PVD). In this process, the solid material to be deposited is first evaporated in a vacuum system using a physical technique. The thin-to-medium thickness film subsequently condensed and deposited on the cooler substrate [51].

Uemiya et al. [52] fabricated asymmetric membranes consisting of palladium, ruthenium, and platinum deposited on the surface or inside the pores (average size, 200 nm) of a tubular alumina membrane, prepared by the chemical vapor deposition (CVD) technique. Sublimation and decomposition temperatures of their acetylacetonato complexes, used as metal sources for CVD, were important factors in preparing the membranes selected for hydrogen separation.

4.2.1.2 Thin-Layer Metallic Membranes

Thin-layer metallic coating has been classified into four types:

1. a thin metal layer (dense or porous) formed on the surface and extraneous to the support;
2. a thin metal layer formed on the walls within a porous support;
3. a microporous ceramic layer formed on the supporting layer by finely distributing metal particles within the pores of the support;
4. a microporous ceramic layer formed on the supporting layer by sintering metal-coated particles onto the surface [53].

Metallic membranes can be prepared as thin layers on various supports such as glasses, ceramics or other metals (porous), in an effort to increase flux (especially hydrogen) while maintaining mechanical strength, thermal stability, and reliability. The current deposition methods are electroless plating, electro-deposition, spray pyrolysis, sol–gel dip coating, physical and chemical vapor deposition (PVD/CVD), or sputtering [45]. Gryaznov et al. [54] deposited thin films of binary and ternary alloys of Pd, Mn, Co, Ru, Sn, and Pb on asymmetric polymeric membranes, porous stainless steel sheets and oxide supports by the sputtering technique.

The spray pyrolysis method is also used for the preparation of metallic membranes. This method involves spraying a solution of metal salts into a heated gas stream where it is pyrolyzed. This method has been successfully applied for the production of fine metals or metal oxide particles [44]. Li et al. [55] obtained a Pd–Ag alloy membrane on the surface of a porous alumina hollow fiber by spray pyrolysis of a $Pd(NO_3)_2$ and $AgNO_3$ solutions on a H_2–O_2 flame. Zhang et al. [56] prepared thin palladium membranes by depositing Pd on α-alumina supports via modified electroless plating technique. It was suggested by Zhang et al. that the supports with smoother surfaces and uniform pore sizes could obtain a dense membrane with a thickness less than 3 μm to improve separation and permeation properties.

4.2.2 Silica Membranes

Silica membranes are generally comprised of three layers, namely, a membrane layer, an intermediate layer, and a support. Silica membranes are synthesized primarily through two different methods—sol–gel modification and chemical vapor deposition (CVD). Sol–gel modification provides good selectivity and permeability, while loss of permeability and enhanced selectivity are provided by CVD.

Silica-gel processing can be done in three ways:

1. Silica polymers method
2. Particulate-sol method
3. Template method

The silica polymers route involves the hydrolysis and condensation of alkoxysilane precursors, such as tetraethoxysilane (TEOS), under controlled conditions. The particulate-sol route is based on the packing of nanoparticles to make a highly porous structure [57]. Asadea and Yamasaki [57] used cylindrical α-alumina porous tubes (O.D., 10 mm; thickness, 2 mm; average pore size, 1 μm; porosity, 0.5; length, 100–250 mm) to pack on the membrane support. Glass tubes were connected to both ends of the cylindrical substrate as shown in Fig. 4.9. Before coating the colloidal sols on the outer surface of the substrate, fine α-alumina particles (average particle diameter; 0.19 μm) were deposited on the surface (binder; sol-A) to make it smooth and homogeneous and to relax the difference in thermal expansion between the porous α-alumina substrate and a silica layer to be coated. The colloid coating was done while the substrate was hot at around 190 °C by contacting the hot substrate with a cloth wet with the sols, for very quick drying, and then it was fired at around 500 °C for more than 10 min. The coating was repeated several times with the colloidal sols diluted to a concentration less than 0.5 wt% of equivalent original TEOS.

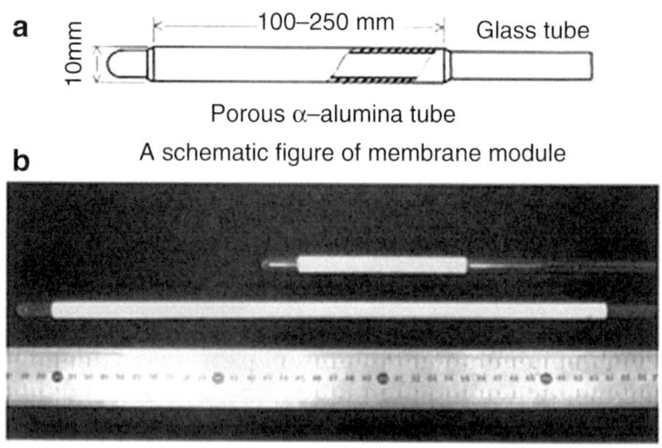

A photograph of membrane modules

Fig. 4.9 Cylindrical porous-alumina substrates connected to glass tubes

Silica particles of different sizes can be packed into the support substrate to process membranes with different pore sizes. The template route uses organic molecules as templates in the sol matrix that are burned out upon calcinations. The organic molecules size and shape can be imprinted in the sol for a tuned porosity. Surfactants, organic ligands, and polymers have been reported as templates.

Brinker at al. [58] fabricated silica-based microporous membranes by sol–gel deposition. Ultramicroporous (pore radius < 10 Å) separation layers with thicknesses in the range of 200–1,200 Å were deposited from polymeric silicate sols onto commercial alumina supports using a dip-coating/casting procedure. The silicate sols were prepared using a two-step acid-catalyzed hydrolysis of tetraethoxysilane under pH conditions, where the condensation rate is low, producing polymers of low fractal dimension which readily interpenetrate during film deposition to provide amorphous layers with extremely small pore sizes.

Plasma-enhanced chemical vapor deposition (PECVD) using organosilanes as starting materials is a promising method for depositing both inorganic SiO_2-like films and polymer-like $SiO_xC_yH_z$ films, depending on the plasma composition [59]. However, this method for the preparation for gas separation membranes has not been extensively studied.

4.3 Composite Membrane Preparation/Mixed Matrix Membranes

In the past two decades, a variety of fabrication methods for composite membranes/MMMs have been in practice, namely, the physical blending method, sol–gel method, infiltration method, in situ polymerization method, interfacial polymerization method, chemical atomic layer deposition method, layer-by-layer assembly method, etc. [60]. The most often used methods are physical blending and sol–gel.

Physical blending method: In this technique, the filler is prepared prior to membrane fabrication, and then physically dispersed into the polymer matrix by solution blending or melt blending, followed by polymer solidification (Fig. 4.10a).

Sol–gel method: In this technique, both polymer and filler precursor are mixed at the molecular level in a casting solution containing a certain amount of water, and the composite membrane is obtained through simultaneous sol–gel reaction and polymer solidification (Fig. 4.10b) [60].

The above mentioned methods are based on two different strategies. In physical blending, the filler is synthesized prior to membrane formation. While in sol–gel, the filler is synthesized during membrane formation. There is a third strategy where the filler is synthesized after membrane formation and this is called the infiltration method [60]. In this technique, the precursor of the filler is allowed to infiltrate into a swollen or nanoporous polymeric membrane, and then the composite membrane is obtained through in situ filler growth and polymer curing (Fig. 4.10c).

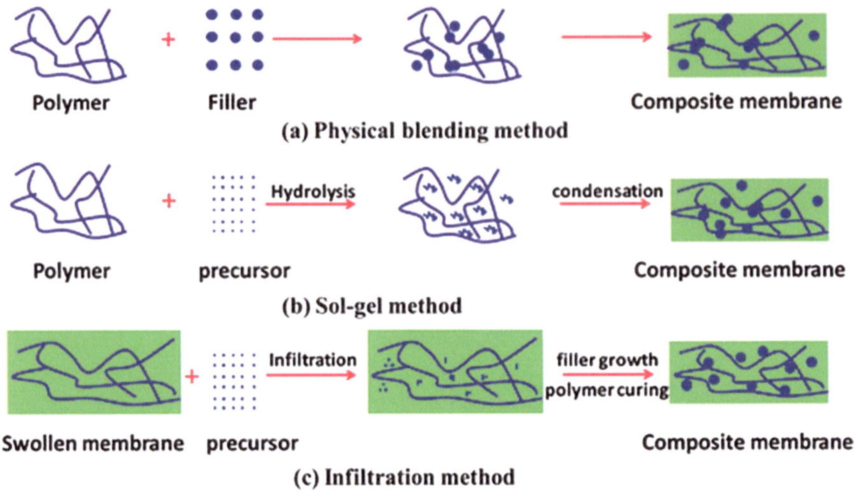

Fig. 4.10 Three typical methods for fabricating composite membranes: (**a**) physical blending method; (**b**) sol–gel method; and (**c**) infiltration method

Preparation of zeolite membranes were discussed in Chap. 3. Homogeneous dispersion of zeolites without side effects is a difficult subject. Many approaches have been made to improve polymer–zeolite contact, such as silanizing, annealing, priming, and Grignard treatment [61]. The most significant work has been performed by creating nanoscale morphologies on the surface of the zeolite particles [62]. Four techniques have been reported—Grignard decomposition reactions, solvothermal depositions, modified solvothermal depositions, and ion exchange—the details of which can be found elsewhere [63].

4.4 Preparation of Metal-Organic Framework Membranes (MOFs)

Metal-organic frameworks (MOFs) are a new class of hybrid organic/inorganic porous materials. MOFs are essentially coordination polymers formed by connecting metal ions with polytropic organic linkers, and often result in fascinating structural topologies. The most direct approach—a simple coating with MOF-particles by dipping suitable substrates into a suspension of MOF powder particles and letting the solvent evaporate—puts no special requirements on the supporting surface. Five different concepts have been employed for the growth or preparation of MOF thin films on the substrate [64]:

1. Growth/deposition from solvothermal mother solutions.
2. Microwave-induced thermal deposition (assembly of preformed, size selected nanocrystals).

3. Stepwise layer-by-layer growth onto the substrate.
4. Electrochemical deposition of thin MOF-films on metal substrates.
5. Deposition of MOF thin films using a gel-layer approach.

The differences between these methods relate first to the type of substrate used and second the deposition procedure itself.

4.4.1 Growth/Deposition from Solvothermal Mother Solutions

Typical MOF synthesis involves solvothermal synthesis or slow interdiffusion of the separate solution of the respective building-blocks [65]. The first successful preparation of well-defined MOF thin films using solvothermal methods was made by Herms et al. [66]. In the case of MOF-5, the conventional scheme was as follows: after mixing the reactants ($Zn(NO_3)_2$-$4H_2O$, terephthalic acid) the solution was kept at elevated temperatures for an extended amount of time (105 °C, 1 atm). After a period of 3 days the reaction product was a powder consisting of particulate precipitates at the bottom of the container. When aiming at a rigid deposition of the MOF particles on a substrate the direct approach would be to simply immerse a substrate in the solution during MOF formation. In order to obtain such a grafting of MOFs onto a solid substrate, first a suitable surface functionalization has to be chosen. Most important is that the termination of functional groups on this surface should allow for a direct binding of the MOF material.

Usually MOFs are prepared under solvothermal or hydrothermal conditions. In this method, substances are crystallized from high temperature aqueous solutions at high vapor pressures. This type of synthesis depends on the solubility of minerals in hot water under high pressures. Crystal growth is performed in an autoclave to provide high vapor pressure. The nutrients are supplied in it along with water. A temperature gradient is maintained at the opposite ends of the chamber, such that the hotter end dissolves the nutrients and the cooler end causes seeds to take additional growth.

Hu et al. [67] developed a facile reactive seeding (RS) method for the preparation of continuous membranes on alumina porous supports, in which the porous support acted as the inorganic source reacting with the organic precursor to grow a seeding layer. An example of the preparation of MIL-53 membrane on alumina support via the RS method is shown in Fig. 4.11.

First, alpha alumina support instead of $Al(NO_3)_3 \cdot 9H_2O$ acts as the aluminum precursor, which reacts with 1,4-benzenedicarboxylic acid (H_2BDC) under mild hydrothermal conditions to produce a seed layer. This is followed by a secondary growth process when $Al(NO_3)_3 \cdot 9H_2O$ and H_2BDC form the MIL-53 membrane under hydrothermal conditions (typical synthesis conditions 220 °C for 12 h).

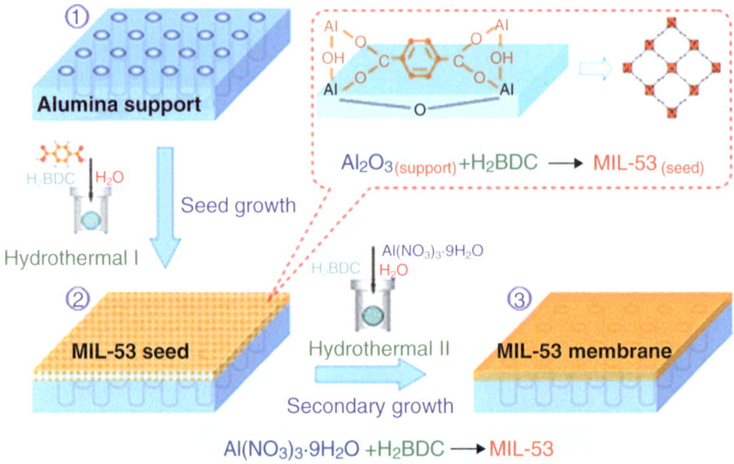

Fig. 4.11 Schematic diagram of preparation of the MIL-53 membrane on alumina support via the RS method

4.4.2 Microwave-Induced Thermal Deposition (MITD)

MITD is another novel method to prepare MOF thin films on porous substrates by the rapid production of MOF crystals in a facile manner. Yoo and Jeong [68] demonstrated a novel method to rapidly fabricate nanoporous MOF thin films and patterns on porous alumina substrates under microwave irradiation. It was noticed that the thin layers of conductive materials such as amorphous carbon, graphite, and other materials such as Au drastically enhanced the kinetics of heterogeneous nucleation and growth of MOF-5 crystals. In brief, a mixture of metal precursor and corresponding spacing ligands was dissolved in N-N-diethylformamide solvent. To create a homogeneous seeding environment, the mixture was thoroughly stirred to get a clear solution [69]. Substrates (nanoporous anodized alumina discs, Anodisc®, Whatman Co.) coated with various conductive thin films were then placed vertically in vials containing the precursor solution and MOF-5 crystals were grown under microwave irradiation in a domestic microwave oven with 500 W power for 5–30 s. Anodisc® substrates were chosen as mechanical supports with negligible transport resistance to gas molecules, which is important for potential applications of MOF films in gas separation. The resulting films and powders were thoroughly washed, dried, and stored in desiccators for analysis. Thin layers of conductive materials such as amorphous carbon, graphite, and other materials such as Au were found to drastically enhance the kinetics of heterogeneous nucleation and growth of MOF-5 crystals.

4.4.3 Stepwise Layer-by-Layer Growth onto the Substrate

Stepwise layer-by-layer growth onto the substrate was first demonstrated by Shekhah et al. [70], using a novel approach for depositing metal-organic open frameworks (MOFs) based on benzenetricarboxylic acid ligands and Cu(II)-ions on a COOH-terminated organic surface. This low-temperature deposition of highly porous, oriented metal organic frameworks exhibiting crystalline order both perpendicular and parallel to the substrate surface, is based on the layer-by-layer (lbl) method first employed by Langmuir and Blodgett (LB) [71] for the fabrication of multilayer organic LB films. Bétard et al. demonstrated the fabrication of MOF membranes by stepwise deposition of reactants [72].

4.4.4 Electrochemical Deposition of Thin MOF-Films on Metal Substrates

This method to prepare thin films of HKUST-1 on copper substrate was introduced by BASF [73]. For this electrochemical synthesis of powdered MOF-5 a copper electrode was immersed in a solution of the organic MOF building block, 1,4-benzenedicarboxylic acid (BDC) in N-N-diethylformamide (DEF). By applying an appropriately biased electric voltage, the electrochemical oxidation of Cu-atoms leads to the dissolution of Cu^{2+} metal ions and, subsequently, to the formation of crystallites in the vicinity of the electrode surface. The continuous supply of more Cu^{+} leads to a continuous growth of the crystallites.

4.4.5 Deposition of MOF Thin Films Using a Gel-Layer Approach

Schoedel et al. [74] have used a gel layer approach to synthesize oriented MOF thin films on modified Au substrates. Modification of the Au substrates was done by growing –COOH or –OH terminated alkanethiolate-based SAM (self assembled monolayer). Figure 4.12 shows the scheme for the gel-layer approach leading to uniquely oriented nanoscale films of metal-organic frameworks.

Subsequently, the substrate was coated with a layer of a poly(ethylene oxide) gel, which served as a storage medium for the metal-containing reactants. The formation of the MOF thin films was then induced by pouring solution of H_2btc (benzenetricarboxylic acid) or NH_2-bdc (amino-benzenedicarboxylic acid) on top of the gel layer. A structural analysis of the MOF films using XRD revealed an orientation growth of HKUST-1 and the amino-functionalized, flexible framework, Fe-MIL-88B-NH_2. The thicknesses of the MOF thin films deposited using this gel layer method could be varied by adjusting the concentration of the metal-containing reactant within the gel.

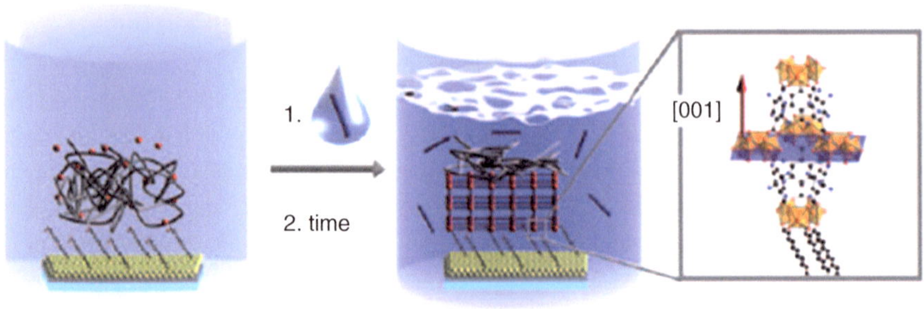

Fig. 4.12 Representation of the gel-layer approach leading to uniquely oriented nanoscale films of metal-organic frameworks. A Sam-functionalized gold slide is loaded with the metal-salt-containing poly(ethylene glycol) gel layer (metal ions in *red*) and covered with a solution containing the linker molecules (*blue*)

4.5 Ultrathin Membranes

Favorable membranes should have high flux and separation capabilities. Either property can be combined by the preparation of a so called composite membrane which consists of a highly porous supporting membrane coated with a thin, homogeneous and dense separating layer. In general, this separating layer is cast from a solution. However, solution casting is limited to the preparation of layers of more than a micrometer in thickness. If a thinner separating layer is need, more complicated and sophisticated preparation techniques have to be used. One of the methods is the Langmuir-Blodgett (LB) technique, based on molecular self assembly such as physiosorption or chemisorption of organic compounds on a solid support [44]. Uniform, defect-free silicone/polycarbonate membranes as thin as 0.015 μm have been formed by spreading solutions of the copolymer on water surfaces. The membranes were readily applied to support materials. Composite membranes formed from several laminations of ultrathin membrane on a support were readily able to be handled and provide the basis for practical gas separation processes [75]. Ackern et al. [76] demonstrated that the layer-by-layer adsorption of oppositely charged polyelectrolytes on porous supporting membranes is a useful method to prepare composite membranes with an ultrathin separation layer. A 60 nm thick polyelectrolyte membrane can be obtained, only by 60 pairs of polyelectrolyte layers.

Sullivan and Bruening [77] described a convenient method for forming ultrathin, gas selective polyimide films at the surface of porous alumina by alternating electrostatic adsorption of poly(amic acids) and poly(allyamine hydrochloride) (PAH) followed by heat-induced imidization. By controlling deposition conditions, membranes can be tailored to contain primarily polyimide, and fully imidized membranes exhibit permeability coefficients and selectivities that are comparable to literature values for the bulk polyimides.

Uniform, defect-free silicone/polycarbonate membranes as thin as 0.015 μm were formed via spreading solutions of the copolymer on water surfaces by Ward et al. [75]. Pinnau et al. [78] fabricated ultrathin asymmetric gas separation membranes by a dry/wet phase inversion process. The Chung group [18] described a method for the ultrathin skin hollow fiber membranes with a skin layer 474 Å using mainly a polymer and one solvent system.

Graphine-based materials have great potential to make ultrathin membranes for gas separation. Li et al. [79] fabricated ultrathin graphine oxide (GO) membranes, with thicknesses approaching 1.8 nm, using a facile filtration process. These membranes showed mixture separation selectivities as high as 3,400 and 900 for H_2–CO_2 and H_2–N_2 mixtures, respectively, through selective structural defects were identified on GO membranes.

Thin layers of the membrane on the surface of a substrate/support can be deposited by Electroless plating and electroplating. Electroless plating is based upon the controlled autocatalyzed decomposition or reduction of metastable metallic salt complexes on target surfaces. In the case of the production of palladium membranes, the substrate should be pre-seeded with Pd nuclei in activation solution to reduce the induction period of the autocatalytic plating reaction. In electroplating, the substrate acts as a cathode. In a plating bath, the metal or an alloy is coated on the substrate [51].

References

1. Loeb S, Sourirajan S (1963) Sea water demineralization by means of an osmotic membrane. Adv Chem Ser 38:117–132
2. Ren J, Wang R (2011) Preparation of polymeric membranes. In: Wang LK, Chen JP, Hung YT, Shammas NK (eds) Membrane and desalination technologies. Hand book of environmental engineering. Springer, Humana Press, Inc., Totowa, NJ
3. Peng N, Widjojo N, Sukitpaneenit P, Teoh MM, Lipscomb GG, Chung TS, Lai JY (2012) Evolution of polymeric hollow fibers as sustainable technologies: past, present, and future. Prog Polym Sci 37:1401–1442
4. Cohen C, Tanny GB, Prager S (1979) Diffusion controlled formation of porous structures in ternary polymer systems. J Polym Sci Polym Phys Ed 17:477–489
5. Yilmaz L, McHugh AJ (1986) Analysis of non-solvent-solvent-polymer phase diagrams and their relevance to membrane formation modeling. J Appl Polym Sci 31:997–1018
6. McHugh AJ, Yilmaz L (1985) The diffusion equations for polymer membrane formation in ternary systems. J Polym Sci 23:1271–1274
7. Reuvers AJ, van der Berg JWA, Smolders CA (1987) Formation of membranes by means immersion precipitation: Part I. A model to describe mass transfer during immersion precipitation. J Membr Sci 34:45–65
8. Reuvers AJ, Smolders CA (1987) Formation of membranes by means immersion precipitation: Part II. The mechanism of formation of membranes prepared from the system cellulose acetate-acetone-water. J Membr Sci 34:67–86
9. Kim YD, Kim JY, Lee HK, Kim SC (2001) A new modeling of asymmetric membrane formation in rapid mass transfer system. J Membr Sci 190:69–77
10. Termonia Y (1994) Monte Carlo diffusion model of polymer coagulation. Phys Rev Lett 72:3678–3681

11. Wang XL, Qian HJ, Chen LJ, Lu ZY, Li ZS (2008) Dissipative particle dynamic simulation on the polymer membrane formation by immersion precipitation. J Membr Sci 311:251–258
12. Twarowska-schmid K, Wlochowicz A (1997) Melt-spun poly(4-methyl-1-pentene) hollow fiber membrane. J Membr Sci 137:55–61
13. Li SG, Koops GH, Mulder MHV, van den Boomgaard T, Smolders CA (1994) Wet spinning of integrally skinned hollow fiber membranes by a modified dual-bath coagulation method using a triple orifice spinneret. J Membr Sci 94:329–340
14. Dong G, Li H, Chen V (2010) Factors affect defect-free Matrimid® hollow fiber gas separation performance in natural gas purification. J Membr Sci 353:17–27
15. Julian H, Wenten IG (2012) Polysulfone membranes for CO_2/CH_4 separation: state of art. IOSR J Eng 2:484–495
16. Ismail AF, Li K (2008) From polymeric precursors to hollow fiber carbon and ceramic membranes. In: Mallada R, Menendez M (eds) Synthesis, characterization and applications, Membrane science and technology series. Elsevier, Amsterdam
17. Ekiner OM, Vassilatos G (2001) Polyaramide hollow fibers for H_2/CH_4 separation: II. Spinning and properties. J Membr Sci 186:71–84
18. Chung TS, Teoh SK, Hu X (1997) Formation of ultrathin high-performance polyethersulfone hollow fiber membrane. J Membr Sci 133:161–175
19. Wallace DW, Staudt-Bickel C, Koros WJ (2006) Efficient development of effective hollow fiber membranes for gas separations from novel polymers. J Membr Sci 278:92–104
20. Cao C, Chung TS, Chen SB, Dong Z (2004) The study of elongation and shear rates in spinning process and in effect on gas separation performance of poly(ether sulfone) (PES) hollowfiber membrane. Chem Eng Sci 59:1053–1062
21. Li DF, Chung TS, Wang R, Liu Y (2002) Fabrication of fluoropolyimide/polyethersulfone (PES) dual layer asymmetric hollow fiber membranes for gas separation. J Membr Sci 198:211–223
22. Pereira CC, Nobrega R, Peinemann KV, Borges CP (2003) Hollow fiber membrane obtained by simultaneous spinning of two polymer solutions: a morphological study. J Membr Sci 226:35–50
23. Matsuyama H, Yuasa M, Kitamura Y, Teramoto M, Lloyd DR (2000) Structural control of anisotropic and asymmetric polypropylene membrane prepared by thermally induced phase separation. J Membr Sci 179:91–100
24. Kim JJ, Hwang JR, Kim UY, Kim SS (1995) Operation parameters of melt spinning of polypropylene hollow fiber membranes. J Membr Sci 108:25–36
25. Kluiters SCA (2004) Status review on membrane systems for hydrogen separation. Intermediate report EU project MIGREYD NNE5-2001-670, ECN-C-04-102
26. Jansen JC, Tasselli F, Tocci E, Drioli E (2006) High-flux composite perfluorinated gas separation membranes of Hyflon® AD on a hollow fibre ultrafiltration membrane support. Desalination 192:207–213
27. Sridhar S, Smitha B, Mayor S, Prathab B, Aminabhavi TM (2007) Gas permeation properties of polyamide membrane prepared by interfacial polymerization. J Mater Sci 42:9392–9401
28. Zhu J, Fan Y, Xu N (2011) Modified dip-coating method for preparation of pinhole-free ceramic membranes. J Membr Sci 367:14–20
29. Abu Seman MN, Khayet M, Hilal N (2010) Nanofiltration thin film composite polyester polyethersulfone-based membranes prepared by interfacial polymerization. J Membr Sci 34:109–116
30. Lau WJ, Ismail AF (2011) Progress in interfacial polymerization technique on composite membrane preparation. In: Second international conference on environmental engineering and applications. IPCBEE, vol 17. IACSIT Press, Singapore
31. Yasuda H (1981) Glow discharge polymerization. J Polym Sci Macromol Rev 16:199–293
32. Shen M, Alexis TB (1979) Plasma polymerization. American Chemical Society, Washington, DC
33. Michelmore A, Martinek P, Sah V, Short RD, Vasilev K (2011) Surface morphology in the early stages of plasma polymer film growth from amine-containing monomers. Plasma Process Polym 8:367–372

34. Chen RT, Muir BW, Thomsen L, Tadich A, Cowie BCC, Such GK, Alamar P, Postma K, McLean KM, Caruso F (2011) New insights into the substrate plasma polymer interface. J Phys Chem B 115:6495–6502
35. Gaur S, Vergason G (2000) Plasma polymerization: theory and practice. In: 43rd annual technical conference proceedings. Denver, 15–20 Apr 2000, ISSN 0737-592
36. Zang Z (2003) Surface modification by plasma polymerization and application of plasma polymers as biomaterials. Ph.D. Thesis, Johannes Gutenberg University of Mainz
37. Adiga SP, Jin C, Curtiss LA, Monteiro-Riviere NA, Narayan RJ (2009) Nanoporous membranes for medical and biological applications. Wiley Interdiscip Rev Nanomed Nanobiotechnol 1:568–581
38. Sanguanruksa R, Rujiravanit R, Supaphol P, Tokura S (2004) Porous polyethylene membranes by template-leaching technique: preparation and characterization. Polym Test 23:91–99
39. Roy-Chowdhury P, Kumar V (2006) Fabrication and evaluation of porous 2,3-dialdehydecellulose membranes as a potential biodegradable tissue engineering scaffold. J Biomed Mater Res A 76:300–309
40. Decher G (1997) Fuzzy nanoassemblies: toward layered polymeric multicomposites. Science 277:1232–1237
41. Zhu Z, Feng X, Penlidis A (2006) Self assembled nano-structured polyelectrolyte composite membranes for pervaporation. Mater Sci Eng C 26:1–8
42. Decher G, Hong JD (1991) Buildup of ultrathin multilayer films by a self-assembly process: II. Consecutive adsorption of anionic and cationic bipolar amphiphiles and polyelectrolytes on charged surfaces. Ber Bunsenges Phys Chem 95:1430–1434
43. Tieke B, van Ackern F, Krasemann L, Toutianoush A (2001) Ultrathin self-assembled polyelectrolyte multilayer membranes. Eur Phys J E 5:29–39
44. Pandey P, Chauhan RS (2001) Membranes for gas separation. Prog Polym Sci 26:853–893
45. Ockwig NW, Nenoff TM (2007) Membranes for hydrogen separation. Chem Rev 107:4078–4110
46. Roark SE, Mackay R, Mundschau MV (2003) Dense, layered membrane for hydrogen separation. US Patent 7,001,446, 21 Feb 2006
47. Benes NE, Bicseuvel PM, Verweij H (1999) Tensile stress in a porous medium due to gas expansion. AIChE J 45:1322–1328
48. Nam SW, Gavalas GR (1989) Stability of H_2-permselective SiO_2 films formed by chemical vapor deposition. AIChE Symp Ser 268:68–74
49. Tsapatsis M, Kim S, Nam SW, Gavalas G (1991) Synthesis of hydrogen permselective silicon dioxide, titanium dioxide, aluminum oxide, and boron oxide membranes from the chloride precursors. Ind Eng Chem Res 30:2152–2159
50. Richardson JT, Paripatyadar SA (1990) Carbon dioxide reforming of methane with supported rhodium. Appl Catal 61:293–309
51. Shu J, Grandjean BPA, Van Neste A, Kaliaguine S (1991) Catalytic palladium-based membrane reactors: a review. Can J Chem Eng 69:1036–1060
52. Uemiya S, Kajiwara M, Kojima T (1997) Composite membranes of group VIII metal supported on porous alumina. AIChE J 43:2715–2723
53. Uemiya S (1999) State-of-the-art of supported metal membranes for gas separation. Sep Purif Rev 28:51–85
54. Gryaznov VM, Seretryannikova OS, Serov YM (1993) Preparation and catalysis over palladium composite membranes. Appl Catal A 96:15–23
55. Li ZY, Maeda H, Kusakabe R, Moroooka S, Anzai H, Akiyama S (1993) Preparation of palladium-silver alloy membranes for hydrogen separation by the spray pyrolysis method. J Membr Sci 78:247–254
56. Zhang K, Gao H, Rui Z, Lin Y, Li Y (2007) Preparation of thin palladium composite membranes and application to hydrogen/nitrogen separation. Chin J Chem Eng 15:643–647
57. Asadea M, Yamasaki S (2001) Separation of inorganic/organic gas mixtures by porous silica membranes. Sep Purif Technol 25:151–159

58. Brinker CJ, Ward TL, Sehgal R, Raman NK, Hietala SL, Smith DM, Hua DW, Headley TJ (1993) "Ultramicroporous" silica-based supported inorganic membranes. J Membr Sci 77:165–179
59. Wavhal DS, Zhang J, Steen ML, Fisher ER (2006) Investigation of gas phase species and deposition of SiO_2 films from HMDSO/O_2 plasmas. Plasma Process Polym 3:276–287
60. Li Y, He G, Wang S, Yu S, Pan F, Wu H, Jiang J (2013) Recent advances in the fabrication of advanced composite membranes. J Mater Chem A 1:10058–10077
61. Bastani D, Esmaeili N, Asadollahi M (2013) Polymeric mixed matrix membranes containing zeolites as a filler for gas separation applications: a review. J Ind Eng Chem 19:375–393
62. Bae TH, Liu J, Lee JS, Koros WJ, Jones CW, Nair S (2009) Facile high-yield solvothermal deposition of inorganic nanostructures on zeolite crystals for mixed matrix membrane fabrication. J Am Chem Soc 131:14662–14663
63. Lydon ME, Unocic KA, Bae TH, Jones CW, Nair S (2012) Structure-property relationships of inorganically surface-modified zeolite molecular sieves for nanocomposite membrane fabrication. J Phys Chem C 116:9636–9645
64. Shekhah O, Liu J, Fischer RA, Wöll C (2011) MOF thin films: existing and future applications. Chem Soc Rev 40:1081–1106
65. Zhuang JL, Lommel K, Ceglarek D, Andrusenko I, Kolb U, Maracke S, Sazama U, Fröba M, Terfort A (2011) Synthesis of a new copper-azobenzene framework in the form of hierarchical bulk solids and thin films without and with patterning. Chem Mater 23:5366–5374
66. Herms S, Schroder F, Chelmowski R, Wöll C, Fischer RA (2005) Selective nucleation and growth of metal-organic open framework thin films on patterned COOH/CF_3-terminated self-assembled monolayers on Au(III). J Am Chem Soc 127:13744–13745
67. Hu Y, Dong X, Nan J, Jin W, Ren X, Xu N, Lee YM (2011) Metal-organic membranes fabricated via reactive seeding. Chem Commun 47:737–739
68. Yoo Y, Jeong HK (2008) Rapid fabrication of metal organic framework thin films using microwave-induced thermal deposition. Chem Commun: 2441–2443
69. Ni Z, Masel RI (2006) Rapid production of metal-organic frameworks via microwave-assisted solvothermal synthesis. J Am Chem Soc 128:12394–12395
70. Shekhah O, Wang H, Kowarik S, Schreiber F, Paulus M, Tolan M, Sternemann C, Evers F, Zacher D, Fischer RA, Wöll C (2007) Step-by-step route for the synthesis of metal-organic frameworks. J Am Chem Soc 129:15118–15119
71. Blodgett KB, Langmuir I (1937) Built-up films of barium stearate and their optical properties. Phys Rev 51:964–983
72. Bétard A, Bux H, Henke S, Zacher D, Caro J, Fischer RA (2012) Fabrication of a CO_2-selective membrane by stepwise liquid-phase deposition of an alkylether functionalized pillard-layered metal-organic framework $[Cu_2L_2P]_n$ on a macroporous support. Microporous Mesoporous Mater 150:76–82
73. Muller U, Putter H, Hesse M, Schubert M, Wessel H, Huff J, Guzmann M (2007) Method for electrochemical production of a crystalline porous metal organic skeleton material. US Patent 20,070,227,898 A1, 4 Oct 2007
74. Schoedel A, Scherb C, Bein T (2010) Oriented nanoscale films of metal-organic frameworks by room temperature gel-layer synthesis. Angew Chem Int Ed 49:7225–7228
75. Ward WJ, Browall WR, Salemme RM (1976) Ultrathin silicone/polycarbonate membranes for gas separation processes. J Membr Sci 1:99–108
76. van Ackern F, Krasemann L, Tieke B (1998) Ultrathin membranes for gas separation and pervaporation prepared upon electrostatic self-assembly of polyelectrolytes. Thin Solid Films 327–329:762–766
77. Sullivan DM, Bruening ML (2003) Ultrathin, gas-selective polyimide membranes prepared from multilayer polyelectrolyte films. Chem Mater 15:281–287
78. Pinnau I, Wind J, Peinemann KV (1990) Ultrathin multicomponent poly(ether sulfone) membranes for gas separation made by dry/wet phase inversion. Ind Eng Chem Res 29:2028–2032
79. Li H, Song Z, Zhang X, Huang Y, Li S, Mao Y (2013) Ultrathin, molecular-sieving graphene oxide membranes for selective hydrogen separation. Science 342:95–98

Chapter 5
Membrane Modules and Process Design

Hundreds of thousands of square meters of membrane are needed to perform the required separation of compounds in industrial plants. There are several efficient and economical ways to create a large surface area in a membrane package for effective compound separation. From an overall cost standpoint, not only the cost of membranes per unit area is crucial but also the cost of the containment vessel into which they are mounted. These packages are called membrane modules. The most important are:

1. Plate-and-frame
2. Tubular
3. Spiral-wound
4. Hollow fiber

Baker et al. [1] discussed some of the factors that affect the design of membranes for the vapor-gas separation process. The design of a membrane separation process depends on the separation to be performed and the properties of the membrane used. The type of membrane structure depends on the nature of the selective materials. Glassy polymers are commonly formed into high-performance anisotropic (skinned) membranes by varying the solution precipitation procedure invented by Loeb and Sourirajan.

Alternatively, rubbery polymer membranes made by this method would collapse under the high pressure of the gas separation processes; thus, rubbery polymer membranes are formed as composite structures consisting of a mechanically strong, highly permeable microporous support layer coated with a thin film on the selective rubbery material. The support layer should be 10–100 times more permeable than the selective layer to ensure that the separation properties of the composite are determined by the rubbery layer and not by the support. Table 5.1 shows some examples of gas separation applications and the type of modules used.

A.F. Ismail et al., *Gas Separation Membranes*,
DOI 10.1007/978-3-319-01095-3_5

Table 5.1 Examples of gas separation applications and the type of module used [2]

Application	Membrane material	Selectivity, α	Module design
O_2/N_2	Polyamide	6–7	Hollow fiber
H_2/N_2	Polysulfone	100	Hollow fiber
CO_2/CH_4	Cellulose acetate	18–20	Spiral wound or hollow fiber
VOC/N_2	Silicon rubber	10–30	Spiral wound
H_2O/air	Hydrophilic rubber	>200	Capillary

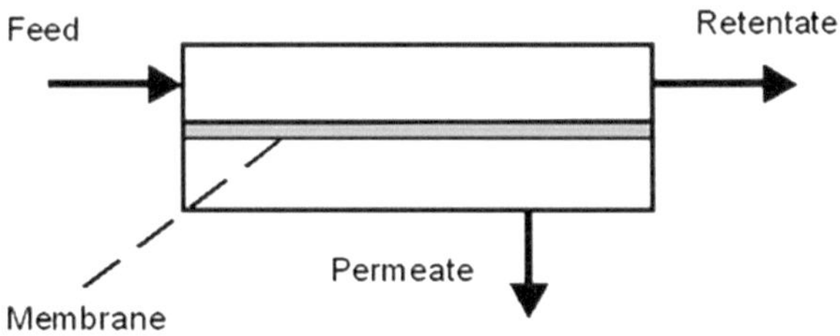

Fig. 5.1 Process principle of gas separation with membranes

Table 5.2 Approximate dimensions of tubular membranes [3]

Configuration	Diameter (mm)
Tubular	>10.0
Capillary	0.5–10.0
Hollow fiber	<0.5

5.1 Membrane Modules

The separation units into which the membranes are fitted are called membrane modules. They must allow for the separate conduction of the feed and permeate currents on either side of the membrane (Fig. 5.1).

The gas mixture, which is introduced into the separation apparatus, is called the feed. Inside the apparatus, this current is divided into two streams; the one that penetrates through the membrane is called the permeate, and the other stream that leaves the unit depleted is called the retentate.

All module types applied for gas separation are based on two types of membrane configurations such as flat and tubular. Modules based on flat membranes are the plate-and-frame and spiral-wound modules. Tubular-type membrane modules are subdivided into tubular, capillary, and hollow fiber. The differences between tubular, capillary, and hollow fiber modules are their tube's dimensions, as shown in Table 5.2.

5.1.1 Plate and Frame

Plate-and-frame modules are the closest to common laboratory setups. Flat membranes can be assembled as plate, bag or spiral wound. Their designs have their origins in the conventional filter press-concept. Each pair of membranes is separated by a

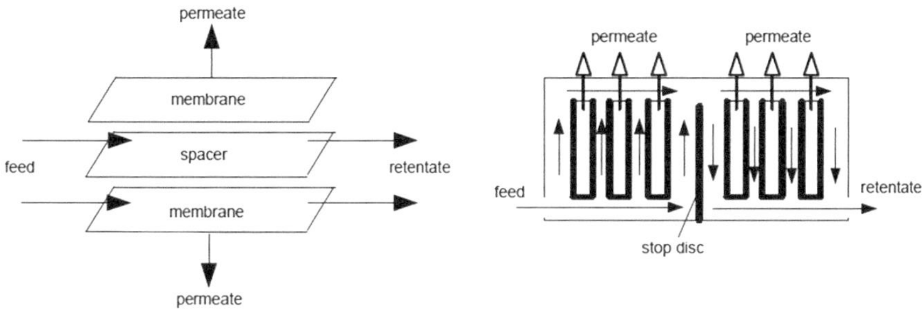

Fig. 5.2 Schematic drawings of plate-and-frame modules

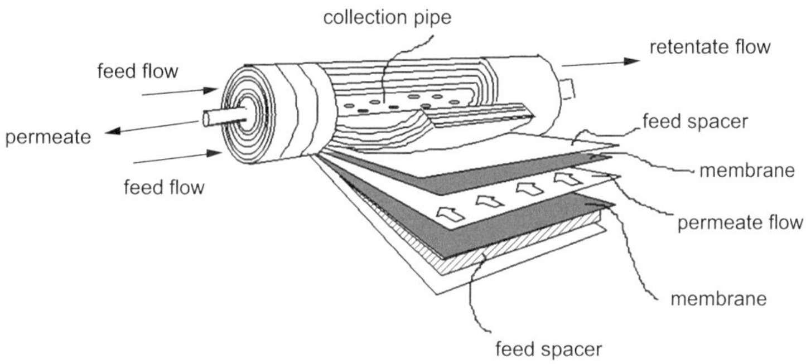

Fig. 5.3 Schematic drawing of a spiral-wound membrane module

spacer (feed spacer) with the separated layers stacked towards each other (like in a sandwich). Between every pair a permeate spacer is inserted (Fig. 5.2). These spacers are incorporated into frames, which simultaneously seal the module and make possible the material conduction through the alternating channels and drilled holes. The spacer plate separates the feed flow running alongside different membranes in the module. The packing density (i.e., membrane surface per module volume) is around 100–400 m^2/m^3. The stop disc in the right of Fig. 5.2 is used to improve the flow pattern in order to use the membrane surface as efficiently as possible (to reduce so-called "channelling", i.e., the tendency of the flow to move along a fixed pathway) [4].

5.1.2 *Spiral Wound*

A variation of the basic plate-and-frame concept is the spiral-wound module, which is widely used today for gas separation application. Its basic design is illustrated in Fig. 5.3.

The spiral-wound format was the first to be commercialized. In this technique, two flat sheet membranes are sealed together with a spacer placed both between the sheets and on one external side. The sandwiched sheets are then rolled into a spiral format around a central permeate collection channel. In this modular design the permeate flow is cross-flow to the feed, an arrangement which is less thermodynamically efficient than countercurrent flow. However, the relatively wide permeate channels ensure a low pressure drop and the feed side spacers act to increase turbulence, thus reducing concentration polarization.

In this design, the feed flows through the channel spacer, the membrane, and the porous membrane support form an envelope which is rolled around a perforated central collection tube and inserted into an outer tubular pressure shell. The feed passes in an axial direction through the feed channel across the membrane surface. The filtrate moves along the permeate channel and is collected in a perforated tube in the center of the roll. Small spiral-wound units consist of just one envelope, which limits the total membrane area that can be installed in a unit to about 1–2 m^2. The main reason for the limitation of the installed surface area in a module containing one single envelope is the pressure drop encountered by the permeate moving down the permeate channel to the central collection tube. Because the channel in a practical unit is very narrow, its length is limited to 2–5 m. A significantly longer path would result in an unacceptable pressure drop in the permeate channel. To install larger membrane surfaces in a spiral-wound module, a multi-leaf arrangement in used.

Another module for flat membranes is a bag module. In these modules, the membranes are welded as bags and arranged around a tube with drilled holes.

5.1.3 Tubular

Tubular membranes consist of a thin selective membrane layer deposited on the inner or outer sides of a tubular support with a diameter generally larger than 10 mm. The number of tubes put together in the module may vary from 4 to 8, but is not limited to this number (Fig. 5.4). The feed flows through the center of the membrane tubes and the permeate crosses the membrane from the inside to the outside, flowing subsequently in the larger tube. Monolithic modules, constructed using a ceramic block and a number of membrane tubes inserted in the block, form a special category of tubular modules [4].

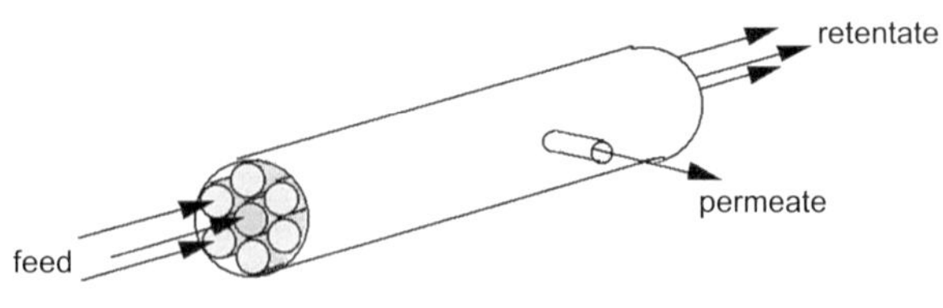

Fig. 5.4 Tubular module with seven individual tubes bundled in a shell tube

In some modules, the membranes are cast directly on the porous pipes and in others they are prepared separately as tubes and then installed into the support pipes.

Usually, 10–30 individual tubes are installed in a larger tube and potted at the end of the tube. The feed solution is fed in parallel through the tubular bundle while permeate of the individual tubes is collected in the outer shell tube as indicated in Fig. 5.4. The main advantage of the tubular module is that concentration polarization effects and membrane fouling can be easily controlled.

5.1.4 *Capillary*

Figure 5.5 shows two types of capillary modules, consisting of a large number of membrane capillaries with an inner diameter of 0.2–3 mm arranged in parallel as a bundle in a shell tube. The capillaries are self-supporting and bound together at the free ends (potted) with agents such as epoxy resins, polyurethanes, or silicone rubber. The feed flow can go through the bores of the capillaries, with the permeate exiting the membrane sideways (left scheme), but the feed can also run through the capillaries on the outside with the permeate exiting through the bores of the membrane. Packing densities are in the range 600–1,200 m^2/m^3 [4].

The capillary membrane module requires membranes in a self-supporting capillary configuration, which, when asymmetrically structured, carry the selective barrier on the inner side of the capillary as indicated in the scanning electron micrograph of Fig. 5.6.

The capillary membrane module provides a high membrane area per module volume. The production costs are very low, and concentration polarization and membrane fouling can effectively be controlled by the proper feed flow and back-flushing of the permeate at certain time intervals. The main disadvantage of the capillary membrane module is the required low operating pressure. Because of the limited stability of the capillary membranes, operating pressures generally cannot exceed 4–6 bars. Therefore, the capillary membrane is used in applications where low transmembrane pressures are applied—i.e., in dialysis, microfiltration, and low pressure ultrafiltration. The most significant application of the capillary membrane

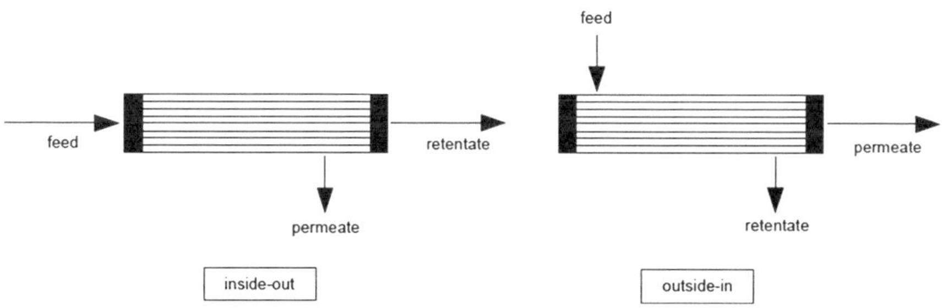

Fig. 5.5 Two schemes of the capillary model

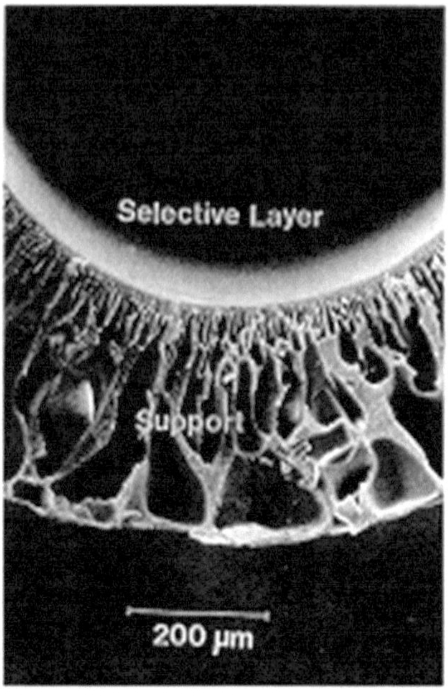

Fig. 5.6 SEM of a capillary membrane with the selective "skin" on the inside of the capillary (ultrafiltration membrane prepared by wet-spinning process)

module is as an artificial kidney. Strathmann et al. [5] discussed gas separation module development for gas separation and pervaporation.

Takaba and Nakao [6] used a computational fluid dynamics (CFD) technique for modeling capillary tube membrane modules without providing detailed descriptions of the models. They used porous ceramic membrane for extracting a H_2 from H_2–CO gas mixture in the steam reforming process.

5.1.5 Hollow Fiber

The hollow fiber module is essentially the same as the capillary module, only the sizes of the tubes are smaller. Hollow fibers are essentially self-supporting and resistant to collapse at high pressures and environmentally difficult situations [4]. Figure 5.7 shows a hollow-fiber module. Hollow-fiber membranes are incorporated in bundles with a synthetic resin at the ends. The hollow fibers are open at the permeate side; the module shown below is a possible variant, closed at one end of the hollow fibers. Another alternative design is that a number of hollow fibers are collected together and "potted" in an epoxy resin at both ends and installed into an outer shell. Both ends of hollow fibers remain open.

Asymmetric hollow fibers are attractive for membrane-based gas separation, since they provide high active surface area-to-volume ratio, low resistance to gas flow,

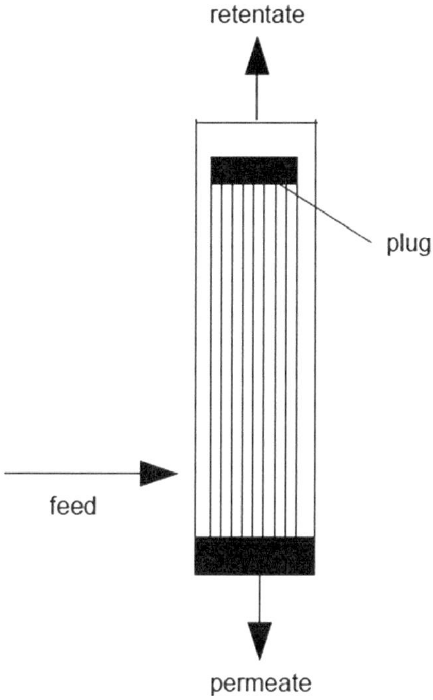

Fig. 5.7 Outside-in hollow fiber configuration for gas separation

and the ability to support high transmembrane pressure drops. Each of these factors contributes to high productivity.

The membrane modules presented here differ in their obtainable packing densities (membrane surface per volume of module) and basic prices, which are relatively expensive for plate and frame modules in comparison to spiral and hollow fiber modules. Hollow-fiber modules are characteristically 4–8 in. (10–20 cm) in diameter and 3–5 ft (1.0–1.6 m) long. These units mostly run with the feed stream on the outside of the fiber.

Katoh et al. [7] developed a simulation model to examine the unsteady-state behaviors of hollow-fiber membrane modules for multi component gas separation. They considered the nonideal mixing flows in the permeate and residue sides by using a tanks-in-series model. The relaxation method was applied to solve the governing simultaneous ordinary differential equations.

TNO of the Netherlands patented a cross-flow membrane module design, which offers good mass-transfer characteristics and scale-up potential [8]. In this module, CO_2 flows in the shell-side perpendicular to the fiber, but overall the two phases flow counter-currently through the module. A schematic diagram of the module is given in Fig. 5.8.

Coker et al. [9] developed a model for multicomponent gas separation using hollow-fiber contactors that permit simulation of co-current, counter current, and cross-flow contacting patterns with the permeate (or sweep). They followed a stage-wise approach to convert the differential equations to a set of coupled, nonlinear

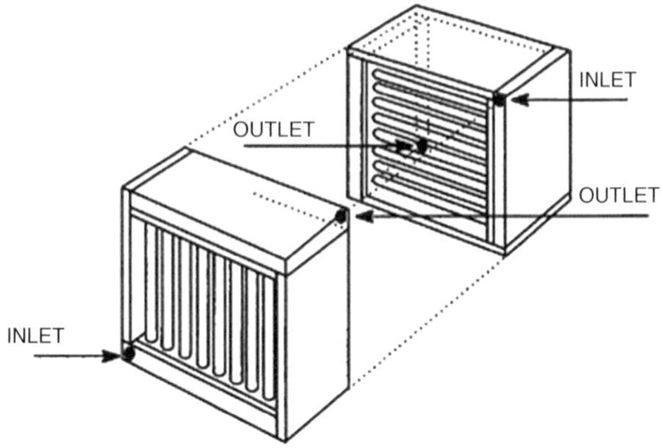

Fig. 5.8 TNO cross-flow module

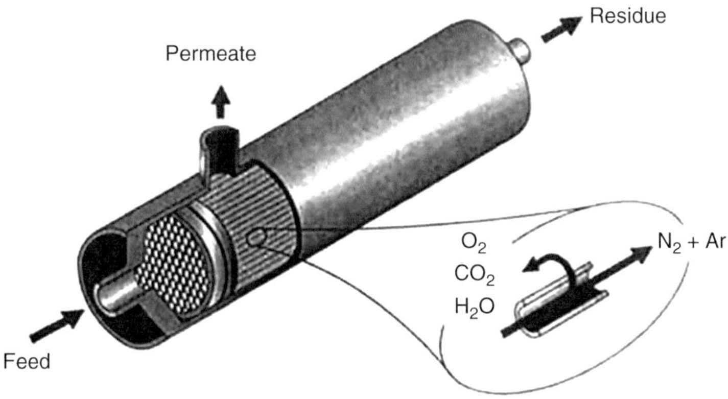

Fig. 5.9 Hollow-fiber module

differential equations. Although they claimed that their methodology could easily incorporate pressure dependence permeability, they assumed constant permeability in their modeling work. Model validation was not verified with experimental data. Figure 5.9 shows the flow configuration and internal structure of a typical hollow-fiber gas separation module. The hollow-fiber bundle is sealed on both ends by epoxy tube sheets and is contained inside a high-pressure housing. Feed gas may be introduced on the bore of the hollow fibers (as shown in Fig. 5.9) or on the shell side of the module. Figure 5.9 shows the schematic diagram of the hollow-fiber cross-section.

The principal assumptions for the model are:

1. Shell side pressure change is negligible.
2. Bore side pressure change is given the Hagen–Poiseuille equation.
3. The hollow fibers consist of a very thin membrane separation layer on a porous support as shown in Fig. 5.10. All mass-transfer resistance is confined to the separation membrane or the total membrane wall.

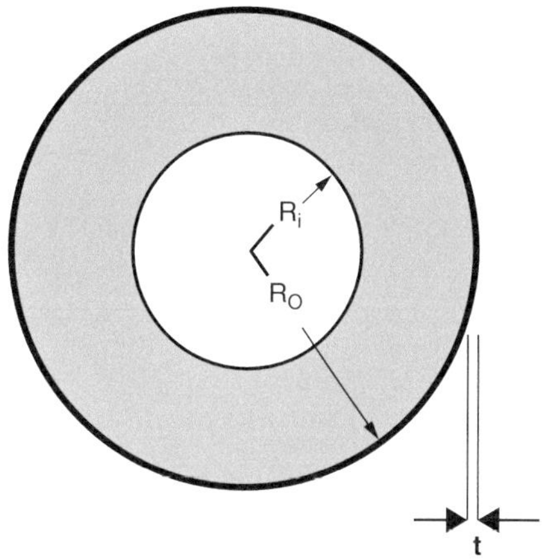

Fig. 5.10 Hollow-fiber cross section of inside diameter 2Ri, outside diameter 2R, and dense separating layer effective thickness *t*

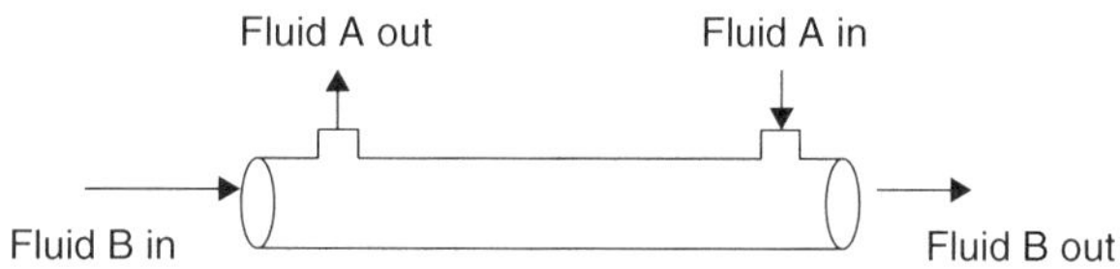

Fig. 5.11 Schematic diagram of a parallel-flow hollow fiber membrane contactor

4. There is no axial mixing of shell or lumen side gases in the direction of bulk gas flow.
5. The gas on the shell side of the hollow fibers and in the lumen is in plug flow.
6. The performance of a single hollow fiber is calculated during the simulation and these results are scaled in proportion to the number of fibers in the module, to account for total gas flow and membrane area.
7. The deformation of the hollow fiber under pressure is negligible.
8. All fibers have uniform inner and outer radiuses as well as uniform membrane thicknesses.
9. The membrane module is operated at steady state.

5.1.6 Membrane Contactors

Thus far, membrane modules for gas separation have been shown, but different kinds of hollow-fiber modules have also been used for membrane contractor applications. The membrane contactors are devices that achieve gas–liquid or liquid–liquid mass transfer without dispersion of one phase within the other. Figure 5.11 shows a simple module of a membrane contactor.

Figure 5.12 shows the module of the hollow-fiber membrane contactor used by Nymeijer et al. for olefin/paraffin separation [10].

Fig. 5.12 Schematic representation of the hollow fiber membrane modules of membrane contactor

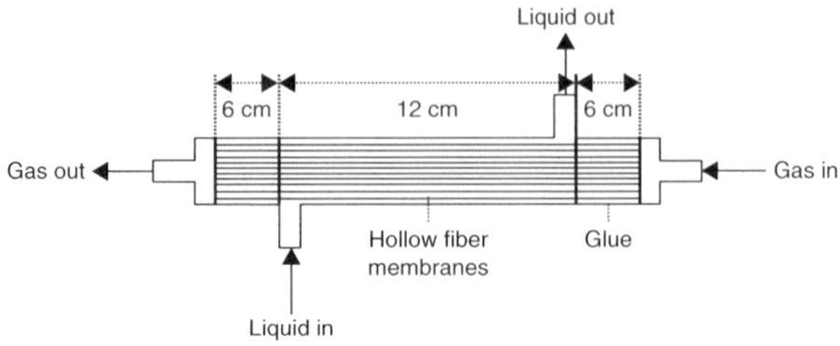

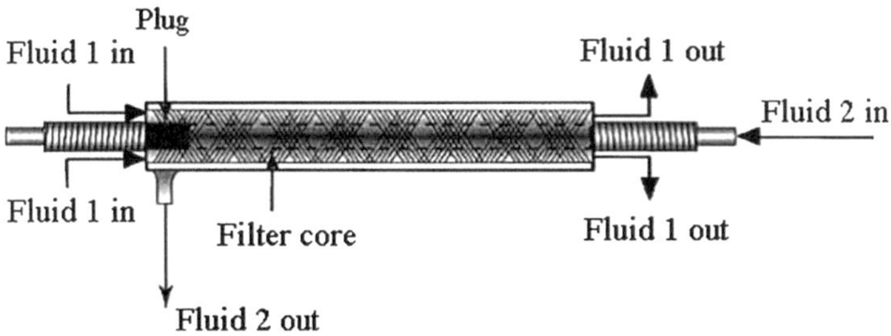

Fig. 5.13 Module containing hollow fibers wound helically around a central core. (The plug forces liquid entering the core radically outward so that the flow is perpendicular to the fibers)

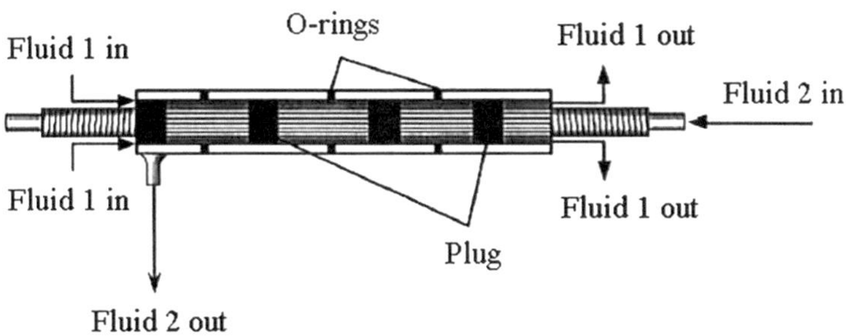

Fig. 5.14 Module containing woven hollow fiber wound helically around a central core. (The plug and O-rings provide multiple shell passes)

Stanojević et al. [11] wrote a general review of membrane contactor designs and operation, especially in applying the contactors for removing dissolved oxygen from aqueous solutions by vacuum degassing. They discussed different type of modules for membrane contactors, illustrated in Figs. 5.13, 5.14, 5.15, 5.16, and 5.17.

Depending on the relative flow directions of the two phases, the membrane module can be classified into two groups [12].

1. Longitudinal-flow module
2. Cross-flow module

Longitudinal-flow module: In such a module, the gas and liquid phases flow in parallel (either countercurrently or cocurrently) to each other on the opposite sides

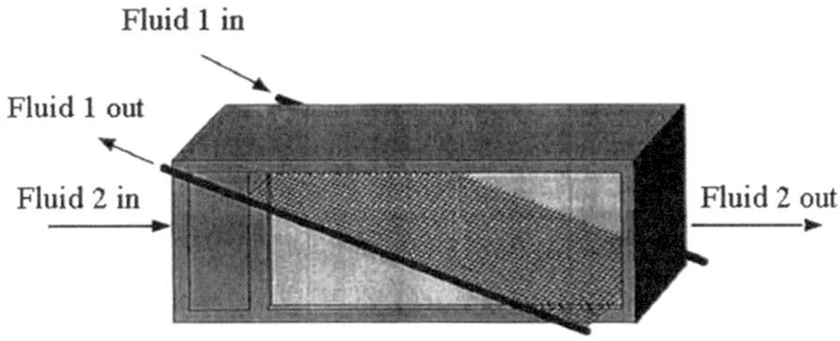

Fig. 5.15 Module containing woven hollow fiber wound-mounted diagonally in an open-ended box

Fig. 5.16 Rectangular module containing two baffles

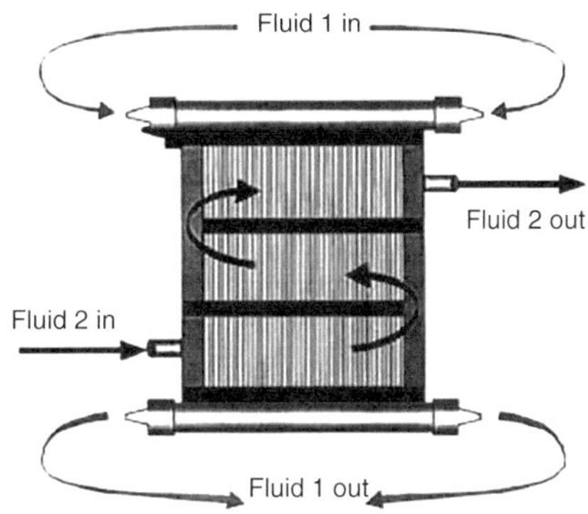

Fig. 5.17 Fully baffled cylindrical module

of the fibers (membrane). A schematic description of this type of module is given in Fig. 5.11. The advantage of this module is its simplicity in manufacturing. In contrast, its advantage is mainly offset by its mediocre efficiency in mass transfer compared with the cross-flow module.

Cross-flow module: Compared to the longitudinal flow module, the characteristic of the cross-flow module is the presence of some baffles in the module design, as is shown in Fig. 5.18. The baffles can improve the mass-transfer efficiency by minimizing shell-side bypass and providing a velocity component normal to the membrane surface. As a result, the cross-flow module can maintain higher mass-transfer coefficient.

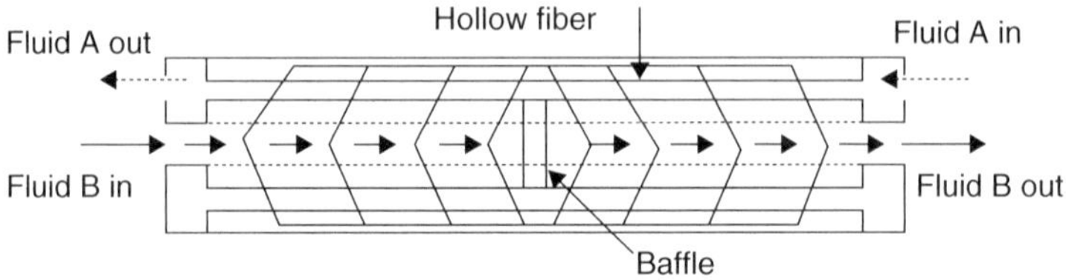

Fig. 5.18 Schematic diagram of a cross-flow hollow fiber membrane contactor (Liqui-Cel®)

Table 5.3 Active area of different type of modules

Type of module	Area per volume (m^2/m^3)
Plate	200–600
Spiral	800–1,000
Hollow fiber	2,000–5,000

Table 5.4 Parameters for membrane module design

Parameter	Hollow fibers	Capillary fibers	Spiral wound	Plate and frame	Tubular
Manufacturing cost ($/m²)	2–10	5–50	5–50	50–200	50–200
Concentration polarization/fouling control	Poor	Good	Moderate	Good	Very good
Permeate-side pressure drop	High	Moderate	Moderate	Low	Low
Suitability for high pressure operation	Yes	No	Yes	Marginal	Marginal
Limitation to specific types of membrane material	Yes	Yes	No	No	No

5.2 Comparison of the Module Configuration

Table 5.3 shows the active area of different type of modules [13].

The plate module format has been chosen for laboratory tests and for the design of membranes because flat membranes are easily produced without defects and the modules are easy to shape. A majority of gas separation membranes are formed into spiral-wound or hollow-fiber modules. The choice of the most suitable membrane module type for a particular membrane separation must balance a number of factors. The principal module design parameters that enter into the decision are summarized in Table 5.4.

The single greatest advantage of hollow-fiber modules is the ability to pack a very large membrane area into a single module. For example, an 8-in. diameter, 40-in. long spiral-wound module contains about 20–40 m² of membrane area. The equivalent hollow-fiber module filled with fibers of 100-mm diameter will contain approximately 600 m² of membrane area.

The membrane contactors have a number of advantages in contrast to conventional dispersed phase contactors. Some of them are:

1. No flooding at high flow rates.
2. The absence of emulsions.
3. No density difference is required between the fluids.

Membrane contactors reduce the volume of equipment and offer an interfacial area in non-dispersive contact across a membrane, leading to a decrease in the height of transfer unit (HTU) values. The membrane should be attentively chosen to enable, as much as possible, higher values of the mass transfer coefficient. Membrane contactors give any wanted shape of fluid–fluid interface, in contrast to conventional separation equipment where the shape of the fluid–fluid contact is an accident of nature [11].

Gottschlich et al. [14] developed computer models to compare the performances of facilitated transport (FT) and conventional solution-diffusion (SD) membrane modules, and supplied sample calculations given for the separation of CO_2–CH_4 mixtures. For the conditions examined, the facilitation effect in the FT membrane module was significant only when the partial pressure of CO_2 was relatively low (<10 psia). For a low CO_2 partial pressure (7.5 psia) in the feed, the FT membrane with an ethylene diamine concentration of 8 M had an area requirement lower by a factor of 2 and 2 % greater methane recovery than for an SD membrane with a selectivity of 30. Above 50 psia CO_2 partial pressure, the SD and FT membrane modules functioned identically.

5.3 System Design

The main focus of system designs for gas separation is choice of suitable separation membranes, membrane modules, and the design of the whole system based on basic engineering principles. Membrane processes are commonplace in industrial applications; they replace evaporation and distilation and have offered improved economics.

Membrane systems have become viable alternatives to conventional gas separation technologies such as pressure swing adsorption and cryogenic distillation. The economics of membrane separation processes depend critically on the process design. The main task for every engineer is to arrange the modules in an optimal design at the lowest product cost. Single-stage systems have low capital costs, but they are appropriate only for moderate product purity and recovery requirements. Multiple separation stages and recycle are required for more demanding applications. The design of a membrane system involves the determination of: (1) the configuration of the permeator network; and (2) the operating conditions of the individual permeators [15].

Membrane systems currently are designed via a sequential procedure in which the permeator configuration is chosen as a priority and the operating conditions are determined using some type of optimization procedure [16]. An essential part in the design of gas separation by membranes is the determination of the separation configuration. A single stage arrangement with no recycle is the most common and simplest design form; however, the demand for higher product purity and recovery ratio of the desired species necessitates the use of recycle streams as well as multistage configurations. Commonly, the multistage systems are designed using two, three or four stages [17].

System design, therefore, is the process of defining the architecture, components, modules, interfaces, and (to reduce the cost of power) improved thermal efficiency. Superior environmental performance and attaining favorable products are also key to systems design. As plant size increases, the enlargement of membrane modules becomes increasingly important. In a number of processes, the goal of the membrane system is to recover and recycle vapor components previously lost with an inert gas purge [18]. Baker discussed process design for different systems, namely, one-stage selective purge systems, multistep and multistage systems, and hybrid system designs [19].

Many design studies for multistage gas membrane systems are based on this approach. Spillman et al. [20] investigated several permeator configurations for the separation of CO_2–CH_4 mixtures encountered in natural gas treatment and enhanced oil recovery. Babcock et al. [21] evaluated the economics of single- and three-stage membrane systems for natural gas treatment by providing comparisons with amine treatment processes. Bhide and Stern proposed a grid search method to design membrane separation systems for natural gas treatment [22, 23] of economic parameters and membrane properties [24]. Xu and Agrawal [25], Agrawal [26], and Agrawal and Xu [27, 28] developed a stepwise procedure for design of membrane cascades using a limited number of recycle compressors.

Qi and Henson proposed an efficient methodology for the preliminary design of multistage membrane (spiral-wound) separation systems for binary gas mixtures [15]. In another article Qi and Henson [16] proposed an optimal design strategy based on a permeator model and mixed-integer nonlinear programming (MINLP) for membrane systems separating multicomponent gas mixtures. It was concluded that the MINLP strategy is an effective tool for preliminary design of multistage, multicomponent gas membrane systems, including those with very small component concentrations.

Khalilpour et al. [29] analyzed the performance of a membrane system over key design/operation parameters. A computation methodology was developed to solve the model of hollow-fiber membrane systems for multicomponent gas feeds. The model represented by a nonlinear differential algebraic equation system was solved via a combination of backward differentiation and Gauss–Seidel methods. A natural gas sweetening problem was investigated as a case study. Model parametric analyses of variables, namely, feed gas quality, pressure, area, selectivity and permeance, resulted in a better understanding of operating and design optima. In particular, high selectivities and/or permeabilities were shown not to be necessary targets for optimal operation. This model-based membrane system engineering approach was proposed for the synthesis of efficient and cost-effective multistage membrane networks.

There could be many different designs of membrane contactors as they have a large number of applications and module configuration. The module is the central part of a membrane installation and can be called a separation unit. The membrane contactors are devices that achieve gas–liquid or liquid–liquid mass transfer without dispersion of one phase within another. The membrane is used to accomplish a particular separation and transport of one component more easily than another because of differences in physical and/or chemical properties between the membrane and the permeating components. In evaluating and describing membrane contactor designs, mass transfer coefficient is very important. The design objective in mass transfer with membrane contactors depends on the application.

5.4 Process Parameter

A membrane can be considered a permselective barrier or interface between two phases—two gas phases in the case of gas separation. Separation occurs because one of the components in the feed gas passes the membrane more easily and quickly than the other components.

Important properties in membrane separation are permeation rate and selectivity. Permeability says something about how easy a given gas will go through the membrane, and selectivity says something about how easy a given gas goes through the membrane compared to another gas. The ease with which a gas goes through a given membrane material is determined by a combination of diffusivity and solubility. Each gaseous component transporting through the membrane has a characteristic permeation rate that is a function of the ability to dissolve and diffuse through the membrane material.

Gas separation is done mostly by using non-porous polymeric membranes. For the polymeric membranes, polymers of both high permeability and selectivity are desirable. Higher permeability decreases the amount of membrane area required to treat a given amount of gas, thereby decreasing the capital cost of membrane units. Higher selectivity results in higher purity of product gas. The most used polymers are polysulfone, polyimide, cellulose acetate and polycarbonates.

Figure 5.19 shows a schematic diagram of the basic membrane gas separation process. There are three different "streams" in the membrane process for gas separation:

1. Feed gas (starting point).
2. Gas that permeated through the membrane.
3. Retentate gas that does not pass the membrane.

According to Fig. 5.19 the overall material balance becomes

$$L_f = L_o + V_p \tag{5.1}$$

where L_f is feed flow rate, L_o is retentate flow rate, and V_p is permeate flow rate.

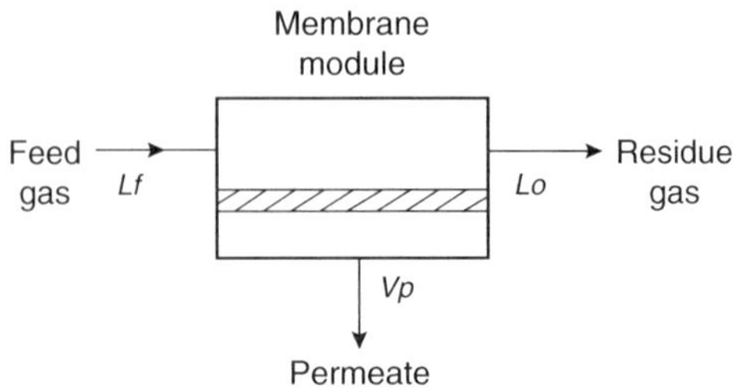

Fig. 5.19 Basic membrane gas separation process

A parameter of the economic importance in commercial membrane processes is stage cut, or cut rate, θ, given by

$$\theta = V_p / L_f \tag{5.2}$$

It is necessary to choose the cut rate in commercial processes because it directly influences the product purity and yield.

Another important parameter is recovery, R, for a given gas component, A. Recovery tells you something about the selectivity of the membrane, how much of your wanted component is present in the product stream compared to the amount in the initial feed stream. It indirectly tells you how much of your wanted component you have lost. When the wanted product is in the retentate stream, recovery, R, is given by

$$R = \left\{ \left(x_{o,A} L_o \right) / \left(x_{f,A} L_f \right) \right\} \cdot 100\% \tag{5.3}$$

where $x_{f,A}$ is mole fraction of component A in the feed and $x_{o,A}$ is mole fraction of component A in the retentate. When the wanted product is in the permeate stream, the corresponding equation using $y_{p,A}$ and V_p, instead of $x_{o,A}$ and L_o, can be used, where $y_{p,A}$ is mole fraction of component A in the permeate.

He et al. [30] established successfully a mathematical model to investigate the effects of both the membrane separation properties and operation parameters on the concentration polarization in a gas separation processes. The influences of membrane performance and operating parameters on concentration polarization were studied in terms of permeation fluxes of the more and less permeable gases and separation factors. Sample calculations were presented for the two typical gas separation applications, hydrogen recovery and air separation, with shell side feed in a hollow-fiber module. The permeation rate was found to be a dominating factor in affecting concentration polarization, while the influence of the separation factor was found to be significant initially and to level off gradually. Increasing feed gas velocity led to a decrease in the concentration polarization. The effect of operating pressure was limited and the composition of feed gas showed no effect. It was concluded that concentration polarization is important for process analysis and design when the permeation rate of the more permeable gas is larger than 1×10^{-4} cm^3 (STP) cm^{-2} s^{-1} cmHg^{-1} (100 GPU).

Zhao et al. described a detailed parameter study of mass and energy balances for a single membrane and also the energetic and economic analyses of a multistage process [31]. In other articles, Zhao et al. [32, 33] suggested that the membrane permeability, selectivity and area are decisive parameters for membranes. Each of these properties should be taken into account by a membrane developer. Jusoh et al. [34] noticed in the purification of natural gas with impurities by membranes, the sorbed concentration and permeability coefficient of a binary mixture were found to be lower as compared to pure gases, which represented the competition between penetrants.

In the separation of a propylene–propane mixture via facilitated transport mechanism, transmembrane pressure and carrier concentration were two important parameters. Better separation performance can be obtained with more transmembrane pressure and more carrier concentration [35].

5.5 Energy Requirements

Energy crises and environmental pollution are an ongoing problem for humanity. For solving these problems, efforts to replace fossil fuels with other energy sources and with clean fuels have been undertaken. Fuel cells, due to their particular properties, are on the verge of creating a vast revolutionary change in the field of electricity.

Energy consumption is a significant problem for gas separation membrane processes. In principle, a multistage membrane system consumes more energy than a single-stage process. Merkel et al. [36] investigated the influences of membrane parameters and process configurations on energy consumption and CO_2 capture costs, and some important suggestions were proposed for future studies. Membranes offer significant opportunity to be low cost, low energy solutions for flue gas CO_2 capture.

Membrane processes using polymeric materials are based on the difference in rates of diffusion of gases through a membrane separating high-pressure and low pressure process streams. A major benefit of membrane separation is the simple, continuous nature of the process at near ambient conditions and cost cutting; however polymers do not have the highest available selectivities but can easily and cost-effectively be processed [37].

Lababidi et al. [17] developed optimization models used for cost optimization of enriching a binary mixture of methane and carbon dioxide. Seppälä [38] discussed irreversible losses in membranes and contactors for gas separation and expressed some optimal parameters and operating conditions based on the second law of thermodynamics. Bhide and Stern [24, 39] demonstrated an economic evaluation of membrane processes for the production of oxygen-enriched air. As a result, a single stage system without permeate recycle was identified as the best module configuration; however, the permeate mole fraction of oxygen was limited to 30 %. Geometric parameters, pressure losses and concentration polarization were also not taken into account since the study was concentrated on economic parameters.

Often, the optimization of processes takes into account capital and operating costs, which are dependent on time and scale. Meriläinen et al. [40] presented the

Table 5.5 The specific energy consumption of air separation processes when pure oxygen is produced

Process	Specific energy consumption (kJ/mol(O_2))
Cryogenic distillation, large-scale (10,000 t(O_2)/d)	23.0
Cryogenic distillation, small-scale (25 t(O_2)/d)	63.4
Pressure swing adsorption	41.5
Ion transfer membranes	20.7
Theoretical minimum	6.1

modeling and optimization of air separation by polymer membrane. A counter-current lumen-feed gas separation model was presented and used to optimize various parameters in order to minimize the specific energy consumption of air separation in membrane systems. Important geometry parameters were recognized. A two-stage configuration, where the permeate of the first module was further enriched in a second stage, was also considered. By optimizing both stages the specific energy consumption could, in some cases, be further reduced. The optimization of the module geometry is not as important as the optimization of the feed parameters—the pressure ratio and stage cut. The results of air separation by polymeric membrane were compared to cryogenic distillation, pressure-swing adsorption, and ion transfer membranes. In their opinion, present-day polymeric membranes modules could compete with traditional techniques in specific energy consumptions when the required oxygen mole fraction is low. A single-stage polymeric membrane module with a selectivity of 100 would be more efficient than other techniques up to an oxygen purity of approximately 92 %. The lower the desired oxygen mole fraction, the more air could be used in the mixing, thus lowering the energy needed for separation. The theoretical minimum work required for the separation was calculated from the change in the Gibbs free energy. Table 5.5 contains a comparison of different air separation processes when pure oxygen is produced [41].

References

1. Baker RW, Wijmans JG, Kaschemekat JH (1998) The design of membrane vapor-gas separation systems. J Membr Sci 151:55–62
2. Baker RW (2012) Membrane technology and applications. Wiley, Hoboken, NJ
3. Mulder M (1996) Basic principles of membrane technology. Springer, Berlin, Heidelberg, NY
4. Kluiters SCA (2004) Status review on membrane systems for hydrogen separation. Intermediate report EU project MIGREYD NNE5-2001-670, ECN-C-04-102
5. Strathmann H, Bell C-M, Kerres J (1990) Gas separation and pervaporation: membrane and module development. Desalination 77:259–278
6. Takaba H, Nakao S (2005) Computational fluid dynamics study on concentration polarization in H_2/CO separation membranes. J Membr Sci 249:83–88
7. Katoh T, Tokumura M, Yoshikawa H, Kawase Y (2011) Dynamic simulation of multicomponent gas separation by hollow-fiber membrane module: nonideal mixing flows in permeate and residue sides using the tanks-in-series model. Sep Purif Technol 76:362–372

8. Feron PHM, Jansen AE (1995) The production of carbon dioxide from flue gas by membrane gas absorption and reuse in the horticultural industry. Energy Convers Manag 36: 411–414
9. Coker DT, Freeman BD, Fleming GK (1998) Modeling multicomponent gas separation using hollow-fiber membrane contactors. AIChE J 44:1289–1302
10. Nymeijer DC, Visser T, Assen R, Wessling M (2004) Composite hollow-fiber gas–liquid membrane contactors for olefin/paraffin separation. Sep Purif Technol 37:209–220
11. Stanojević M, Lazarević B, Radić D (2003) Review of membrane contactors designs and applications of different modules in industry. FME Trans 31:91–98
12. Li JL, Chen BH (2005) Review of CO_2 absorption using chemical solvents in hollow fiber membrane contactors. Sep Purif Technol 41:109–122
13. Sengbusch GV (1994) Future of membranes, technological and economical aspects. Vortrag: Synthetic Membranes in Science and Industry, Tüebingen
14. Gottschlich DE, Roberts DL, Way JD (1968) A theoretical comparison of facilitated transport and solution-diffusion membrane modules for gas separation. Gas Sep Purif 2:65–71
15. Qi R, Henson MA (1998) Optimal design of spiral-wound membrane networks for gas separations. J Membr Sci 148:71–89
16. Qi R, Henson MA (2000) Membrane system design for multicomponent gas mixtures via mixed-integer nonlinear programming. Comput Chem Eng 24:2719–2737
17. Lababidi H, Al-Enezi GA, Ettouney HM (1996) Optimization of module configuration in membrane gas separation. J Membr Sci 112:185–197
18. Baker RW, Lokhandwala KA, Jacobs ML, Gottschlich DE (2000) Recover feed stock and product recovery from reactor vent streams. Chem Eng Prog 96:51–57
19. Baker RW (2001) Membranes for vapor/gas separations. MTR Inc., Menlo Park, CA
20. Spillman RW, Barrett MG, Cooley TE (1988) Gas membrane process optimization. In: AIChE spring annual meeting, New Orleans, LA
21. Babcock RE, Spillman RW, Goddin CS, Cooley TE (1988) Natural gas cleanup: a comparison of membrane and amine treatment processes. Energy Progress 8:135–142
22. Bhide BD, Stern SA (1993) Membrane processes for the removal of acid gases from natural gas. II. Effect of operating conditions, economic parameters, and membrane properties. J Membr Sci 81:239–252
23. Bhide BD, Stern SA (1993) Membrane processes for the removal of acid gases from natural gas. I. Process configuration and optimization of operating conditions. J Membr Sci 81:209–237
24. Bhide BD, Stern SA (1991) A new evaluation of membrane processes for the oxygen-enrichment of air. II. Effects of economic parameters and membrane properties. J Membr Sci 62:37–58
25. Xu J, Agrawal R (1996) Gas separation membrane cascades. I. One compressor cascades with minimal energy losses due to mixing. J Membr Sci 112:115–128
26. Agrawal R (1997) A simplified method for synthesis of gas separation membranes cascades with limited number of compressors. Chem Eng Sci 52:1029–1044
27. Agrawal R, Xu J (1996) Gas separation membrane cascades. II. Two compressor cascades. J Membr Sci 112:129–146
28. Agrawal R, Xu J (1996) Gas separation membrane cascades utilizing limited numbers of compressors. AIChE J 42:2141–2154
29. Khalilpour R, Abbas A, Lai Z, Pinnau I (2013) Analysis of hollow fibre membrane systems for multicomponent gas separation. Chem Eng Res Des 91:332–347
30. He G, Mi Y, Yue PL, Chen G (1999) Theoretical study on concentration polarization in gas separation membrane processes. J Membr Sci 153:243–258
31. Zhao L, Riensche E, Menzer R, Blum L, Stolten D (2008) A parametric study of CO_2/N_2 gas separation membrane processes for post-combustion capture. J Membr Sci 325:284–294
32. Zhao L, Riensche E, Menzer R, Blum L, Stolten D (2010) Multi-stage gas separation membrane processes used in post-combustion capture: energetic and economic analyses. J Membr Sci 325:160–172
33. Zhao L, Riensche E, Menzer R, Blum L, Stolten D (2011) How gas separation membrane competes with chemical absorption in postcombustion capture. Energy Procedia 4:629–636

34. Jusoh NW, Lau KK, Shariff AM (2012) Purification of natural gas with impurities using membrane processes: parameter estimation. AJEAS 5:78–83
35. Ravanchi MT, Kaghazchi T, Kargari A, Soleimani M (2009) A novel separation process for olefin gas purification: effect of operating parameters on separation performance and process optimization. J Taiwan Inst Chem E 40:511–517
36. Merkel TC, Lin H, Wei X, Baker R (2010) Power plant post-combustion carbon dioxide capture: an opportunity for membranes. J Membr Sci 359:126–139
37. Smith AR, Klosek J (2001) A review of air separation technologies and their integration with energy conversion processes. Fuel Process Technol 70:115–134
38. Seppälä A (2010) Irreversibility in a gas separation membrane medium. Chem Phys 367:99–109
39. Bhide BD, Stern SA (1991) A new evaluation of membrane processes for the oxygen enrichment of air. I. Identification of optimum operating-conditions and process configuration. J Membr Sci 62:13–35
40. Meriläinen A, Seppälä A, Kauranen P (2012) Minimizing specific energy consumption of oxygen enrichment in polymeric hollow fiber membrane modules. Appl Energy 94:285–294
41. Baker RW (1991) Membrane separation systems—recent developments and future directions. Noyes Data Corporation, NJ

Chapter 6
Application of Gas Separation Membranes

Membrane-based gas separation (GS) systems are today widely accepted and, in some cases, used as unit operations for generation, separation, and purification of gases in gas, chemical, petroleum, and allied industries. There are several fields of application of membrane GS, and several membrane materials and membrane modular solutions are available today for the various fields of interest. However, the growth of large-scale industrial applications for GS is still far from reaching the real potential this technology offers. Together with the investigation of new materials with improved properties, a key component for widespread use of this technology is a better understanding and utilization of the unit operations already available on the market in integrated membrane systems, combining various membrane operations in industrial processes. The role of membrane engineering is crucial to overcome this hurdle [1].

Membranes are currently employed in:

1. Removal of *nitrogen* or *oxygen* from air.
2. Separation of *hydrogen* from gases like *nitrogen* and *methane*.
3. Recovery of hydrogen from product streams of *ammonia* plants.
4. Recovery of hydrogen in *oil refinery* processes.
5. Separation of methane from the other components of *biogas*.
6. *Enrichment of oxygen* from air for medical or metallurgical purposes
7. Removal of CO_2 from natural gas.
8. Removal of H_2S from natural gas.
9. Removal of *volatile* organic liquids (VOL) from air of exhaust streams.

In 1980, Permea, which is now a division of Air Products, started its hydrogen separation by Prism membranes [2]. This was the first large industrial application of gas separation membranes. Since then, membrane-based gas separation has grown. Within a few years, Permea systems were installed in many separation plants. By the mid-1980s, Cynara (now part of Natco), Separex (now part of UOP), and GMS (now part of Kavaerner) were using cellulose acetate membranes to remove carbon dioxide from natural gas. Generon (now part of MG) introduced a membrane system to separate nitrogen from air.

A.F. Ismail et al., *Gas Separation Membranes*,
DOI 10.1007/978-3-319-01095-3_6

6.1 Large-Scale Applications

The main industrial applications developed for membrane gas separation are summarized in Table 6.1.

MTR's (Membrane Technology and Research, Inc.) Gas Separation Process Development Group develops membrane technology for natural gas production, refinery operations, and petrochemical processes.

Recent development activities include the following [4]:

1. Recovering hydrogen from refinery off-gases and steam reformer waste gas systems.
2. Separating carbon dioxide from hydrogen in steam reforming and coal gasification plants.
3. Removing natural gas liquids (NGLs), nitrogen, carbon dioxide, and water from natural gas.
4. Separating carbon dioxide from coal plant flue gas.
5. Separating olefin–paraffin mixtures.
6. Recovering hydrocarbon feedstock's from oxidation reactor purge streams.
7. Upgrading associated gas from oil production and biogas from landfills and animal feed lots.

6.1.1 Air Separation (Nitrogen and Oxygen Production)

The modern air separation industry was started at the beginning of the twentieth century with the development of cryogenic air processing, at that time driven by a desire to produce calcium carbide, required for the manufacture of acetylene.

Table 6.1 Main industrial applications of membrane gas separation [3]

Separation	Process
H_2/N_2	Ammonia purge gas
H_2/CO	Syngas ratio adjustment
H_2/hydrocarbon	Hydrogen recovery in refineries
O_2/N_2	Nitrogen generation, oxygen–enriched air production
CO_2–hydrocarbons (CH_4)	Natural gas sweetening, land fill gas upgrading
H_2O–hydrocarbons (CH_4)	Natural gas dehydration
H_2S/hydrocarbons	Sour gas treating
He/hydrocarbons	Helium separation
He/N_2	Helium recovery
Hydrocarbons/air	Hydrocarbons recovery, pollution control, air dehumidification
Volatile organic species (e.g., ethylene or propylene)/light gases (e.g., nitrogen)	Polyolefin purge gas purification

Oxygen demand received a further boost, and an industry was born when it was realized that combining oxygen and an acetylene torch resulted in a very high temperature flame, able to melt and therefore weld a wide variety of metals.

Demand for oxygen increased in the 1950s as new blown steel furnaces (LD or BOP/BOS processes) replaced air blown Bessemer furnaces. This new demand for large volumes of dedicated oxygen allowed the air separation industry to develop large-scale "tonnage" oxygen supply plants, now capable of processing thousands of tons of air per day to produce high purity oxygen, nitrogen and argon. Tonnage air plants now find use in a growing range of applications including: onsite oxygen supply to the chemical industry (e.g., ethylene oxide, methanol, and reformer processes), onsite oxygen supply for the developing gas-to-liquid (GTL) sector and nitrogen supply (generally by pipeline) to pressurize oil fields and maintain crude oil production.

Currently, three branches of technologies exist to separate oxygen from the air: (1) distillation, (2) adsorption, and (3) membranes [5]. Membrane technology is the most recent of the three and includes polymeric and high temperature ion transport membranes. While polymeric membranes can produce oxygen-enriched air of various concentrations, ion transport membranes can produce up to 100 % pure oxygen. Membrane air separation is based on the principle that different gases have different permeation rates through the polymer film and separation can be done at ambient temperature. The use of selective membranes for partial air separation progressed rapidly in the 1980s as a promising alternative to cryogenics and adsorption [6]. The basic technology choice between cryogenic or non-cryogenic is largely determined by the number of products that must be supplied (e.g., oxygen or nitrogen or both), the required production rates of each gas and/or liquid product, and required product purities.

The separation of oxygen and nitrogen in polymeric membranes is also based on differences in their diffusivity and solubility in the membrane material. Currently, polymeric materials are dominant because they can easily and cost-effectively be processed into membranes even though the selectivities of polymers are not necessarily the highest. Among many polymeric materials, only a few polymers cover over 90 % of all the gas separation membranes that are commercially installed. In contrast to vast-oxygen-producing plants, membrane systems are relatively simple. They can be operated continuously in ambient conditions [7].

The first air separation system of Generon (now part of MG) was based on poly(4-methyl-1-pentene) (TPX) membranes with an oxygen–nitrogen selectivity of about 4. These membranes were only competitive in a few niche areas, requiring 95 % nitrogen; but by 1990, Generon, Praxair, and Medal had all produced custom polymers with oxygen–nitrogen selectivities of 6–8. The membranes fabricated from these polymers could produce more than 99 % nitrogen and offered a cost competitive process to deliver liquid nitrogen for many small users [2]. This application has grown to represent about one-third of new nitrogen production capacity; to date, more than 10,000 nitrogen systems have been installed worldwide.

A simplified schematic diagram of nitrogen from air separation system is shown in Fig. 6.1 [2]. The feed air is compressed to 8–10 atm with a low cost screw compressor and then passed through a bore side hollow-fiber module.

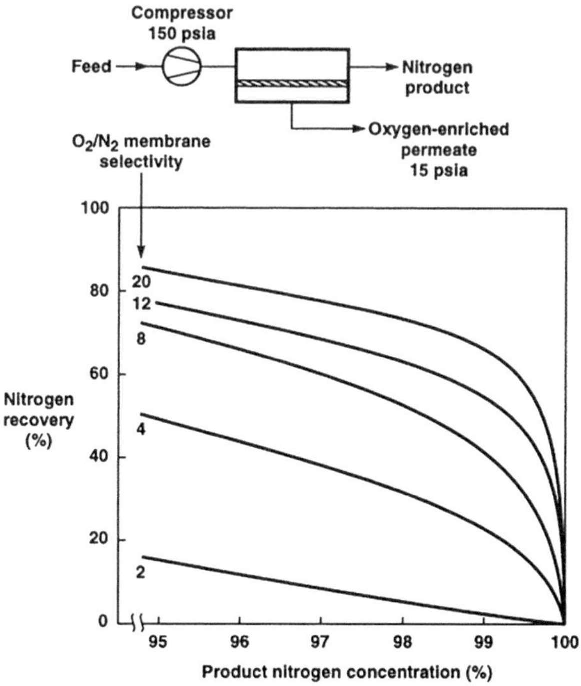

Fig. 6.1 Nitrogen recovery as a function of product nitrogen concentration for membranes with selectivities between 2 and 20

The module operates in counter flow mode. The first membranes used for nitrogen separation had an oxygen–nitrogen selectivity of about 4. As Fig. 6.1 shows, this membrane can produce 95 % nitrogen at a nitrogen recovery of about 50 %.

A simplified flow schematic of a membrane separation process for oxygen-enriched air is shown in Fig. 6.2a. Feed air containing 21 % oxygen is passed across the surface of a membrane that preferentially permeates oxygen. In the schematic, the pressure differential across the membrane required to drive the process is maintained by drawing a vacuum on the permeate gas. Depending on the properties of the membrane and pressure differential, a permeate gas containing 30–60 % oxygen is produced. Pure oxygen can be produced by adding a second separation stage, as shown in Fig. 6.2b.

Compared to the membrane process, cryogenic air separation is a process by which highly purified gases or liquids are produced. The cryogenic distillation process can be very complicated in practice. First precooled air is compressed in a multistage process to 650 kPa. It is subsequently throttled, causing it to cool to low temperatures and liquefy. Oxygen and nitrogen are then separated by phase in a fractional distillation column. While the oxygen settles to the bottom of the column as liquid, the nitrogen boils and is taken out of the top of the column. The contents are then sent to a low pressure distillation column to increase the oxygen purity. The end result is the production of oxygen with purity greater than 99 %. One configuration of an actual cryogenic distillation process is shown in Fig. 6.3 [8].

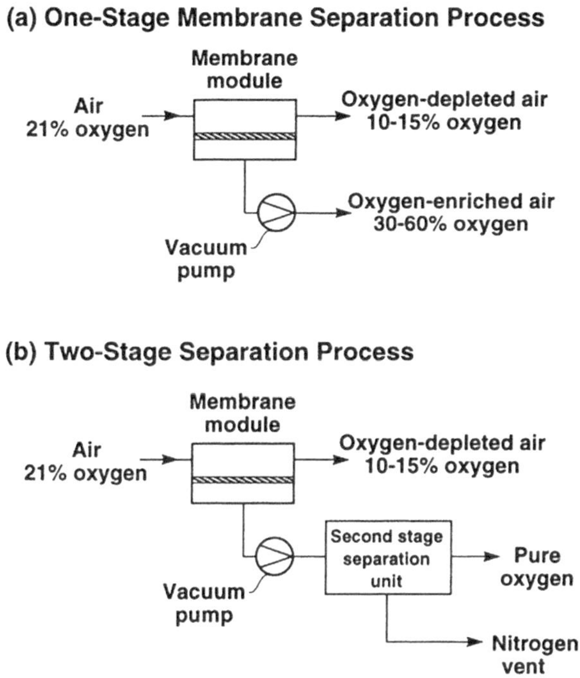

Fig. 6.2 Membrane process flow schematics for the production of (**a**) oxygen enriched air and (**b**) pure oxygen

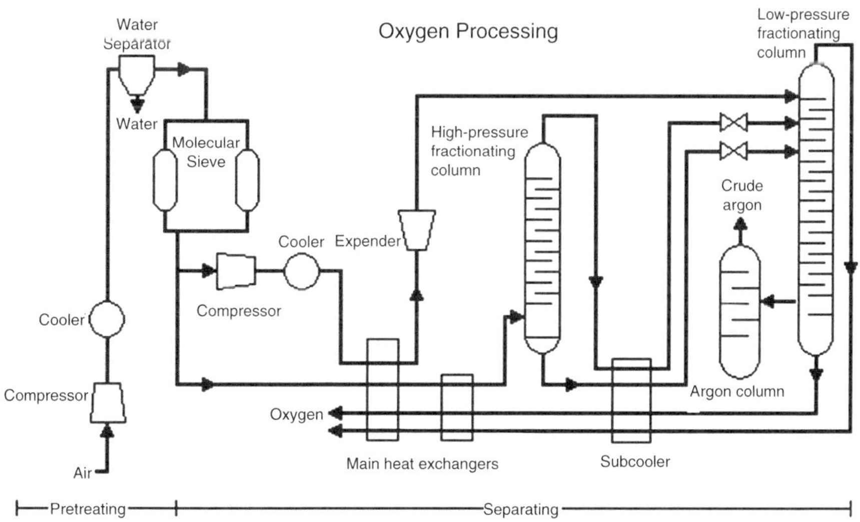

Fig. 6.3 Actual cryogenic distillation system using the Linde process [8, 9]

In cryogenic gas processing, various equipment is used such as distillation columns, heat exchangers, cold inter-connecting piping etc. These operate at very low temperatures and therefore must be well insulated and located inside sealed "*cold boxes*." Another process named adsorption separation (pressure swing adsorption or PSA) relies on the fact that, under pressure, gases tend to be attracted to solid surfaces or *adsorbed*. The higher the pressure, the more gas is adsorbed. When pressure is reduced, the gas is released, or desorbed. PSA processes can be used to separate gases in a mixture because different gases tend to be attracted to different solid surfaces more or less strongly. There is another way called vacuum swing adsorption (VSA), which is different from pressure swing adsorption (PSA) techniques due to the fact that it operates at near-ambient temperatures and pressures.

Meriläinen et al. [7] studied the modeling and optimization of air separation by polymer membranes. They concluded that if air separation is optimized in order to minimize the specific energy consumption, membranes can compete with other more established technologies to produce oxygen-enriched air. On the other hand, choosing the feed parameters poorly will increase the specific energy consumption substantially. Present-day membranes are most suitable in applications where the desired oxygen purity is fairly low. If the desired molar fraction of oxygen is 50 %, an optimized membrane module with selectivity above 8 can compete with even large-scale cryogenic distillation and ion transfer membranes. A polymeric membrane with a selectivity of 100 would be more efficient than other techniques up to an oxygen purity of approximately 92 %.

6.1.2 Hydrogen Recovery

Hydrogen requirements are growing worldwide due to the increased use of hydrotreating and hydrocracking for cleaner and higher-value fuels, as well as demand for chemical products and production of electronics. The process using and/or generating hydrogen typically creates residual gas streams or by-products that still contain a significant amount of valuable pressurized hydrogen. Hydrogen is a valuable commodity in refining and petrochemical processes, which must be economically generated, purchased, or recovered from numerous process streams. The demand for hydrogen is increasing rapidly due to changes in government regulations affecting refiners. The worldwide demand for hydrogen is expected to increase, especially as a key feedstock for leading fuel cell technologies, and in industries such as petroleum refining where regulatory and economic trends will require more hydrogen. This technology will provide economic justification for recovery/separation processes in industries where hydrogen is currently lost or burned as fuel. The separation of hydrogen from light hydrocarbons in oil refineries is an issue of major importance.

For separation of hydrogen from gaseous streams, membranes can provide an attractive alternative to PSA and cryogenic separation, depending on the scale and purity of the product streams required. Hydrogen-selective membranes can be broadly separated into four categories: polymeric (organic), metallic, carbon and ceramic (the latter three jointly called inorganic). For a long time, considerably more effort has been put into development of polymeric membranes than into inorganic membranes. Consequently, polymeric membranes have wide ranging applications and can be bought at relatively low cost. However, interest in inorganic membranes has started to grow in the last decade. Inorganic materials can operate under higher temperatures than polymeric materials and generally possess superior chemical stability relative to polymeric materials. Ceramic-formed membrane is the main class of inorganic membranes [10].

Kaldis et al. [11] tried the modeling of membrane for multicomponent separations for the first time. Membrane modeling was approached by the orthogonal method to ensure solution stability with low computational time and effort. A model prediction was validated by experimental results obtained from the separation of a multi-component mixture in a polyimide membrane unit and the general comparison was very satisfactory. The mathematical model was applied further to various other refinery streams. It was revealed that even for a one-stage membrane unit, high hydrogen purity (up to 99+ %) and significant total recovery (up to 90 %) can be achieved for moderate (0.2–0.6) stage cuts. The residue stream, rich in hydrocarbons and containing 10 % of the initial hydrogen, can be further separated in a second stage or used as a gas fuel.

Table 6.2 compares, in general, the relative operational performances of five membrane types. Each membrane type has advantages and disadvantages [10, 12].

Palladium thin films are known to selectively transmit hydrogen via an adsorption–desorption mechanism. Mixed proton- and electron-conducting materials consisting of barium cerate, doped with rare-earth ions (that is, $BaCe_{1-x}M_xO_{3-\delta}$, where $M = Nd^{3+}$, La^{3+}, Y^{3+}, or Gd^{3+}) have been found to be of potential interest for hydrogen separation [13]. Research into H_2 separation membranes is being conducted

Table 6.2 Properties of five hydrogen-selective membranes

	Dense polymer	Microporous ceramic	Dense ceramic	Porous carbon	Dense metallic
Temperature range	<100 °C	200–600 °C	600–900 °C	500–900 °C	300–600 °C
H_2 selectivity	Low	Moderate	Very high	Low	Very high
H_2 flux	Low	High	Moderate	Moderate	High
Known poisoning issues	HCl, SO_2, CO_2		H_2S	Strong vapors, organics	H_2S, HCl, CO
Example materials	Polymers	Silica, alumina, zirconia, titania, zeolites	$SrCeO_{3-\delta}$ $BaCeO_{3-\delta}$	Carbon	Palladium alloys, Pd-Cu, Pd-Au
Transport mechanism	Solution/ diffusion	Molecular sieving	Solution/ diffusion	Surface diffusion, molecular sieving	Solution/ diffusion

within the framework of the USDOE Vision 21 program [14] aiming at a hydrogen separation rate >10 ml/min cm^2 (74 × 10^{-8} mol/m^2 Pa · s at 1 bar pressure difference) in thin films (100–300 μm) of perovskites [15].

Grainger and Hägg [16] demonstrated that carbon molecular sieves can recover 90 % of the hydrogen from a feed stream containing 5 mol% hydrogen. It was claimed that carbon molecular sieve membranes (CMSMs) have great potential for hydrogen recovery from hydrocarbons. The authors produced pure hydrogen from a leaner stream with lower energy consumption.

Hydrogen membranes are an economical method to recover and purify hydrogen from a refinery's own waste gases and reactor purges. MTR's hydrogen-permeable VaporSep-H_2 membranes can provide 90–99 % hydrogen purity and greater than 90 % recovery. Refinery hydrogen requirements are growing due to the increased use of hydrotreating (to remove sulfur) and hydrocracking (to convert heavy hydrocarbons to lighter, higher value fuels). Residual gas from these processes contains a significant amount of unused hydrogen at pressure, and membranes provide an economical recovery method. MTR's hydrogen-permeable VaporSep-H_2™ polymer-based membranes can provide 90–99 % pure hydrogen and greater than 90 % recovery [17].

VaporSep-H_2™ offers a simple method for recovering hydrogen from refinery streams (Fig. 6.4) [17]. Hydrogen permeates preferentially through the membrane, producing a purified hydrogen "permeate" stream and a hydrocarbon-enriched "residue" stream. The available pressure for the purified hydrogen depends on the feed conditions, but can be as high as 1,500 psi. The hydrocarbon-enriched "residue" is recovered at close to the feed pressure, and can be sent directly to fuel, or first treated for liquefied petroleum gas (LPG) recovery if these components have value.

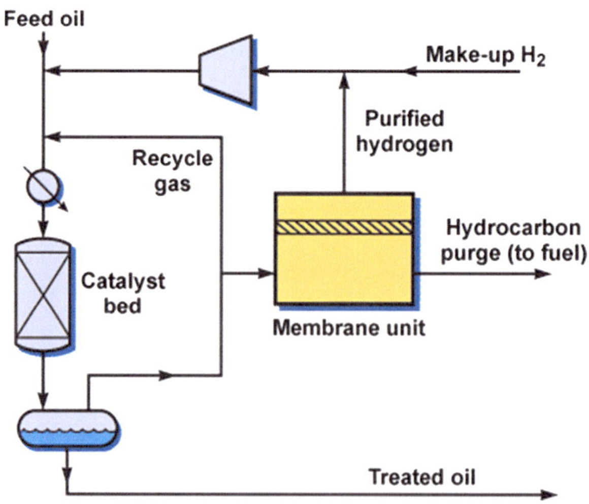

Fig. 6.4 Schematic diagram of the equipment for recovering hydrogen from refinery streams

Larger applications of hydrogen permeable membranes are used in refineries for hydrogen recovery. Hydrogen can be recovered from fuel gas streams, PSA tail gas, FCCU gas, and hydrocracker/hydrotreater off-gas by using membrane technology. These gas streams contain 30–80 % hydrogen mixed with light hydrocarbons (C_1–C_5). In refineries the reliability of membranes is poor, due to fouling, plasticization, and condensation of hydrocarbon vapors on the membrane surface. A typical application, illustrated in Fig. 6.5, is the recovery and reuse of hydrogen from an oil hydrocracker purge gas [2, 18].

Hydrocrackers are used in refineries to break down high molecular-weight compounds, to remove impurities, and to hydrogenate aromatics. Heavy oil is cracked to C_{5+} hydrocarbons, while some methane, ethane, and propane are formed as by-products. The oil–gas mixture from the hydrocracker is sent to a lower pressure separator from which the C_{5+} product is removed. Untreated hydrogen is recirculated

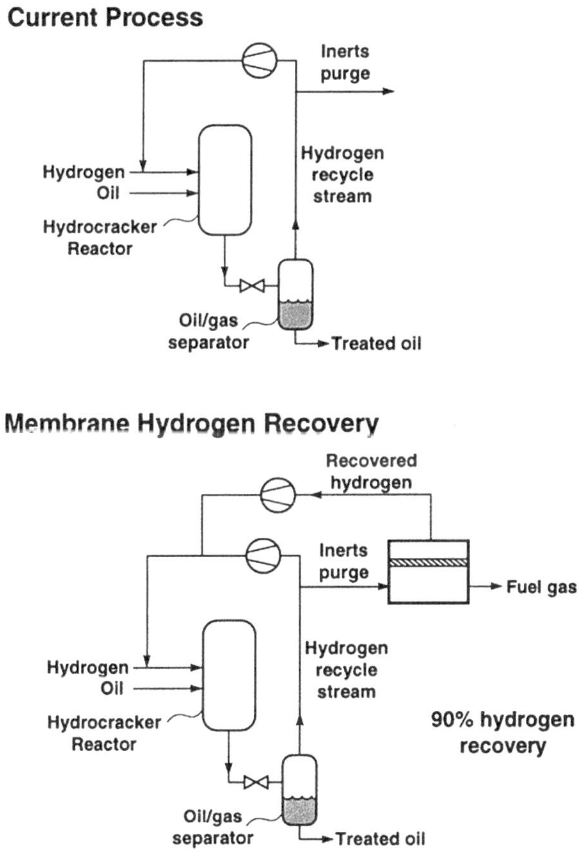

Fig. 6.5 Use of hydrogen-permeable membranes to recover and recycle hydrogen from hydrotreater purge gas streams

back to the reactor. Methane, ethane, and propane accumulate in the recycle stream and must be removed as an inert purge.

The feed gas, off-gas from a separator with a dew point of 31 °C, contains 75 % hydrogen; 22 % methane, ethane, and propane; and 3 % C_{4+} hydrocarbons. As hydrogen is removed through the membrane, the remaining gas becomes enriched in hydrocarbons, and the dew point increases to 64 °C. To minimize plasticization of the membrane, the gas must be heated to 15–20 °C above the expected residue gas dew point. Even heating the gas does not provide absolute membrane protection. High free volume polyacetylene polymers such as poly(1-trimethyl-1-propyne) (PTMS) and poly(4-methyl-2-pentyne) (PMP), the microporous absorbent carbon membranes developed by Air Products, are used for the abovementioned method., as well as silicon rubber.

The commercial success in the mid-1970s of the Permea hollow-fiber prism system for in-process recycling of hydrogen from ammonia purge gases was the starting point of the introduction of membrane technology in large-scale manufacturing [3]. This technology has been applied for recovery of hydrogen from gas mixtures (H_2–CO_2 or H_2–CH_4 ratios adjustment for *syn* gas production) and has been successfully competing with cryogenic distillation and the pressure swing adsorption (PSA) process. H_2 recovery from refinery streams is an emerging field of gas separation in the petrochemical industry. It is a key approach to meet the increased demand for hydrogen (for hydrotreating, hydrocracking, or hydrodesulfurization processes) owing to new environmental regulations. China started its first hydrogen recovery unit based on membrane technology, for hydrocracking dry gas and PSA resolving gas production in Sinopec's Zhenhai Plant. The membrane in the Prism system is a PSf hollow fiber with a thin silicon film on it. Polyimide membranes are successfully applied for hydrogen recovery in refineries due to their stability and interesting separation factor (H_2/N_2 of 100–200).

Commercial polymeric membrane modules (e.g., MEDAL) can operate at high pressures (120 bar) with flow rates up to 330,000 Nm^3/h; hydrogen recovery can reach 98 % in volume, with hydrogen purity as high as 99.9 % [19]. Concerning the development of novel membrane materials, Adams [20] demonstrated that commercially available metallic glass membranes can be used for hydrogen purification or separation. These membranes proved to be efficient and effective for separation of hydrogen from a mixed gas feed stream. Compared with the conventional zeolite membranes, the copper net supported $Cu_3(BTC)_2$ membranes exhibit a higher permeation flux and excellent permeation selectivity for H_2. Such characteristics of copper net-supported $Cu_3(BTC)_2$ membranes offer great potential toward applications such as separating, recycling, and reusing H_2 exhausted from steam reforming natural gas [21].

Proton conductors based on ion transport membranes can be applied for hydrogen separation from fossil fuel. Phair and Badwal [22] have discussed H_2 separation via proton conductors. They describe various classes of proton-conducting materials with specific emphasis on their potential use as hydrogen separation membranes in the industrial processes of coal gasification, natural gas reforming and the water-gas shift (WGS) reaction.

6.1.3 Acid Gas Removal from Natural Gas and Syn Gas

Acid gas is natural gas or any other gas mixture containing significant quantities of hydrogen sulfide (H_2S), carbon dioxide (CO_2), or similar acidic gases. The terms *acid gas* and *sour gas* are often incorrectly treated as synonyms. Strictly speaking, a sour gas is any gas that specifically contains hydrogen sulfide in significant amounts; an acid gas is any gas that contains significant amounts of acidic gases such as carbon dioxide (CO_2) or hydrogen sulfide. Thus, carbon dioxide by itself is an acid gas but not a sour gas. Untreated natural gas is not a clean fluid, and it can contain oil mist, glycol, methanol, drilling fluids from earlier operations, and ultra-fine iron sulfide particulates (formed by the reaction of hydrogen sulfide with iron pipes). Before a raw natural gas containing hydrogen sulfide and/or carbon dioxide can be used, the raw gas must be treated to reduce impurities to acceptable levels and this is commonly done with an amine gas treating process. The removed H_2S is most often converted to by-product elemental sulfur in a Claus process or, alternatively, converted to valuable sulfuric acid in a WSA process unit. Hydrogen sulfide is a toxic gas. It also restricts the materials that can be used for piping and other equipment, as many metals are sensitive to sulfide stress cracking.

Syngas, or synthesis gas, is a fuel-gas mixture consisting primarily of hydrogen, carbon monoxide, and very often some carbon dioxide. The name comes from its use as intermediates in creating synthetic natural gas (SNG). Syngas is used when hydrocarbons like coal are gasified and the gas is then burned to create electricity. Syngas is also used as an intermediate in producing synthetic petroleum for use as a fuel or lubricant via the Fischer–Tropsch process. Syngas is combustible and often used as a fuel for internal combustion engines.

The emission of greenhouse gases (GHGs) from fossil fuel combustion and other human economic and social activities has been escalating over the last century. The atmospheric concentration of CO_2 has been increasing since the mid-nineteenth century, and the annual rate of increase is greater than ever, which is believed to be largely associated with global warning [23]. Today, a large number of carbon sources such as fossil fuels, biomass energy facilities, chemical industries, natural gas processing, synthetic fuel plants, and fossil fuel-based hydrogen production plants result in the emission of megatons of CO_2 per day, which increases atmospheric CO_2, affecting climate and the natural environment. The percentage of carbon dioxide in the air between 7 and 11 % can cause dizziness, headaches, visual and hearing dysfunction, and unconsciousness within a few minutes to an hour. Concentrations of CO_2 above 17 % are lethal for humans and animals when they are exposed for more than 1 min.

Unprocessed natural gas contains undesirable gases such as CO_2, H_2S, and N_2. These must be removed from natural gas prior to its use or sale to meet standards of less than 4 % inert gases. Acid gases such as CO_2 and H_2S are corrosive and water molecules can be a major source of maintenance problem in pipelines. Acid gas removal plants are essential to natural gas processing. Table 6.3 shows fossil fuel emission levels per year globally [24].

Table 6.3 Fossil fuel emission levels (pounds per billion BTU of energy input)

Fuel sources/pollutant (pounds/billion BTU)	Natural gas	Oil	Coal
Carbon dioxide	117,000	164,000	208,000
Carbon monoxide	40	33	208
Nitrogen oxides	92	448	457
Sulfur dioxide	1	1,122	2,591
Particulates	7	84	2744
Mercury	9.0	0.007	0.016

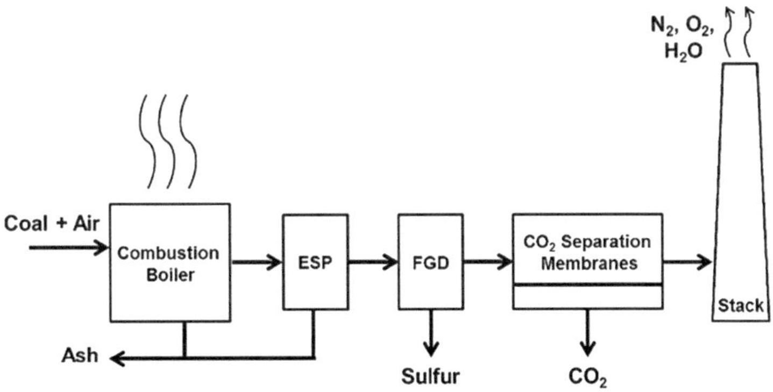

Fig. 6.6 Schematic of a flue gas cleanup train for coal fired power plant [27]

Acid gases must be removed from natural gas in order to

1. Increase the heating value of natural gas;
2. Decrease the volume of gas transported in pipelines;
3. Reduce corrosion during the transport and distribution of natural gas; and
4. Prevent atmospheric pollution by SO_2, which is generated during the combustion of natural gas containing H_2S.

The CO_2 separation process is the most challenging separation process, known to be about 70 % of the total cost of the carbon capture and storage (CCS) process [25]. Capture costs can be reduced by developing efficient CO_2 separation processes [26] in post-combustion CO_2 capture. The CO_2 separation process is applied at the final stage of the power generation process after desulfurization (Fig. 6.6). Post-combustion flue gas from coal-fired plants mainly consists of carbon dioxide (around 17 %) and nitrogen (around 80 %) with a small amount of oxygen and water vapor.

Membranes for the separation of CO_2 from natural gas, and other compounds, are also classified into polymeric and inorganic. Polymeric membranes are relatively easy to manufacture and are suited at low temperature application. The polymer morphology and mobility determine the gas permeability and selectivity. Asymmetric hollow fibers are used for the separation. Figure 6.7 shows an asymmetric hollow-fiber membrane. A thin layer of functional cardopolyimide material supported by a porous structure allows high permeability [28].

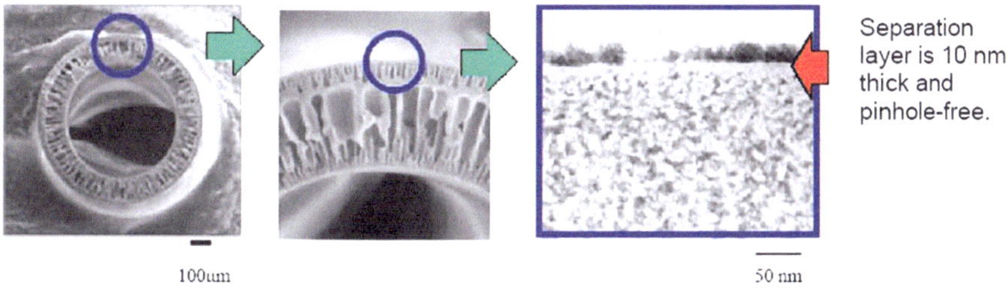

Fig. 6.7 Cardopolyimide hollow-fiber membrane with a thin, functional outer layer

There are three main types of organic (polymeric) membranes commercially available for CO_2 removal: cellulose acetate, polyimides, and perfluoropolymers. These membranes are easy to process through to membrane modules, are mechanically strong, and have thermal and chemical resistance to ensure long term viability [29].

The search for novel polymeric materials continues today. Polymer membranes highly permeable to CO_2 and having good selectivity should be developed for the membrane process to be viable. Du et al.'s [30] article (review) on "Advances in high permeability polymeric membrane materials for CO_2 separations" summarizes natural gas processing with membranes. Noteworthy advances are provided in polymeric materials having very high permeability and good CO_2–N_2 selectivity that largely surpass the separation performance of conventional polymer materials. Although the focus is on CO_2–N_2 separation, the membranes have potential to be applied in CO_2 separation from natural gas. Five important classes of polymer membrane materials are highlighted, including: polyimides, thermally rearranged polymers (TRs), substituted oxyacetylene, polymers with intrinsic microporosity (PIM) and polyethers; insights are provided into polymer design suitable for CO_2 separation from, for example, the flue gases coming from coal-fired power plants. The majority of these polymers exhibit excellent characteristics for CO_2 separation application. Their chemical structures are given in Fig. 6.8. Recently NOTT-220 has been invented, which has a bright future for the removal of CO_2 from the atmosphere. It is in very early stage development, as discussed in Chapter 3.

Du et al. [31] introduced a new class of membranes called PIMs, incorporating tetrazoles (TZPIM)—described on Chapter 3 of this book—for CO_2–N_2 separation. The presence of the tetrazole groups leads to favorable CO_2 sorption and selective pore blocking by presorbed CO_2 molecules, thus limiting access by other light gas molecules such as nitrogen (Fig. 6.9). The introduction of tetrazoles into PIM is the first example of {2+3} cycloaddition of a polymer containing aromatic nitrile groups with an azide. This strategy of incorporating nitrogen heterocycles into PIMs provides new directions in the design of other polymeric membrane materials for important CO_2 separation process.

Lee et al. [32] described novel supported liquid membranes, which were prepared by incorporating room-temperature ionic liquids into PVDF matrix via a phase separation technique. It consisted mainly of two processes; the low temperature

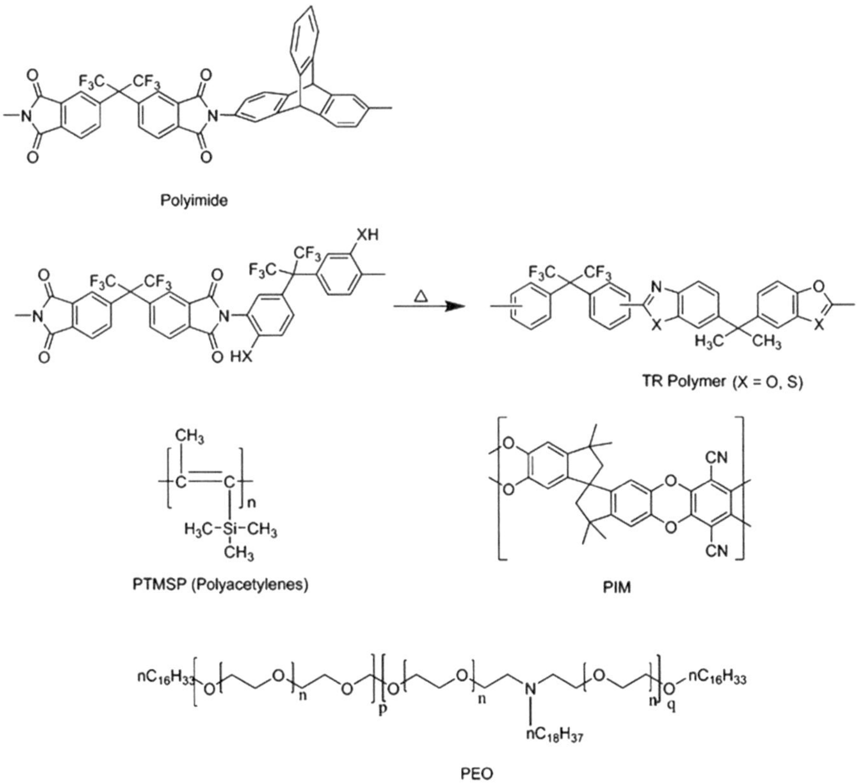

Fig. 6.8 Representative chemical structures of polymers with high CO_2 permeability

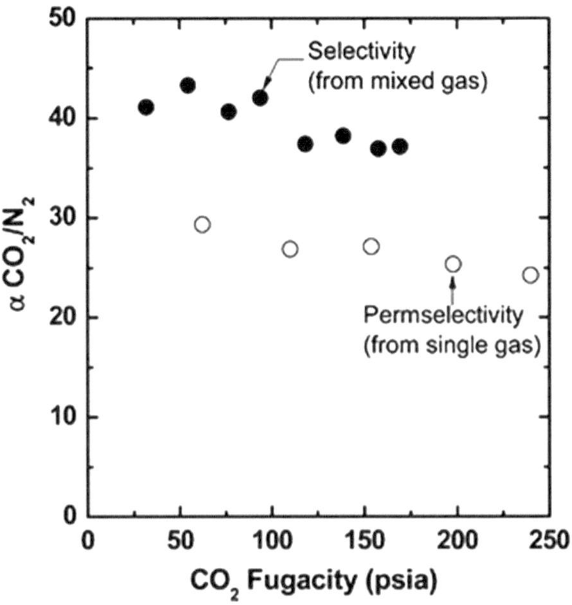

Fig. 6.9 Effect of CO_2 partial pressure on mixed-gas CO_2–N_2 selectivity in TZPIM-2 at 25 °C. Mixed gas composition (in mol% CO_2 : mol% N_2) was 50:50. Adapted from the ref. [31]

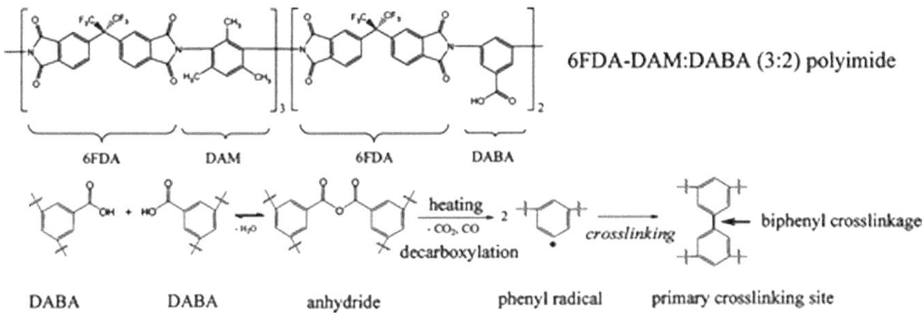

Fig. 6.10 Chemical structure of 6FDA-DAM:DABA (3:2) and the cross-linking mechanism of decarboxylation, radical induced cross-linking (DRIC) [38, 39] (cross-linking bond is noted with an *arrow*)

phase separation and high temperature quenching process. The membranes had excellent stability under severe operating conditions. The permeability coefficients of H_2S and CO_2 through the membranes were very high compared to that of CH_4 which is one of the essential elements of natural gas. Hence, the membrane exhibited very high H_2S–CH_4 and CO_2–CH_4 selectivities, ranging from 200 to 600 and from 50 to 100 depending on volume fraction of ionic liquid in the membranes, respectively. From these permeation results, the authors suggested that the novel-supported liquid membrane is an attractive alternative to conventional processes in the upgrading of crude natural gas.

Nano-engineered thermally rearranged polymer (TOR) membranes show high gas permeability as well as good separation properties, especially in CO_2 separation processes such as post-combustion flue gas and natural gas sweetening [26, 33–36].

Koros group [37] suggested that thermally cross-linkable hollow-fiber membranes made of polyimide were effective for the removal of CO_2 from natural gas. The structures of the cross-linkable polyimide used to make hollow fibers are shown in Fig. 6.10. The results demonstrated that cross-linking improved the membranes' selectivities and effectively eliminated swelling-induced hydrocarbon loss at high pressure.

The 6FDA moiety and DAM moiety provide high permeability due to their bulky groups, which inhibit chain packing. The DABA moiety provides a carboxylic acid pendant group as a cross-linking site. The average molecular weight of the polyimide was 229,400.

Due to poor quality, raw natural gas often cannot be used to fuel gas engines or gas turbines driving pipeline compressors. Significant concentrations of H_2S and/or C_{3+} will cause corrosion and carbon buildup in the gas engine. CO_2 and nitrogen will lower the BTU value. Any of these impurities can compromise engine operation, increase downtime or, at a minimum, put emissions out of compliance. Their presence in the raw gas can even render the gas unusable as fuel so expensive diesel has to be trucked in. Membranes made of rubbery polymers can be used for the treatment of raw gases.

MTR's FuelSep™ systems purify raw gas side streams to premium quality fuel gas (Fig. 6.11). MTR membranes easily remove H_2S, C_{3+}, CO_2, N_2, and water from

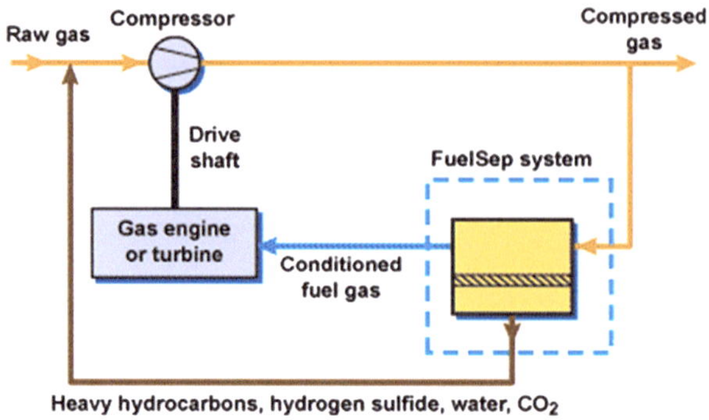

Fig. 6.11 Schematic diagram of MTR's FuelSep™ systems to purify raw gas side streams to premium quality gas

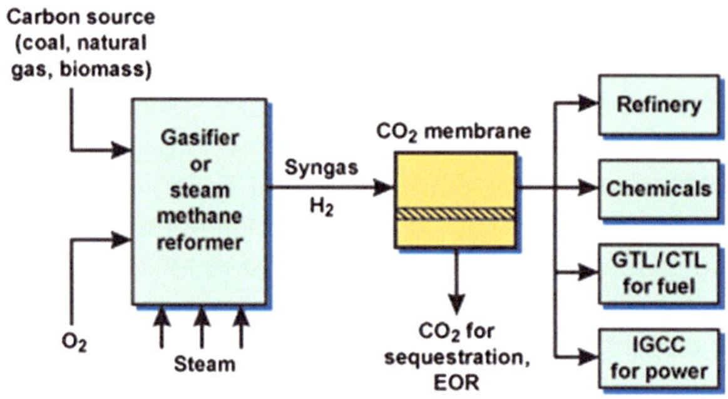

Fig 6.12 CO_2 removal from syngas using Polaris™

fuel gas at moderate pressure. Because these impurities are taken out and returned to the compressor suction, there are no effluent streams to be disposed of. Any C_{3+} removed from the fuel gas goes back into the main gas stream, so all NGL in the raw gas stream is available for downstream recovery, if desired.

MTR's unique Polaris™ membrane (based on polymers) is the first commercially available membrane that separates CO_2 from syngas. The Polaris™ membrane is much more permeable to CO_2 than to other syngas constituents and can be used to recover and purify CO_2 for sequestration, enhanced oil recovery (EOR), or for use in chemical and industrial applications. The resulting CO_2-enriched stream can be produced in gas or liquid form, depending on the final use for CO_2. Figure 6.12 shows the schematic diagram of the system using MTR's unique Polaris™ membrane.

Bhide et al. [40] designed a hybrid process combining membrane separation and gas adsorption in diethanolamine for the removal of acid gases from natural gas up to 40 mol% and up to 1 mol % H_2S from crude natural gas. Membrane separation is used first for the bulk removal of acid gases, in particular CO_2, from the crude natural gas feed, while final purification to US pipeline standards ($\leq$2 mol% CO_2, $\leq$4 ppm H_2S) is performed by gas absorption.

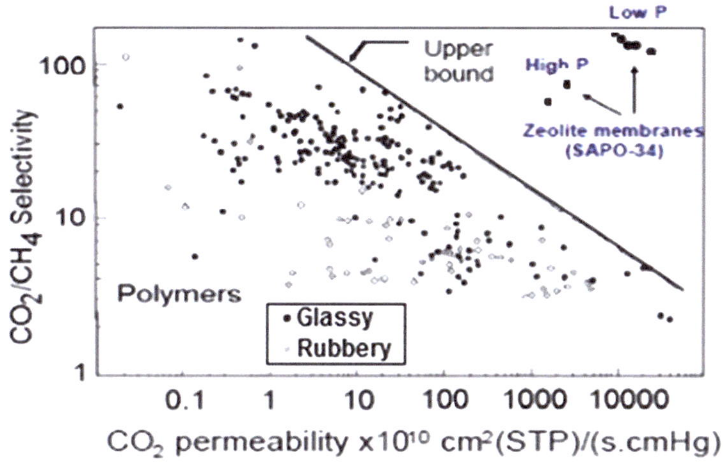

Fig. 6.13 Zeolite (SAPO-34) membrane performances in Robeson's plot

Zeolite membrane such as SAPO-34 is a typical inorganic membrane that could give higher separation performance compared to the polymeric membrane, but the separation performance is inversely proportional to the pressure loaded. The performance of both organic and inorganic membranes is summarized in Fig. 6.13 [24, 41].

In general, the permeability of gas in a specific gas mixture varies inversely with its separation factor. The tighter molecular spacing shows the higher the separation characteristic of the polymer. However, as the operating pressure increases, the permeability decreases due to the lower diffusion coefficient. Generally, as the permeability of the gas increases, the permselectivity has tended to decrease in most cases of polymeric membranes [42]; however, for inorganic membranes like SAPO-34, higher separation performances occur compared to polymeric membranes. The separation performance is inversely proportional to the pressure loaded.

6.1.4 Hydrocarbon/Carbon Dioxide Separation

Various techniques are used to separate hydrocarbons/carbon dioxide during oil shale retorting, coal gasification, oxygen fire flooding, and carbon dioxide miscible flooding in enhanced oil recovery. Some examples of these separation techniques include refrigerated distillation, extractive distillation, amine scrubbing, and semi-permeable membrane separation. These techniques have various degrees of success in producing substantially pure co-products of carbon dioxide and sulfur compounds, along with fuel gas compounds (including methane and ethane), and heavier hydrocarbons (including propane and higher alkanes).

Membrane technology competes most directly with absorption for carbon dioxide removal. The current technology to separate heavy hydrocarbons from natural gas is cooling and condensation, or lean oil absorption. The condensed heavy hydrocarbons separated from the gas stream are then subjected to fractional distillation to recover the individual components. Because refrigeration is costly and uses large amounts of energy, there is interest in alternative techniques such as membrane gas separation. Membrane separation can be efficiently and successfully used for the purification of natural gas replacing the amine absorption process, which dominates the natural gas industry. Membrane technology can be effectively applied for the removal of acid gases such as CO_2, H_2S and water vapor present in the natural gas streams.

Various membrane techniques are presently described in the literature for separating carbon dioxide and hydrocarbons by differential rates of permeation through the membrane relative to other gas constituents. The carbon dioxide is recovered at low pressure and must be recompressed for reinjection into a carbon dioxide utilizing process, such as the enhanced oil recovery operations utilizing carbon dioxide miscible flooding. Both the compressor equipment and the membrane are high capital cost items and staging of the membranes is frequently required for carbon dioxide recovery at high purity. Typical membranes reported in the literature are as described below [43–47].

In general, commercial membranes used for CO_2 removal are polymer based, for example, cellulose acetate, polyimides, polyamides, polysulfone, polycarbonates, and polyetherimide. The membrane does not operate as a filter, where small molecules are separated from larger ones through a medium with pores, rather, the membrane separates based on how well different compounds dissolve into the membrane and diffuse through it. Yoshino et al. [48] used asymmetric hollow-fiber membranes of copolyimide from equimolecular portions of 2,2-bis(3,4-dicarboxyphenyl)hexafluoropropane dianhydride (6FDA) and 3,3′,4,4′-biphenyltetracarboxylic dianhydride (BPDA) with 3,7-diamino-2,8(6)-dimethyldibenzothiophene sulfone (DDBT) to investigate the permeation properties of single-component light gases, olefins, and paraffins and for mixed components of C_3H_6–C_3H_8 and C_4H_6–C_4H_{10}. The evaluation of membrane quality based on the resistance model suggested that the extremely small surface porosity of defect pores significantly reduced the selectivity for the larger gas pairs. The permeance of C_3H_6 and C_4H_6 of 3.6 and 7.4 GPU, respectively, and separation factors of 15 and 69 for C_3H_6–C_3H_8 and C_4H_6–C_4H_{10} (50/50 mol% in feed) were achieved at 373 K and 1 atm.

Lucadamo [49] invented a process for the low temperature distillative separation of a carbonaceous off-gas into a heavy hydrocarbon stream, a high purity fuel gas product and a high purity carbon dioxide–sulfur stream product wherein the fuel gas product was enriched in fuel gas components and the carbon dioxide–sulfur stream product was enriched in carbon dioxide by further separation in a semipermeable membrane.

MTR is developing a series of membranes especially for the natural gas market. The cross section of one of these membranes is shown in Fig. 6.14. The membrane consists of three layers: a nonwoven fabric that serves as the membrane substrate (the support web); a tough, durable, solvent-resistant microporous layer that provides

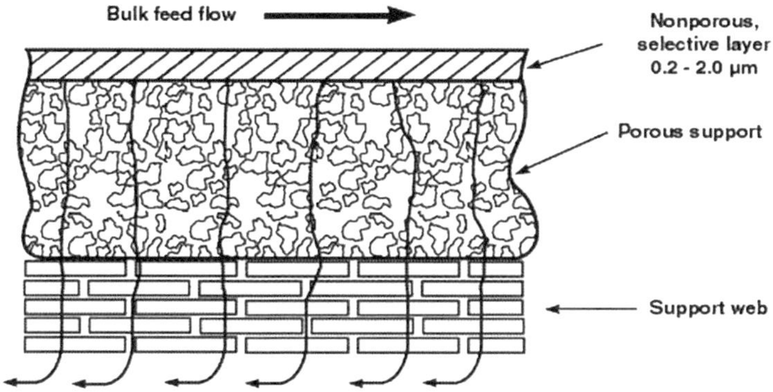

Fig. 6.14 Cross section of an MTR composite membrane

mechanical support without mass transfer resistance; and a nonporous, defect-free selective layer that performs the separation.

This configuration allows each of the layers to be independently chosen to optimize their function: the selective layer for high flux and selectivity, and the support layers for mechanical and chemical stability that will not influence the separation. Specific selective layers may be chosen for specific separations, opening up exciting new applications, such as separation of heavy hydrocarbons from light hydrocarbons and nitrogen, separation of carbon dioxide and H_2S from methane, and dehydration. The selective layer is cross-linked to the microporous layer, thus preventing any delamination or separation of the two layers. The flat membrane sheets are packaged in a module that is spiral wound.

6.1.5 *Vapor Permeation/Pervaporation Gas Separation*

Vapor permeation is similar in principle to PV. The only difference concerns the feed, which is a mixture of vapors or vapors and gases [50]. Vapor permeation differs from membrane gas separation processes, because it requires an extra feature other than high permeability and good selectivity, that is, membrane stability against vapor attack. In terms of its permeability requirement, the membrane of course has to be much more permeable to hydrocarbons than air. The factors affecting the design of membrane gas separation systems have been discussed in detail by Baker et al. [51]

In general, most membranes used for GS can be categorized into two types:

1. Porous ceramic membranes: highly voided structures with randomly distributed interconnected pores. Separation is a function of the permeate character and membrane properties. Different transport mechanisms can be involved.
2. Non-porous polymeric membranes: gas transport is described by the solution-diffusion mechanism, determined by molecular interactions between the gas molecules and the membrane.

For polymeric membranes, a distinction can be made between rubbery and glassy polymers. With glassy polymers, diffusivity is dominant due to the rigid nature of the polymer chains; therefore, small molecules permeate more easily. In rubbery polymers, solubility is dominant. Permeability increases with increasing permeate size. In general, rubbery polymers exhibit high permeabilities and low selectivities, whereas glassy polymers show higher selectivities but much lower permeabilities.

Small gas mixtures and organic isomers can be separated by using MFI zeolite membranes via gas–vapor permeation [52]. Gas mixtures whose components have different adsorption affinities with zeolite pores can be separated by MFI zeolite membranes, because the component with strong adsorption on zeolite pores blocks the pore channels that allow the permeation of the weakly adsorbing component through the membrane, resulting in a high selectivity for the strongly adsorbing component over the weakly adsorbing component [53]. The presence of water vapor does not affect the H_2/CO_2 separation factor for membranes ZSM-5 and silicate membranes [52]. However, these two membranes adsorb water vapor even at temperatures in the range of 300–350 °C and reduce the permeability of hydrogen and carbon dioxide in the zeolitic pores. H_2O/H_2 permselectivity for the ZSM-5 membrane was much higher than that for the silicate membrane due to the strong adsorption capability of the ZSM-5 membrane for water vapor [54].

Yeow et al. coated PVDF hollow-fiber membranes with cross-linkable divinyl-terminated silicon rubber, and divinyl-polydimethylsiloxane (divinyl-PDMS). It was noted that the oxygen–nitrogen selectivity was over 2.24 [55]. The membrane selectivity separating the BTX from nitrogen demonstrated the recovery was greater than 95 %.

6.2 Present and Emerging Large-Scale Applications of Membrane Technology

Currently, only eight or nine polymers have been used to make at least 90 % of the total installed gas separation membrane base. Several hundred new polymer materials have been reported in the past few years. Baker [2] predicted the following advancements for membrane technology for gas separation:

1. The membrane share of the nitrogen-from-air market will grow in the next 20 years.
2. The future growth of membrane gas separation technology will be in the refinery, petrochemical and natural gas industries.
3. As plant size increases, the enlargement of membrane modules becomes increasingly important.
4. Success in refining/petrochemical/natural gas will require more robust membranes and modules than those used today.
5. Custom made, high cost polymers will gradually displace off-the-shelf polymers such as cellulose acetate, polysulfone, and polyaramide as selective membrane materials. The multilayer composite membranes will gradually displace simple Loeb–Sourirajan membranes.

Table 6.4 Commercial membrane success

No.	Success
1	Nitrogen enrichment of air for commercial blanketing and engine applications
2	Hydrogen recovery from ammonia purge
3	Low CO_2 level biogas and natural gas separation
4	Monomer recovery from storage vessels and recycling
5	Nitrogen separation membranes

Table 6.5 Emerging opportunities for membrane technology for gas–gas systems

No.	Opportunities
1.	Air separation for internal combustion engines (e.g., intake, combustion equipment, pre-treatment)
2	Nitrogen separation from natural gas
3	Carbon separation
4	Hydrogen sulfide separation
5	Carbon capture
6	Hydrogen separation from refinery gas streams and coal gas
7	High temperature gas streams (e.g., water-gas shift)
8	Next-generation air separation membranes (oxygen/nitrogen) with better selectivity

6. The production of oxygen from air will slowly become a significant membrane market.
7. Recovery of hydrogen in refineries will become a significant growth area in the next decade.
8. Large opportunities for membranes in the separation of light hydrocarbons exist. The development of facilitated-transport membranes for olefin separation or ceramic membranes for hydrocarbon isomer separation could open up large new membrane markets.
9. The use of membranes to separate natural gas mixtures could easily grow to become the largest single gas separation application.

Though significant time and funds have been invested in membrane technology development, industrial implementation has been somewhat limited. Table 6.4 shows the commercial membrane successes for gas separation.

Emerging opportunities for gas separation membranes are given in Table 6.5 [56].

Membrane dehydration technologies are currently commercially available; however, these technologies are hampered by performance degradation, high energy consumption, hydrocarbon losses, equipment complexity, and high capital costs. Air Products and Petreco seek to resolve these imperfections, developing hollow-fiber membranes for natural gas dehydration [57]. Susceptibility of the membrane material (polymeric) to plasticization by water is also a main problem [58]. Generon® is manufacturing the dehydration membrane modules at Generon® Manufacturing Facilities in Houston TX, Chengdu, China and Grosseto, Italy.

6.3 Dew Pointing of Natural Gas

This book has already discussed that polymeric membranes can remove hydrocarbons from natural gas. The hydrocarbon dew point is the temperature (at a given pressure) at which the hydrocarbon components of any hydrocarbon-rich gas mixture, such as natural gas, will start to condense out of the gaseous phase. Dew point is often also referred to as the HDP or the HCDP. The maximum temperature and the pressure at which such condensation takes place are called the "cricondentherm." The hydrocarbon dew point is a function of the gas composition as well as the pressure.

If the hydrocarbon dew point of pipelined natural gas is too high, some liquids may condense out in the pipeline. This not only degrades the heating value of the remaining gas, it increases the potential for problems in pipeline transmission systems and causes problems for the end users of the gas such as industrial combustion equipment and household gas appliances. It is necessary to remove hydrocarbons from the stream to reduce the dew point of the natural gas; therefore, the hydrocarbon dew point is universally used in the natural gas industry as an important quality parameter, stipulated in contractual specifications, and enforced throughout the natural gas supply train—from producers through processing, transmission, and distribution companies to the final end users.

Gas transportation companies have come to the realization that managing hydrocarbon dew point reduces system liabilities, opens up new gas markets, and generates operating revenue. By managing hydrocarbon dew point, hydrocarbon condensation can be prevented. Potential problems exist in cold spots and liquids collect in low areas, moving as a slug through the pipeline system. This causes over pressuring of the pipe, which overpowers liquid handling facilities, allowing the undesirable liquids to flow into compressors and end user sales points. Also, removing pipeline liquids helps prevent pipe corrosion in the low areas where water is trapped under the hydrocarbon liquid layer, slowly destroying the pipe integrity. Proper managing of gas dew point can also prevent liquids from forming as the gas cools while flowing through pressure reduction stations (e.g., citygates) that feed end user supply systems. Controlling dew point is also necessary when the pipeline company is seeking to have its product qualify as a product for high efficiency gas turbine end users that require a dry and consistent quality of fuel. Membrane technology can contribute to lower the dew point of natural gas.

6.4 Olefin–Paraffin Separations

Light olefins such as ethylene and propylene are very important to petrochemical industries because they are utilized as main raw materials for many essential chemicals and products for industrial and domestic consumption. Approximately 80 % of ethylene produced in the USA and Europe is used to create ethylene oxide, ethylene dichloride, and polyethylene. Ethylene oxide is a key raw material in the production of surfactants, detergents, and automotive anti freeze. Propylene is a raw material for a large variety of products including polypropylene, which is considered a versatile polymer used in packing and other important applications such as textiles, laboratory equipment, and automotive components. Propylene is the second highest volume petrochemical feedstock after ethylene. The production of polymers and other special chemicals from mono-olefins such as propylene requires the olefin to be extremely high purity (>99.9 %), and since light olefins are commonly produced together with paraffin hydrocarbons, i.e., ethane and propane, the techniques for separating both hydrocarbons are of primary importance to the petrochemical industry [59].

In general, a conventional process for olefin–paraffin separation is distillation. Over 2000 articles and patents (which constitute about one-third of this total) have been published on the use of membranes for hydrocarbon separation and removal over the span of the last 30 years [60]. This figure indicates that researchers are still looking into the enhancement of the available membrane technology for olefin–paraffin separation before possible commercialization. The membranes used for this technology can be classified into three main groups: (1) polymeric, (2) inorganic, and (3) facilitated transport membranes [59, 60].

6.4.1 Polymeric Membranes

A large number of articles are reported in the literature on the use of polymeric membranes—without a carrier for olefin–paraffin separation from air mixtures—as well as pervaporation processes such as hydro separation. These membranes include various materials such as glassy, cellulosic, and rubbery polymers. The permeation and separation characteristics of these membranes are governed by the properties of both permeant molecules and polymers; these properties include molecular size and shape and polymeric structure such as packing and rigidity. Among these materials, glassy polymers have been studied intensively for olefin–paraffin separation as well as the separation of aromatic, alicyclic, and aliphatic hydrocarbons, whereas rubbery polymers have mainly been used in gas–vapor separation applications such as hydrocarbon extraction from aqueous solutions. Asymmetric hollow-fiber membranes are more attractive than other membranes configuration due to their high surface area per unit volume.

Ito and Hwang [61] studied the permeation of propylene and propane through commercial glassy, cellulosic, and rubbery hollow-fiber membranes prepared from polysulfone (PS), cellulose acetate (CA), and silicon rubber, respectively. Table 6.6 shows the permeation and separation characteristics through glassy, rubbery, and cellulosic hollow fibers. The results indicated that CA was the best among all those membrane materials for propylene–propane separation.

Chan et al. [62] studied the transport of olefins and paraffins in aromatic poly (1,5-naphthalene-2-2′-bis(3,4-phthalic) hexafluoropropane) diimide (6FDA-1,5NDA) dense membranes. The gas permeability coefficients were measured at pressures

Table 6.6 Permeation and separation characteristics through glassy, rubbery, and cellulosic hollow fibers [61]

Polymer	Temperature (°C)	Permeability (Barrer)		Selectivity
		C_3H_8	C_3H_6	$\alpha_{C3H6/C3H8}$
CA	−17	4.5	4	0.884
	20			2.7
	40	3.2	12.1	3.78
	80			4.8
Ps	−17	31.6	34.1	1.08
	20			1.3
	40	26.3	37.5	1.43
	70			2.3
Silicon rubber	−17	29,500	15,200	0.52
	20			0.9
	40	6,490		0.992
	60			1.1
EC	20	4[a]	27[a]	6.75[a]
	20	17[b]	70[b]	4.1[b]
	40	5.5[a]	32[a]	5.8[a]
	40	18[b]	72[b]	4[b]
	60	8[a]	35[a]	4.38[a]
	60	20[b]	80[b]	4[b]

All data are presented for pure gases
[a]Cast dense EC (ethyl cellulose) membrane
[b]Coated EC hollow-fiber membrane

from 2.5 to 16 atm for the C_2 hydrocarbon gases and pressures up to 87.4 atm for C_3 system at 35 °C. This particular membrane had permeabilities of 0.15, 0.87, 0.023, and 0.24 Barrer with respect to pure ethane, ethylene, propane, and propylene, and showed an ideal selectivity of 5.8 for the separation of ethylene–ethane, 10 for propylene–propane, 7.6 for nitrogen–ethane, and 250 for nitrogen–propane. The olefins showed a preferred permeability to paraffins. The plasticization effect was also found for propane and propylene, respectively, although it was not detected in the saturated C_2 hydrocarbons at pressures up to 16 atm. The diffusivity coefficient decreased with the increase in carbon amount from methane to ethane and propane, and from ethylene to propylene, because of the increasing penetrant size and different molecular configuration.

6.4.2 *Inorganic Membranes*

Carbon molecular sieve (CMS) membranes open a new avenue for the separation of olefins/paraffins. These membranes are usually prepared by the pyrolysis of polymeric precursors for which polyimide materials are frequently used. CMS composite membranes composed of a selective carbonized layer on top of an inorganic support has excellent mechanical strength but suffers from complicated preparation procedures.

6.4.3 *Facilitated Transport Membranes*

Due to the similar physical and chemical properties and molecular sizes of propylene and propane (olefin and paraffin), conventional polymeric membranes are not competitive for the separation of these mixtures based only on the sorption/diffusion mechanism. A simultaneous increase of permeability and selectivity by incorporating specific agents in the polymer matrix—agents that interact reversibly with propylene but not with propane—is one possible way to overcome this constraint. In this way, propylene permeation would occur by a facilitated transport mechanism. Transition metal cations are used as carriers for propylene transport due to their capacity to react specifically and reversibly with unsaturated hydrocarbons.

Olefins are capable of forming reversible chemical bonds with transition metal ions due to the specific interaction between the olefin's hybrid molecular orbitals and the metal's atomic orbitals, commonly known as the π-bonding. The bonds formed between transitional metal ions and olefins are stronger than those formed by Van der Waals forces alone, so it is possible to achieve high selectivity and high capacity for the component to be bound. At the same time, the bonds are still weak enough to be broken by using simple operating techniques such as increasing the temperature or decreasing the pressure. These facilitated transport membranes can be operated in both the liquid and solid states such as in liquid membranes [63].

Facilitated transport has been receiving tremendous attention as a potential energy-saving separation technology, because it can simultaneously improve permeability and selectivity [64, 65]. The separation of olefin–paraffin mixtures by facilitated transport membranes containing silver salts is promising and a potential alternative to energy intensive distillation processes [66, 67]. It has also been reported that silver polymer electrolyte membranes, which are composed of silver ions dissolved in a polar polymer containing oxygen atoms, exhibit high separation performances (around 60 propylene–propane selectivity) [68].

Nymeijer et al. [63] fabricated a composite hollow-fiber membrane with an olefin-selective stabilizing layer of sulfonated poly(ether ether ketone). The membrane was used in a gas–liquid membrane contactor for the separation of olefin and paraffins. Continuous contact between the absorption silver nitrate solution and the SPEEK layer prevents the layer from drying out and a subsequent loss of selectivity. Previously, unknown high ethylene–ethane selectivities (>2,700) were obtained in combination with reasonable ethylene productivities (7.6×10^{-10} cm^3/cm^2 s Pa) (1×10^{-6} cm^3/cm^2 s cm Hg).

Kim et al. [68] fabricated a novel facilitated transport membrane containing silver salts in an inert PDMS matrix. The physically dispersed silver salt carrier was not initially active in facilitated transport, but became active with exposure to propylene–propane mixtures. The PDMS/silver salt membranes exhibit extremely high separation performance—the highest selectivity ever recorded of 200—when tested on propylene–propane mixtures. Pollo et al. [69] fabricated polyurethane (PU) membranes using two different silver salts (triflate and hexafluorantimonate). The membranes were structurally characterized and their performance evaluated for the separation of propylene–propane. The results of the characterization analysis indicated that the triflate salt was the most efficient carrier agent. These membranes containing triflate salt showed the best performance, reaching an ideal selectivity of 10 and propylene permeability of 188 Barrer.

Chilukuri et al. [70] described a conceptual design for the separation of propylene and propane, based on equilibrium calculations. The authors proved that separation using a gas–liquid membrane contactor with silver salt solution is technically feasible. A structural design method was applied, step-by-step. The process flow sheet for the required separation is given in Fig. 6.15. All calculations were based on mass and heat balances. Gas and liquid phases were separated by a composite membrane consisting of a dense polymeric top layer on top of a porous support (Fig. 6.16). Propylene and propane diffused through the membrane, where propylene selectively reacted with silver ions, resulting in a silver–propylene complex. The reaction is based on the ability of silver ions to reversibly form complexes with propylene via a combination of a π and a σ bond between silver ions and propylene. Propane is only physically absorbed in the silver salt solution. The propylene-rich silver salt solution leaves the absorber at the bottom. By changing the temperature and/or pressure, the equilibrium of complexation can be influenced, allowing desorption of propylene in a desorber. The lean silver solution is cooled and recycled back to the absorber. The propane-rich gas stream leaves the absorber at the top. Figure 6.17 shows the functional diagram of the propylene–propane separation process based on a gas–liquid membrane absorber using a silver salt solution.

The above conceptual design resulted in the use of a 5.5 M $AgBF_4$ solution at an absorber temperature of 308 K in combination with a desorber temperature of

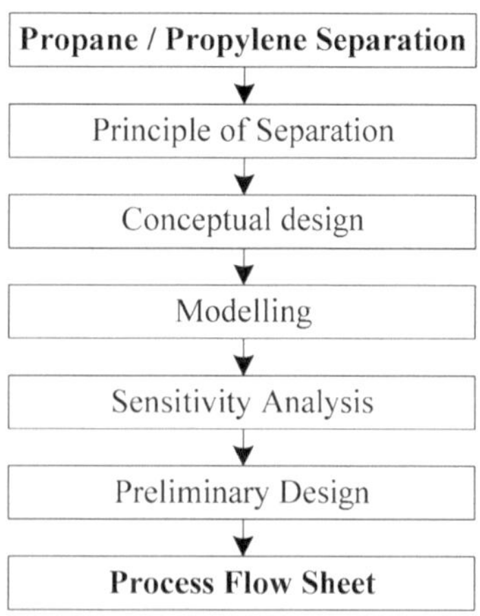

Fig. 6.15 Steps to create process flow sheet

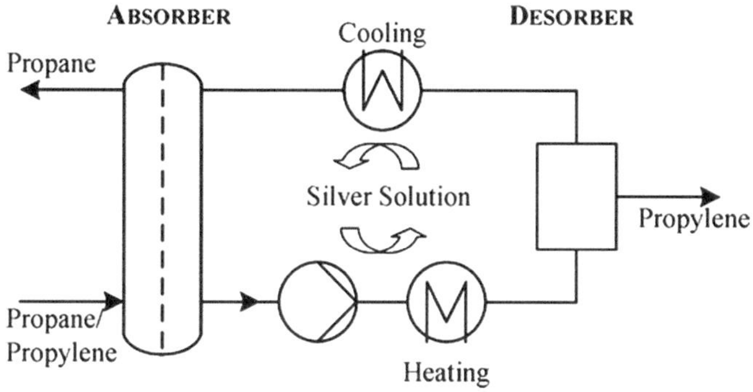

Fig. 6.16 Scheme of propylene–propane separation using a silver salt solution and a gas–liquid membrane contactor as absorber

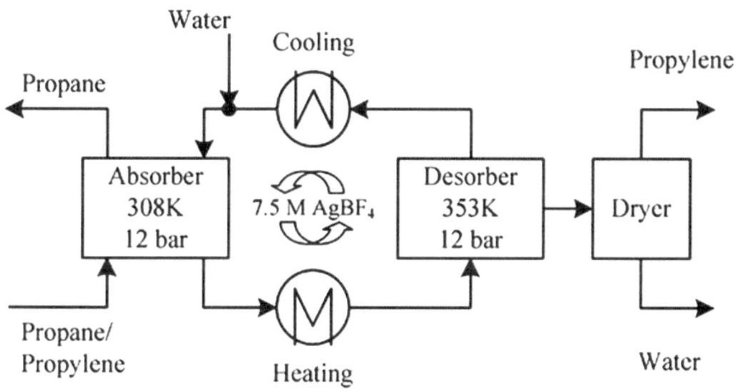

Fig. 6.17 Functional diagram of propylene–propane separation process based on a gas–liquid membrane absorber using a silver salt solution [70]

353 K. This gave the best separation of propylene and propane regarding cost and purity. Modeling of the gas–liquid membrane contactor showed that a membrane area of 80,000 m^2 is required to achieve the desired propane and propylene purities. Kang et al. [71] used the silver nanoparticles to separate propylene–propane mixtures by applying a facilitated transport technique using ionic liquid $BMIM^+BF^-$ (1-butyl-3-methylimidazolium tetrafluoroborate). In membranes containing $BMIM^+BF^-$/Ag nanoparticles, mixed-gas selectivity was 17 for a 50/50 (vol %) propylene–propane mixture with stability performance of 100 h.

Faiz and Li [59] concluded that, in general, glassy and elastic polymers are unsuitable materials for olefin–paraffin separation due to low permeation rates, separation factors, or a combination of both as olefins and paraffin have similar molecular sizes and solubilities. Conventional additive-free membranes, including glassy, rubbery and cellulosic materials, are not good as they do not contain any active groups towards the permeation of olefins. There are, however, some exceptions of glassy polymeric membranes such as polyphenylene oxide (PPO) and its copolymers. Moderate permeation rates and acceptable selectivities are achieved for propylene–propane separation by these membranes. Rubbery polymeric membranes show high permeation rates but very low selectivities. Therefore, they are unsuitable membrane materials. Cellulosic membranes such as ethyl celluslose (EC) have shown moderate permeation rates and ideal selectivities towards propylene–propane separation. These results suggest that more stable materials are needed before practical modules can be anticipated for olefin–paraffin separation.

6.5 Membrane/Pressure Swing Adsorption Process

Membrane permeation and pressure swing adsorption (PSA) are two widely used processes for gas separation. They are often considered to be alternative or compliments to the more conventional cryogenic separation processes. Membrane separation is based on the difference in the rate of permeation through a membrane, while adsorption separation depends on the difference in either the rate or the equilibrium of adsorption on an adsorbent. Membrane gas separation is pressure-driven and normally operates continuously, whereas PSA is a cyclic process in which the adsorbers undergo adsorption at a high pressure and desorption at a reduced pressure, thereby making it suitable for processing a gas mixture in a continuous fashion. PSA is suitable for producing gases of higher purities. Several membrane/PSA combination schemes have been proposed in the patent literature for various applications, including helium recovery, hydrogen purification, acid gas removal, and nitrogen production [72].

A novel process for gas separation, called pressure swing permeation was introduced by Feng et al. [73] to evaluate the relatively low permeate pressure by pressurization with high-pressure feed gas. The pressure swing permeation process is analogous to the pressure swing adsorption. Although the pressure swing permeation process can be run as a batch process, it is more efficient to operate the membrane system in a continuous fashion. In a continuous pressure swing permeation process, the membrane system, in a simple form, comprises two membrane modules A and B to perform a gas separation in a steady-state cyclic fashion. Each module contains two distinct void spaces separated by the membrane for the admission and removal of gas streams, respectively. After the high-pressure feed gas is introduced to membrane module A, permeation takes place, and the permeate from module A, which is at a relatively low pressure level as compared to the feed pressure, is received by module B. As the permeation in module A proceeds with time, the pressure of permeate collected in module B increases. Then the gas feeding to module A is stopped, and the permeate stored in module B is pressurized with the high-pressure feed gas, during which period the permeate product at a desired elevated pressure is displaced out of module B. In the meantime, the residue in module A is released. The two modules are arranged such that the void space in one module used for receiving the permeate from the other module will receive the feed gas when the two modules are switched, to reverse their functions in terms of feed admission and permeate reception. Thus, the membrane system is ready for the next cycle of operation. The membrane modules should be properly designed and the pressurization and displacement steps should be carried out in a sufficiently short period of time to prevent any breakthrough of feed gas into the permeate product. Hollow-fiber membranes are preferred because the lengthwise gas mixing in the membrane module is very limited. The two membrane modules can be synchronized, and each undergoes five basic operating steps:

1. Collection of the low-pressure permeate from the other module.
2. Pressurization of permeate by the high-pressure feed gas.
3. Release of permeate product at an elevated pressure.
4. Admission of feed to carry on permeation.
5. Withdrawal of residue stream.

Clearly, step 4 is responsible for the separation, and step 2 is responsible for producing a high-pressure permeate product. Engineering designs directed to minimizing back mixing of the gas during the pressurization step are necessary to ensure product purity. Figure 6.18 represents the cycle sequence of a two-module process. A brief description of the cycle steps is shown in Table 6.7.

Esteves and Mota [74] presented a new hybrid gas separation process combining permeation and pressure swing adsorption (PSA). An integrated model was formulated which successfully predicted all process characteristics. The coupled process increased the efficiency of the pressurization and high-pressure adsorption steps, and thus improved the separation performance as compared to stand-alone PSA. The new process was applied successfully to the bulk separation of a mixture of H_2–CH_4 (50/50) and preliminary results were obtained for CO_2, CH_4 and H_2–CO_2–CH_4 mixtures.

Cryogenic Process distillation is a technology that has been around since the early 1900s when the Linde process was developed. Cryogenic distillation is considered the most economical large-scale process and can produce very pure gas and liquid products. Gases such as oxygen and nitrogen, the primary constituents of air, decrease in temperature and condense when throttled because they have a positive Joules–Thompson coefficient. By controlling how much the temperature decreases through pressurizing and throttling the air, oxygen and nitrogen can be separated by phases because they have different boiling temperatures. By combining cryogenic distillation with membrane gas separation, the hybrid system is more productive in small-to-medium-scale applications than in large-scale application [8].

A typical cryogenic distillation plant (Fig. 6.19) has a two- or three-stage main compressor (MAC), an air pre-cooler and pre-purification unit, large heat exchanger, an expander for refrigeration, and a multi-effect cryogenic distillation unit [75]. Air is pre-purified in a temperature-swing adsorption (TSA) bed that typically contains layers of activated alumina and zeolite.

Wankat and Kostroski [75] developed a hybrid membrane/cryogenic distillation system (see Fig. 6.20).

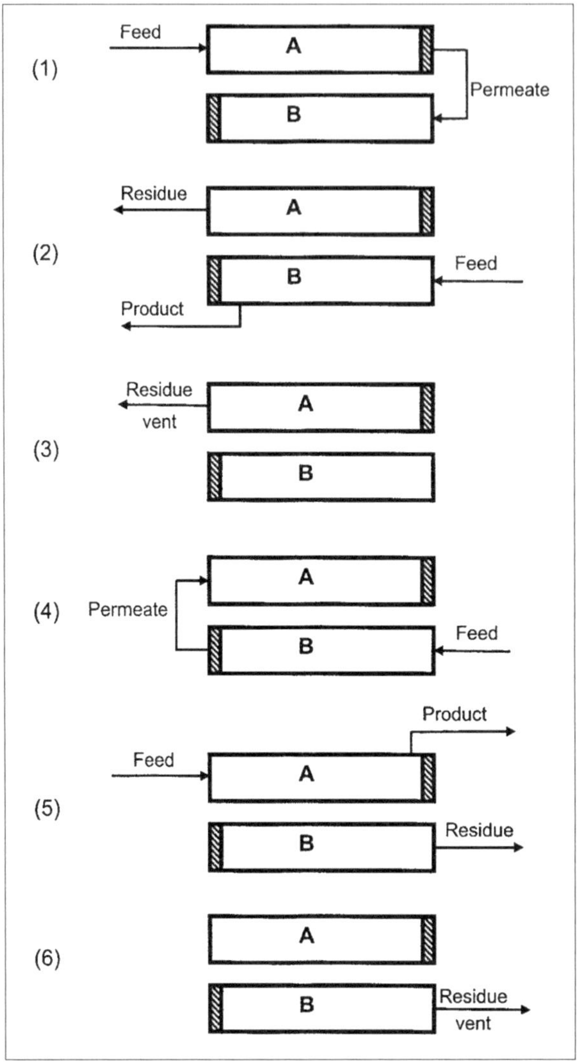

Fig. 6.18 Cycle sequence of a two-permeator pressure swing permeation process

Table 6.7 Cycles steps in a two-module pressure Swing permeation process

Step	Module A	Module B
1	Admission of feed gas and permeation	Reception of permeate from module A
2	Withdrawal of residue from module	Pressurization with feed gas and product release
3	Removal of residue remaining in gas line	Gas feeding stopped
4	Reception of permeate from module B	Admission of feed gas and permeation
5	Pressurization with feed gas and product release	Withdrawal of residue from module
6	Gas feeding stopped	Removal of residue remaining in gas line

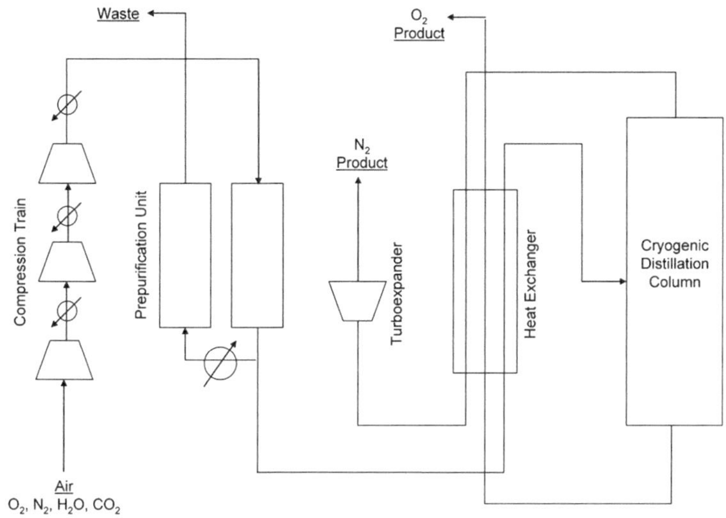

Fig. 6.19 Schematic of a conventional cryogenic air separation plant

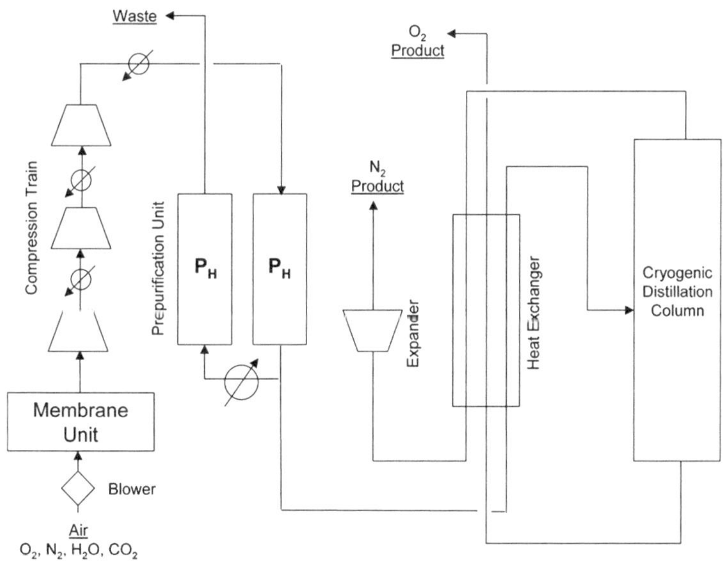

Fig. 6.20 Hybrid membrane/cryogenic distillation system

Wankat and Kostroski [75] concluded that a high recovery of pressurized oxygen is possible in the membrane/cryogenic distillation unit. The resulting hybrid system reduces flow rate in downstream units by 11 %. Agrawal et al. [76] developed a novel membrane/cryogenic hybrid scheme wherein crude argon from a cryogenic air separation unit was fed to an oxygen selective membrane unit to remove a substantial portion of oxygen. The oxygen-enriched permeate from the membrane unit was returned to the crude argon distillation column of the cryogenic air separation process. The non-permeate stream was enriched in argon and could be further purified in a catalytic unit to produce an oxygen-free argon stream.

6.6 Membrane/Distillation Process

Phillips Petroleum Company [77] invented a membrane separation process incorporated in a distillation cycle for efficient recovery of CO_2 from a stream containing natural gas along with carbon dioxide. In this process methane and carbon dioxide were separated from a feed stream in a first distillation to produce a stream containing essentially methane and carbon dioxide. The stream was substantially free from ethane and higher molecular weight hydrocarbons and was subjected to further distillation to produce a carbon dioxide rich product stream and a process stream enriched in methane. The methane-enriched process stream was then passed to a membrane separation unit for separating methane and carbon dioxide to obtain high purity methane.

6.7 Membrane Contactor

Membrane contactors have become the preferred technology for controlled removal of oxygen and carbon dioxide from water for corrosion control. Membrane contactors are devices that allow a gaseous phase and a liquid phase to come into direct contact with each other, for the purpose of mass transfer between the phases, without dispersing one phase into the other. There are two types of membrane contactors. One is to remove gas from liquid (mostly from water). In the other gas is absorbed preferentially through membrane pores. The latter process can be used for gas separation.

For removal of dissolved gases from an aqueous stream, membrane contactors are operated with the aqueous fluid flow on one side of a hydrophobic membrane and a sweep gas and/or a vacuum applied to other side. Since the microporous membrane is hydrophobic, the membrane will not allow liquid water to pass through the pore into the gas side of the membrane. This type of membrane contactors are usually used for the removal of oxygen and carbon dioxide from water for corrosion control.

Usually, large scrubbing towers are used where one of the phases is dispersed in the other either as a drop, bubble, or a thin film. In membrane contactors, gas and liquid absorbent solution are contacted via porous membranes. The principle for the separation of gases based on the absorption of a particular gas in the liquid phase in a hollow-fiber membrane is shown in Fig. 6.21. By adjusting the partial pressure of the gas with the water, gases can be selectively removed.

The essential elements in a membrane absorber are porosity, water repellence, and a polymeric membrane. The gas remains separated from the liquid absorbent as a

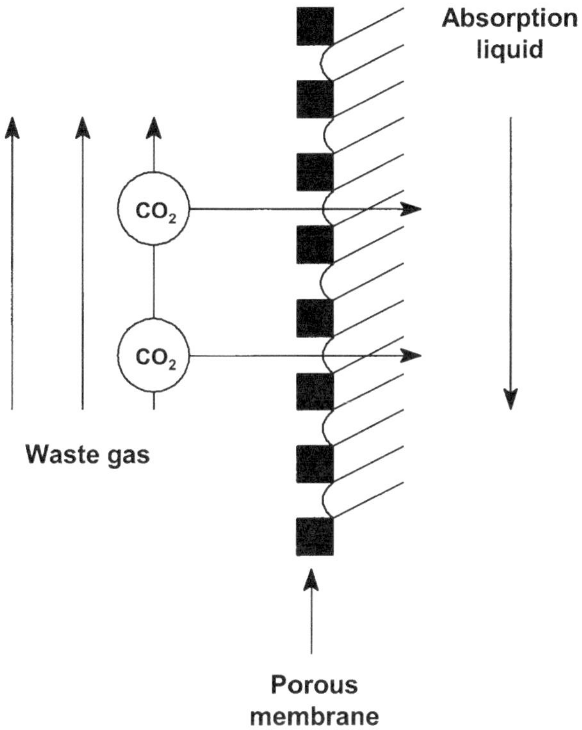

Fig. 6.21 Membrane gas absorption principle

result of the hydrophobicity of the membrane. In a membrane contactor the membrane acts as an interface between the feed gas and the absorption liquid. For example, in the case of CO_2–CH_4 separation, CO_2 diffuses from the feed gas side through the membrane and is then absorbed in the selective absorption liquid. The loaded liquid circulates from the absorber to the desorber, which can be a traditional stripper or a second membrane contactor in which desorption of CO_2 occurs. The selectivity of the process is not only determined by the absorption liquid, but also the membrane plays a significant role and contributes to the selectivity, depending on whether selective or non-selective membranes are used. Figure 6.22 shows the schematic representation of a membrane contactor for the separation of CO_2 and CH_4.

The laboratory scale model contactors may be classified into two categories:

1. Absorbers for which the fluid dynamics of the liquid phase are well understood.
2. Absorbers which reproduce, on a laboratory scale, the characteristics of industrial absorbers.

The first category model contactors are used to determine the reaction kinetics and physical properties of gas liquid systems and the second category contactors are usually used to simulate the industrial contactors. The details on these types of contactors are discussed by Danckwerts [78]. The Sirkar group [79] employed a porous hydrophobic membrane absorber/contactor where the gas and the absorbent

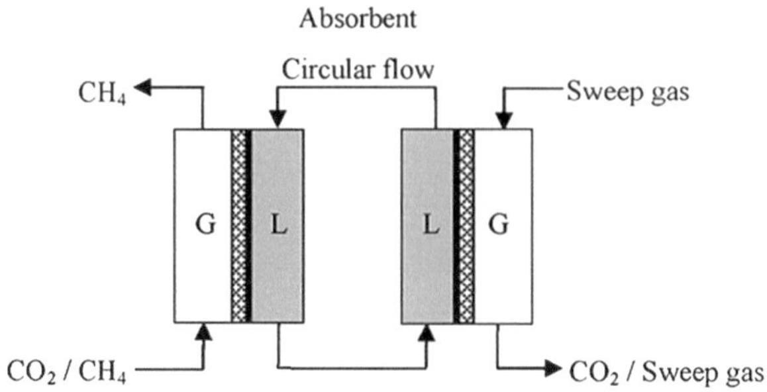

Fig. 6.22 Schematic representation of a membrane contactor for the separation of CO_2 and CH_4

solution flow on two sides of a porous gas-filled membrane without being dispersed as bubbles, thin films, drops, etc. As long as the correct phase pressure difference is maintained, the membrane pores do not become wetted and neither the gas nor the liquid is dispersed into the other phases. Hollow fibers were used to provide a high contacting surface area per unit equipment volume. The shell-side cross flow yielded a high liquid-side mass transfer coefficient.

The components to be removed from the gas stream diffuse through the gas filled pores of the membrane. On the other side of the membrane they will be absorbed into the absorption liquid, such as aqueous solutions of sodium hydroxide, sodium carbonate, monoethanolamine (MEA), or diethanolamine. The uses of a membrane absorber have the following advantages [80]:

1. Gas and liquid flow are independent resulting in avoidance of problems encountered in packed/tray columns such as flooding, foaming, channeling and entrainment.
2. There is no need to have a wash section after the absorber to recover absorption liquid, which is carried over.
3. Operation is not influenced by the orientation of the absorber, which is important in off-shore application and zero gravity applications.
4. Equipment will be compact through the use of hollow-fiber membranes.

There are disadvantage to the membrane absorber, including the additional resistance for the gas permeation through the membrane.

In this process the choice of a suitable combination of absorption liquid and membrane is very important. The hydrophobicity and pore characteristics of the membrane material should be such that the pores remain gas-filled at an excess pressure of preferably more than 1 bar on the liquid side. The overall flow sheet of a membrane gas absorption process is shown in Fig. 6.23.

Feron and Jansen [80] discussed membrane gas absorption (MGA) using dedicated absorption liquids (CORAL) in conjunction with porous polypropylene hollow-fiber membranes for carbon dioxide removal from a CO_2–air mixture. The impact of carbon dioxide partial pressure, liquid loading, and liquid temperature on the carbon dioxide membrane flux was discussed. The novel absorption liquids showed an excellent performance in terms of system stability and mass transfer.

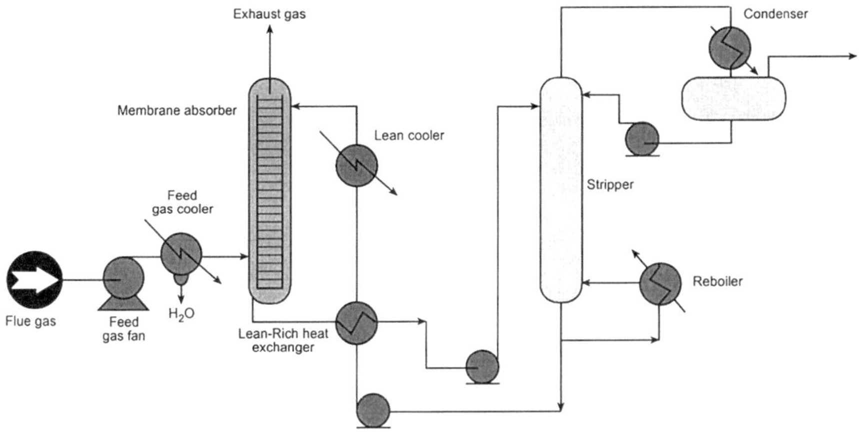

Fig. 6.23 Flow sheet of the membrane gas absorption process

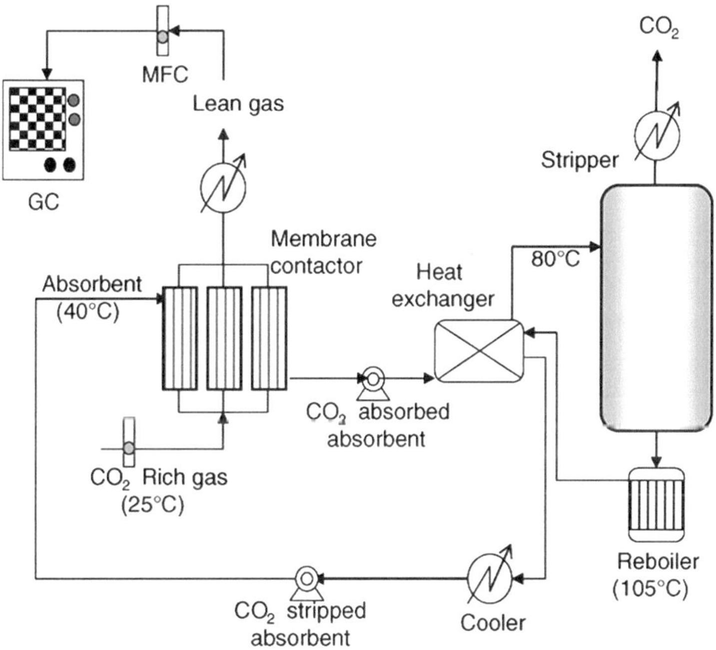

Fig. 6.24 Schematic illustration of membrane contactor hybrid system for CO_2 recovery

The CO_2 membrane mass flux decreased as the cyclic liquid loading was increased. The CO_2 mass flux increased with the increase in temperature from 25 to 40 °C.

Yeon et al. described a pilot-scale membrane contactor hybrid process to recover CO_2 from flue gas [81]. Figure 6.24 shows the schematic of the membrane contactor hybrid system for CO_2 recovery. A porous PVDF hollow-fiber module was used as membrane contactor and its performance was compared with a conventional packed column. Monoethanolamine and triethanolamine solutions were used as the absorbent. The CO_2 was recovered with steam from the absorbent in the thermal stripping tower and the stripped absorbent was recycled. The membrane contactor increased the available gas–liquid contacting area and mass transfer coefficient, and thus, the hybrid process showed a higher CO_2 removal efficiency than the conventional absorption tower. The PVDF module with an asymmetric pore structure showed a stable gas–liquid interface, and the CO_2 absorption rate per unit volume of the membrane contactor was 2.7 times higher than that of the packed column. In addition, the membrane contactor hybrid process was successfully operated for 80 h maintaining a greater CO_2 removal efficiency of above 90 %.

For the evaluation of membrane contactors during polluting gas removal, a hollow-fiber membrane absorber is used. The component to be removed from the gas stream flowing along one side of the membrane is forced by the concentration difference to diffuse through the membrane pores and to be absorbed into the liquid on the other side of the membrane.

Compared to the conventional technologies for CO_2 capture, one of the membrane-based technologies for gas separation—a gas–liquid membrane contactor—offers two major advantages:

1. Hollow-fiber contactors are about 30 times more efficient for gas absorption than conventional equipment.
2. Membrane contactors may reduce the size of the absorber and stripper units by 65 %.

Boucif et al. [82] developed a device for mathematical and numerical investigations on gas–liquid absorption of CO_2 in monoethanolamine solutions in hollow-fiber membrane contactors. The reactive absorption mechanism was built based on momentum and mass transport conservation laws in all three compartments involved in the process—i.e., the gas phase, the membrane barrier, and the liquid phase. The outlet gas and liquid concentrations, the reactive absorption flux, and the gas removal efficiencies were parametrically simulated with operational parameters such as gas flow rate, fresh inlet amine concentrations, and fiber geometrical characteristics.

Castro-Domínguez et al. [83] reported the implementation of perfluorotributylamine (PFTBA) (Fig. 6.24) imbued in porous alumina tubes as a supported liquid membrane to carry out the separation of O_2 and N_2. A conventional ionic liquid membrane [emin][BF_4] (Fig. 6.25) gave maximum O_2/N_2 selectivity of around 6 with an O_2 permeance of 3.4×10^{-12} mol m^{-2} s^{-1} Pa^{-1}. The permeance of the PFTBA

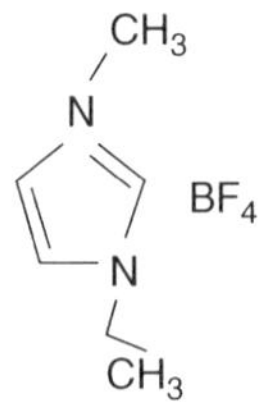

Fig. 6.25 Structure of ionic liquid [emin][BF_4]

membrane correlated with the gas molecular size as $H_2 > O_2 > N_2$, suggesting that the liquid forms pockets for accommodating these gases. Although the PFTBA SLM presented an excellent separation performance, the poor stability of the membrane is a serious issue that requires significant improvements.

Younas et al. [84] used hollow-fiber membrane contactors (HFMCs) for the recovery of aroma compounds from an aqueous feed phase with the help of a conventional solvent. Membrane contactors are typically fabricated with hydrophobic hollow-fiber microporous membranes. As mentioned earlier, since the membranes are hydrophobic and have small pores, water cannot easily pass through the pores. The pressure required to force water to enter the pore can be calculated by the Young-Laplace equation modified for use with hydrophobic membranes [85, 86]. The pressure is often called the liquid entry pressure of water (Eq. (6.1))

$$P = -2\sigma \cos\theta / r \tag{6.1}$$

where P is the liquid entry pressure of water, θ is the contact angle, σ is the surface tension of water, and r is the radius of pore in microporous membrane.

To explain the contactor's ability to remove dissolved gas from water, it is important to discuss the driving force or mass transfer. Henry's law states that the amount of gas that will dissolve into water at equilibrium is proportional to its partial pressure in the gas phase when in contact with water

$$p = Hx \tag{6.2}$$

where p is the gas partial pressure, H is the Henry's law coefficient, a function of water temperature, and x is the concentration of dissolved gas at equilibrium.

Membrane contactors give any wanted shape of fluid–fluid interface in contrast to conventional separation equipment where the shape of the fluid/fluid contact is an accident of nature [87]. There are different membrane configurations like hollow-fiber, flat-sheet, rotating annular and spiral-wound.

By careful control of the pressure difference between the fluids, one is immobilized in the pores of the membrane so that fluid/fluid interface is located at the mouth of each pore. This approach offers a number of important advantages over conventional dispersed phase contactors, including absence of emulsions, no flooding at high flow rates, no unloading at low flow rates, no density difference between fluid required, and surprisingly high interfacial area. The features of the hollow-fiber membrane contactor are very suitable as a gas–liquid model contactor and offer numerous advantages over the conventional model contactors. In the last decade, hollow-fiber membrane contactors have received growing attention in a variety of fields, such as liquid–liquid extraction, gas absorption and stripping, osmotic distillation, and waste water treatment.

Dindore et al. [88] used hollow-fiber membrane contactors as a gas–liquid model contactor for the determination of physical and kinetic parameters for gas–liquid systems. They concluded that a hollow-fiber membrane contactor can be successfully used as a model contactor for the determination of various gas–liquid physicochemical properties.

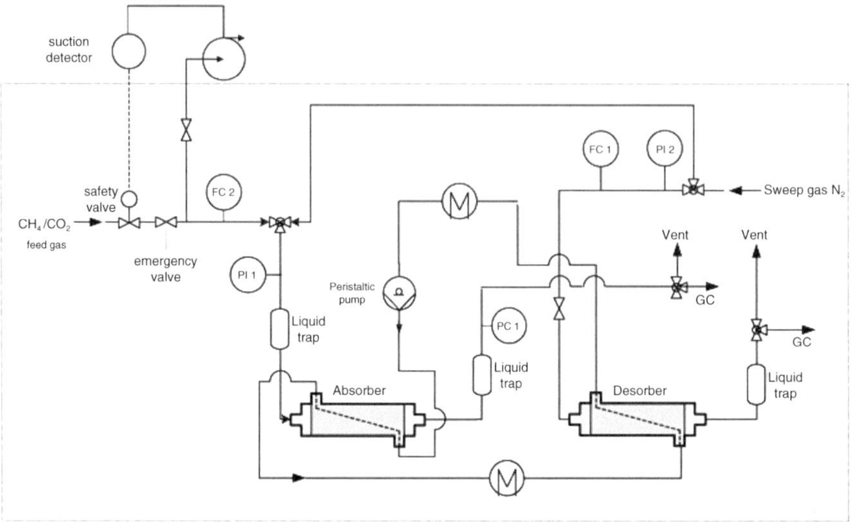

Fig. 6.26 Schematic representation of the experimental membrane contactor setup used for CO_2–CH_4 separation

The early development of membrane contactors was aimed at the separation of CO_2. Simons et al. [89] used both commercially available porous PP hollow-fiber membranes and asymmetric PPO hollow-fiber membranes as membrane contactors for the separation of CO_2 and CH_4 (CO_2–CH_4 20/80 % vol. %). Monoethanolamine (MEA, 10 wt% aqueous solution) was used as an absorption liquid. Figure 6.26 shows the experimental membrane contactor setup used for CO_2–CH_4 separation. The influence of the different parameters on productivity was evaluated.

The performance of the PP membranes outperforms the performance of the PPO membranes in terms of productivity and selectivity. However, the PP hollow fibers were extremely sensitive to only small variations in the feed pressure, resulting in severe performance loss. Besides this, extremely high liquid losses were observed for the PP fibers, especially at elevated temperatures.

Figure 6.27 shows the CO_2 permeance (a) and CO_2–CH_4 selectivity (b) as a function of the absorbent liquid flow rate for the two different membranes investigated at (ΔT (temperature difference between absorption and desorption) = 35 °C; T_{abs} (temperature of the absorber) = 29 °C; Δp = 0.2 bar) [89].

Wetting is the problem for an efficient absorption process with membrane contactors. If the liquid absorbent is water or aqueous solutions with inorganic solutes, the liquid has a high surface tension and usually cannot wet the common hydrophobic membranes such as polypropylene (PP) and polytetrafluoroethylene (PTFE). In the case where liquid contains organic compounds instead of inorganic material, even at low temperatures, its surface tension will drop rapidly. When the concentration of

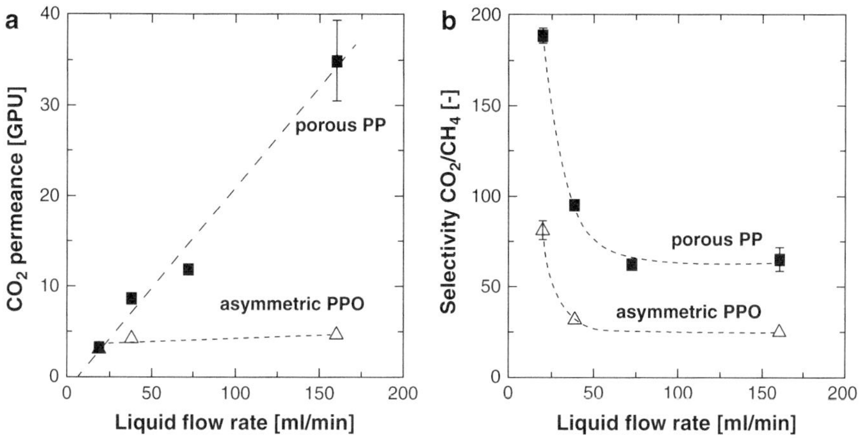

Fig. 6.27 CO_2 permeance (**a**) and CO_2–CH_4 selectivity (**b**) as a function of the absorbent liquid flow rate for the two different membranes investigated

the organic compounds exceeds a critical value, the contact angle between the liquid and membrane surface will decrease to a value less than 90°, and the result will be wetting the membrane surface. For microporous membranes the water can penetrate into the pores. The concentration of the organic component, at which the liquid penetrates into the membrane, is called "the maximum allowable concentration" (MAC). This concentration can be determined by the penetration drop method, in which a liquid drop, only under the gravitational force, is brought into contact with the membrane [90].

Rongwong et al. [91], in order to understand the role of absorbents on membrane wetting, used three absorbent solutions: (1) monoethanolamine (MEA primary amine), (2) diethanolamine (DEA secondary amine), and (3) 2-amino-2-methyl-1-propanol (AMP, sterically hindered amine) as absorbent solutions, and used PVDF hollow fibers. The results revealed that the use of MEA solution absorbent gave the highest CO_2 flux. The overall mass transfer coefficients obtained from the experiments also showed the same trend as CO_2 flux, i.e., the values were in the following order: MEA>AMP>DEA.

Rajabzadeh et al. [92, 93] studied seven kinds of asymmetric PVDF hollow-fiber membrane contactors with considerably different structures at the outer surfaces. These membranes were prepared by thermally induced phase separation (TIPS) and applied for CO_2 absorption as gas–liquid membrane contactors. The authors also studied propylene absorption by using these membranes in a similar way [92, 93]. A commercial microporous poly(tetrafluoroethylene) (PTFE) hollow-fiber membrane was also used as a highly hydrophobic membrane. MEA solutions were used as absorbents. A mathematical model for pure propylene and pure CO_2 absorption in a membrane contactor, assuming the membrane resistance was negligibly small and the total areas were effective for gas absorption, was proposed to simulate propylene and CO_2 absorption rates. It was the first experimental and theoretical study on propylene absorption in membrane contactors [94].

Kartohardjono et al. [94] used diethanolamine (DEA) solution to absorb CO_2 from the gas flow through a hollow-fiber membrane (PP) contactor. DEA solution reduced the mass transfer resistance in the liquid phase, and on the other side, the large contact area of the membrane surface covered the disadvantage of membrane contactors and additional mass transfer resistance in the membrane phase. Mass transfer coefficients and CO_2 fluxes using DEA solution can achieve 28,000 and 7.6 million times greater separation than using water as a solvent, respectively.

Faiz et al. [95] demonstrated that the use of ceramic hollow-fiber membrane contactors would extend operating periods without the need to replace the membrane, as is commonly found with polymeric membranes. These longer operating periods could easily encounter structural damages due to the weak chemical stability with most solvents. Marjani and Shirazin [96] developed a model to describe the transport of SO_2 through the membrane contactor. The simulation results indicated that the removal of SO_2 increased when either increasing the liquid velocity or decreasing the gas velocity in the membrane contactor. The model predictions also revealed that main mass transfer resistances for transport of SO_2 were located in the membrane and gas phase. A ceramic membrane has the highest resistance because of its high tortuosity.

A mathematical model based on the effective permeability of the gaseous mixtures was used to assess the performance of both porous and nonporous hollow-fiber membranes, and a correlation for the shell-side mass transfer was developed by Al-Saffer et al. [97]. It was concluded that overall mass transfer in the microporous and non-porous membranes is dominated by the liquid film coefficient and a combination of liquid film and membrane resistance.

Carbon dioxide (CO_2) is the main component of greenhouse gases in the atmosphere. An effective and economical technology for CO_2 capture is necessary to help offset climate change. Conventional gas absorption processes for removal of CO_2, including chemical absorption by reactive absorbents, are normally carried out by packed towers, spray towers, venture scrubbers, and bubble columns [91]. Qi and Clussler [98] used microporous polypropylene membranes for CO_2 absorption in a NaOH solution. Karoor and Sirkar [99] conducted comprehensive studies on the gas separation of CO_2, SO_2, CO_2–N_2, and CO_2–air mixtures using distilled water in a microporous hydrophobic hollow-fiber device. They utilized microporous polypropylene hollow fibers as contactors.

Zhikang et al. [100] demonstrated that polypropylene hollow-fiber membrane contactors can be effectively used to separate and capture CO_2 from gas mixtures. The separation performance of this membrane technology can be adjusted flexibly by changing the kind, concentration, and flow of the absorbent solution. Keshavarz et al. [101] developed a steady-state model for a microporous hollow-fiber membrane contactor operated under partially wetted conditions, accompanied by a chemical reaction, to analyze CO_2 absorption into aqueous solution of diethanolamine (DEA). The proposed diffusion-reaction model contained reversible chemical reactions in the liquid bulk as well as the wetted parts of the membrane pores. A numerical scheme was applied to solve the simultaneous nonlinear mathematical expressions, and the results were validated with experimental data in the literature. The results of the model and proposed numerical scheme showed that the membrane wetting, even in very low fractions, could decrease the absorption flux significantly. The wetting fraction of membrane was predicted both with and without consideration of chemical reactions inside the wetted pores. The results indicated that the chemical reactions inside the wetted pores, which have been disregarded in the literature, have considerable effects on the prediction of the membrane wetting fraction.

Ghasem et al. [102] reported that hollow fibers prepared with PVDF/triacetin/glycerol (non-solvent additives) can be used to remove CO_2 from a CO_2–CH_4 gas mixture by membrane contactor techniques. The additive glycerol in the dope solution effects the separation of carbon dioxide. It was noticed that as the percentage of glycerol increased in the dope solution the removal rate of CO_2 increased where complete removal of CO_2 is achieved at an equal ratio of gas-to-liquid volumetric flow rate. It was concluded that additions of glycerol to dope solution is efficient to increase the effective surface porosity and pore radius.

Membrane contactor system can be used for gas dehumidification [103]. Usachov et al. [103] developed a pilot active membrane contactor system (AMCS) for air dehumidification. Polydimethylsiloxane based membranes were used in a spiral-wound membrane contactor. The developed contactor system achieves air dehumidification up to $T_{\text{dew point}} = -20°C$.

Bottino et al. [104] studied the performances of polypropylene (PP) and polyethylene (PE) capillary membranes with different module dimensions and configuration (linear and loop). Monoethanolamine (MEA) and NaOH were used as scrubbing solvents. Even at low absorbent concentration, high CO_2 removal efficiency could be reached (over 97 %). On increasing the MEA concentration, the specific flow of transferred CO_2 also increased. At 3 M MEA solution, the efficiency was higher than 25 % with a specific flow of CO_2 7.8×10^{-3} mol/m² s.

Nymeijer et al. [105] fabricated ethylene propylene dieneterpolymer (EPDM) coated composite hollow-fiber membranes and applied them in a bench-scale membrane contactor process for the separation of olefins and paraffins using $AgNO_3$ solutions as absorption liquid. The membrane modules prepared had an effective surface area of 101.8 cm², containing 10 composite hollow-fiber membranes with an 8 μm thick, defect-free EPDM top layer, and were used as absorber and desorber modules. Ethylene productivities were found to be in the range of 2.1×10^{-6} to 6.1×10^{-6} cm³/cm² s cmHg and gas mixture selectivities in the range of 72.5–13.7.

Table 6.8 gives a short review of membrane contactors and their characteristics, as well as the properties of membrane materials and offers an introduction to membrane contactor applications in gas separation [87].

Table 6.9 shows the polymeric and ceramic materials used for membrane contactors for gas separation and the liquids used for absorption.

Table 6.8 Summary of membrane contactors

Membranes	Porous (hydrophobic or hydrophilic), nonporous, or composite
Thicknesses	20–100 μm
Pore sizes	nonporous or 0.05–1.0 μm
Driving force	Concentration or vapor pressure difference
Separation principle	Distribution coefficient
Membrane material	Hydrophobic (polytetrafluoroethylene, polypropylene, silicon rubber), polyolefin
Application	G-L contactors
	– SO_2, CO_2, CO, NO_x from flue gases
	– CO_2 and H_2S from natural gas
	– VOC from off-gas
	– Saturated/unsaturated (ethane/ethylene)
	– LG contactors
	– O_2 removal from water

Table 6.9 Different polymers including ceramic used for the membrane contactors for gas separation and absorption liquids

Polymer/ceramic	Separation/or other	Absorption solution	References
SPEEK	Olefin–paraffin	Water	[63]
PVDF	CO_2 from flue gas	MEA, TEA	[81]
PEO, PBT	Olefin–paraffin	Silver nitrate	[89]
PVDF, PTFE	Propylene	Silver nitrate	[94]
PVDF	CO_2	MEA, DEA, AMP	[93]
PP	CO_2	DEA	[95]
Ceramic	Olefin and paraffin separation	Silver nitrate	[96]
Ceramic	SO_2	*N-N*-dimethylaniline	[97]
PP, PE	CO_2 separation from flue gases	NaOH, MEA	[104]
EPDM	Olefin–paraffin	Silver nitrate	[105]
PDMS	Olefin–paraffin separation	Silver salts	[71]
PAMAM, PMP	CO_2	MEA	[79]
PU	Propylene–propane	Silver salts	[69]
PP	CO_2	MEA, MDEA	[106]
PP	CO_2 from gas mixture	MEA	[107]
PP	CO_2	DEA	[108]

References

1. Brunetti A, Barbieri G, Drioli E (2013) Gas separation—applications. In: Encyclopedia of membrane science and technology. Wiley, Hoboken, NJ
2. Baker RW (2002) Future direction of membrane gas separation technology. Ind Eng Chem Res 41:1303–1411
3. Bernardo P, Drioli E, Golemme G (2009) Membrane gas separation: a review/state of the art. Ind Eng Chem Res 48:4638–4663

4. http://www.mtrinc.com/gas_separation.html
5. Koros WJ, Mahajan R (2000) Pushing the limits on possibilities for large scale gas separation: which strategies? J Membr Sci 175:181–196
6. Pandey P, Chauhan RS (2001) Membranes for gas separation. Prog Polym Sci 26:853–893
7. Meriläinen A, Seppälä A, Kauranen P (2012) Minimizing specific energy consumption of oxygen enrichment in polymeric hollow-fiber membrane modules. App Energy 94:285–294
8. Burdny T, Struchtrup H (2010) Hybrid membrane/cryogenic separation of oxygen from air for use in oxy-fuel process. Energy 35:1884–1897
9. Allen JB (1995) Making oxygen on the moon. In: Popular science. Bonnier Corporation, USA
10. Kluiters SCA (2004) Status review on membrane systems for hydrogen separation: intermediate report for EU project MIGREYD NNE5-2001670
11. Kaldis SP, Kapantaidakis GC, Sakellaropoulos GP (2000) Simulation of multicomponent gas separation in a hollow fiber membrane by orthogonal collocation-hydrogen recovery from refinery gases. J Membrane Sci 173:61–71
12. Ockwig NW, Nenoff TM (2010) Membranes for hydrogen separation. Chem Rev 107:4078–4110
13. Gopalan S (2002) Using ceramic mixed ionic and electronic conductors for gas separation. JOM—J Min Met Mat S 54:26–29
14. DOE: project facts vision 21, June 2000, http://www.doe.gov
15. Meinema HA, Dirrix RWJ, Brinkman HW, Terpstra RA, Jekerle J, Kösters PH (2005) Ceramic membranes for gas separation: recent developments and state of the art. Interceram 54:86–91
16. Grainger D, Hägg MB (2008) The recovery by carbon molecular sieve membranes of hydrogen transmitted in natural gas networks. Int J Hydrogen Energy 33:2379–2388
17. http://www.mtrinc.com/hydrogen_purification_in_refineries.html. Accessed March 2014
18. Henis JMS (1994) Commercial and practical aspects of gas separation membranes. In: Paul DR, Yampol'skii YP (eds) Polymeric gas separation Membranes. CRC Press, Boca Raton, FL
19. w.w.w.medal.airliquide.com/en/hydrogen-membrane-gas-separation/hydrogen-membrane-refineries.html
20. Adams T (2012) II.C.1 membrane separation—bulk amorphous hydrogen purification/separation membranes. FY 2008 annual progress report, DOE hydrogen program, 2012, 113–116
21. Guo H, Zhu G, Hewitt IJ, Qiu S (2009) "Twin copper source" growth of metal-organic framework membranes: $Cu_3(BTC)_2$ with high permeability and selectivity for recycling H_2. J Am Chem Soc 131:1646–1647
22. Phair JW, Badwal SPS (2006) Review of proton conductors for hydrogen separation. Ionics 12:103–115
23. Metz B, Davidson O, de Coninck H, Loos M, Mayer L (eds) (2005) IPCC special report on carbon dioxide capture and storage. Cambridge University Press, Cambridge, UK
24. Mohshim DF, Mukhtar HB, Man Z, Nasir R (2013) Latest development on membrane fabrication for natural gas. J Eng 101746:7
25. Pires JCM, Martins FG, Alvim MCM, Simões M (2011) Recent developments on carbon capture and storage: an overview. Chem Eng Res Des 89:1446–1460
26. Kim S, Lee YM (2012) Thermally rearranged (TR) polymer membranes with nanoengineered cavities tuned for CO_2 separation. J Nanopart Res 14:949–960
27. Merkel TC, Lin H, Wei X, Baker R (2010) Power plant post-combustion carbon dioxide capture: an opportunity for membranes. J Membr Sci 359:126–139
28. Yama K, Kazama S, Yogo K (2008) Development of innovative gas separation membranes through sub-nanoscale materials control. GCEP, Stanford University
29. Scholes CA, Stevens GW, Kentish SE (2012) Membrane gas separation applications in natural gas processing. Fuel 96:15–28
30. Du N, Park HB, Dal-Cin MM, Guiver MD (2012) Advances in high permeability polymeric membrane materials for CO_2 separations. Energy Environ Sci 5:7306–7322
31. Du N, Park HB, Robertson GP, Dal-Cin MM, Visser T, Scoles L, Guiver MD (2011) Polymer nanosieve membranes for CO_2-capture applications. Nat Mater 10:372–375

32. Lee SH, Kim BS, Lee EW, Park YI, Lee JM (2006) The removal of acid gases from crude natural gas by using novel supported liquid membranes. Desalination 200:21–22
33. Park HB, Jung CH, Lee YM, Hill AJ, Pas SJ, Mudie ST, Van Wagner E, Freeman BD, Cookson DJ (2007) Polymers with cavities tuned for fast selective transport of small molecules and ions. Science 318:254–258
34. Park HB, Han SH, Jung CH, Lee YM, Hill AJ (2010) Thermally rearranged (TR) polymer membranes for CO_2 separation. J Membr Sci 359:11–24
35. Choi JI, Jung CH, Han SH, Park HB, Lee YM (2010) Thermally rearranged (TR) poly(benzoxazole-co-pyrrolone) membranes tuned for high gas permeability and selectivity. J Membr Sci 349:358–368
36. Kim S, Han SH, Lee YM (2012) Thermally rearranged (TR) polybenzoxazole hollow fiber membranes for CO_2 capture. J Membr Sci 403–404:169–178
37. Chen CC, Miller JS, Koros WJ (2013) Characterization of thermally cross-linkable hollow fiber membranes for natural gas separation. Ind Eng Chem Res 52:1015–1022
38. Kratochvil AM, Koros WJ (2008) Decarboxylation-induced cross-linking of a polyimide for enhanced CO_2 plasticization resistance. Macromolecules 41:7920–7927
39. Qiu WL, Chen CC, Xu L, Cui L, Paul DR, Koros WJ (2011) Sub-Tg cross-linking of a polyimide membrane for enhanced CO_2 plasticization resistance for natural gas separation. Macromolecules 44:6046–6056
40. Bhide BD, Voskericyan A, Stern SA (1998) Hybrid process for the removal of acid gases from natural gas. J Membr Sci 140:27–49
41. Carreon MA (2011) Novel membranes for efficient CO_2 separation. In: Proceedings of the 22nd national NSF EPSCoR conference. Coeur d'Alene Idaho, 27 Oct 2011
42. Aitken CL, Koros WJ, Paul DR (1992) Gas transport properties of biphenol polysulfones. Macromolecules 25:3651–3658
43. Cnop T, Dortmundt D, Schott M (2007) Continued development of gas separation membranes for highly sour service. Sour oil and gas advanced technology conference
44. Kulkarni SS, Funk EW, Li NN (1983) Membrane separation processes for acid gases. AICHE summer national meeting, Denver, 28–31 Aug 1983
45. Mazur WH, Chan MC (1982) Membranes for natural gas sweetening and CO_2 enrichment. Chem Eng Prog 78:38–43
46. Cooley ET (1984) Spiral wound membranes carbon dioxide removal process for the well head. AIChE spring national meeting, California, 21 May 1984
47. Grey NR, Mazur WH (1984) Membrane separation of carbon dioxide and hydrogen sulfide from natural gas-field experience. American Institute of Chemical Engineers, Spring national meeting, California, 20–23 May 1984
48. Yoshino M, Nakamura S, Okamoto KH, K-I TN, Kusuki Y (2003) Olefin/paraffin separation performance of asymmetric hollow fiber membranes of 6FDA/BPDA-DDBT system. J Membr Sci 212:13–27
49. Lucadamo GA (1986) Membrane-aided distillation for carbon dioxide and hydrocarbon separation. US patent 4602477, 29 Jul 1986
50. Kujawski W (2000) Application of pervaporation and vapor permeation in environmental protection. Pol J Environ Stud 9:13–26
51. Baker RW, Kaschemekat JG, Wijmans JH (1998) The design of membrane vapor-gas separation systems. J Membr Sci 151:55–62
52. Wang H, Lin YS (2012) Effects of water vapor on gas permeation and separation properties of MFI zeolite membranes at high temperatures. AIChE J 58:153–162
53. Baker WJW, Kapteijn F, Poppe J, Moulijn JA (1996) Permeation characteristics of a metal-supported silicate-1 zeolite membrane. J Membr Sci 117:57–78
54. Noack M, Kolsch P, Caro J, Schneider M, Toussaint P, Sieber I (2000) MFI membranes of different Si/Al ratios for pervaporation and steam permeation. Micro Meso Mater 35–36:253–265
55. Yeow ML, Field RW, Li K, Teo WK (2002) Preparation of divinyl-PDMS/PVDF composite hollow fiber membrane for BTX removal. J Membr Sci 203:137–143

56. U.S. Department of Energy, Membrane technology workshop summary report Nov. 2012, Workshop held on July 24, 2012, Rosemont, Illinois
57. U.S. Department of Energy, Energy efficiency & renewable energy. Advanced membrane technology for hydrocarbon separations: new membrane technology for natural gas dehydration promises improved separation efficiency, May 2009, Rosemont, Illinois
58. Feng H, Zhang H, Xu L (2007) Polymeric membranes for natural gas conditioning. Energy Source Part A 29:1269–1278
59. Faiz R, Li K (2012) Polymer membranes for light olefin/paraffin separation. Desalination 287:62–97
60. Semenova SI (2004) Polymer membranes for hydrocarbon separation and removal. J Membr Sci 231:189–207
61. Ito A, Hwang ST (1989) Permeation of propane and propylene through cellulosic poymer membranes. J Appl Polym Sci 38:483–490
62. Chan SS, Wang R, Chung TS, Liu Y (2002) C_2 and C_3 hydrocarbon separations in poly(1,5-naphthalene-2-2'-bis(3,4-phthalic) hexafluoropropane) diimide (6FDA-1,5NDA) dense membranes. J Membr Sci 210:55–64
63. Nymeijer K, Visser T, Assen R, Wessling M (2004) Super selective membranes in gas-liquid membrane contactors for olefin/paraffin separation. J Membr Sci 232:107–114
64. Suzuki Y, Nishide H, Tsuchida E (2000) Membranes of the picket fence cobalt porphyrin complexed with poly(vinylimidazole and pyridine): selective optical response to oxygen. Macromolecules 33:2530–2534
65. Nishide H, Tsukahara Y, Tsuchida E (1998) Highly selective oxygen permeation through a poly(vinylidene dichloride)–cobalt porphyrin membrane: hopping transport of oxygen via the fixed cobalt porphyrin carrier. J Phys Chem B 102:8766–8770
66. Sungpet A, Way JD, Koval CA, Eberhart ME (2001) Silver doped Nafion-poly(pyrrole) membranes for facilitated permeation of liquid-phase olefins. J Membr Sci 189:271–279
67. Müller J, Peinemann K-V, Müller J (2002) Development of facilitated transport membranes for the separation of olefins from gas streams. Desalination 145:339–345
68. Kim JH, Won J, Kang YS (2004) Olefin-induced dissolution of silver salts physically dispersed in inert polymers and their applications to olefin/paraffin separation. J Membr Sci 241:403–407
69. Pollo LD, Durate LT, Anacleo M, Habert AC, Borges CP (2012) Polymeric membranes containing silver salts for propylene/propane separation. Braz J Chem Eng 29:307–314
70. Chilukuri P, Rademakers K, Nymeijer K, vab der Ham L, Van dewn Berg H (2007) Propylene/propane separation with a gas/liquid membrane contactor using a silver salt solution. Ind Eng Chem Res 46:8701–8709
71. Kang SW, Char K, Kang YS (2008) Novel application of partially charged silver nanoparticles for facilitated transport in olefin/paraffin separation membranes. Chem Mater 20:1308–1311
72. Feng X, Pan CY, Ivory J, Ghosh D (1998) Integrated membrane/adsorption process for gas separation. Chem Eng Sci 53:1689–1698
73. Feng X, Pan CY, Ivory J (2000) Pressure swing permeation: novel process for gas separation by membranes. AIChE J 46:724–733
74. Esteves IAAC, Mota JPB (2002) Simulation of a new hybrid membrane/pressure swing adsorption process for gas separation. Desalination 148:275–280
75. Wankat PC, Kostroski KP (2011) Hybrid membrane-cryogenic distillation air separation process for oxygen production. Sep Sci Technol 46:1439–1545
76. Agrawal R, Auvil SR, Choe JS, Woodward DW (1990) Membrane/cryogenic hybrid scheme for argon production from air. Gas Sep Purif 4:75–80
77. Burkinshaw JR, Waldo RA (1998) Distillation plus membrane processing of gas streams. US patent 4936887 A
78. Danckwerts PV (1970) Gas-liquid reactions. McGraw Hill, New York
79. Kosaraju P, Kovvali AS, Korikov A, Sirkar KK (2005) Hollow fiber membrane contactor based CO_2 absorption-stripping using novel solvents and membranes. Ind Eng Chem Res 44:1250–1258

80. Feron PHM, Jansen AE (2002) CO_2 separation with polyolefin membrane contactors and dedicated absorption liquids: performance and prospects. Sep Purif Technol 27:231–242
81. Yeon SH, Lee KS, Sea B, Park YI, Lee KH (2005) Application of pilot-scale membrane contactor hybrid system for removal of carbon dioxide from flue gas. J Membr Sci 257:156–160
82. Boucif N, Corriou JP, Roizard D, Favre E (2011) Carbon dioxide absorption by monoethanolamine in hollow fiber membrane contactors: a parametric investigation. AIChE J 58:2843–2855
83. Castro-Domínguez B, Leelachaikul P, Takagaki A, Sugawara T, Kikuchi R, Oyama ST (2013) Perfluorocarbon-based supported liquid membranes for O_2/N_2 separation. Sep Purif Technol 116:19–24
84. Younas M, Druon Bocquet S, Sanchez J (2008) Extraction of aroma compounds in HFMC: dynamic modeling and simulation. J Membr Sci 323:386–394
85. Goel V, Mauro AA, DiLeo AJ, Meiser AP, Pluskai M (1992) Deadend microfiltration: applications, design and cost. In: Ho WSW, Sirkar KK (eds) Membrane handbook. Van Vostrand Reinhold, New York
86. Kim B, Harriot P (1987) Critical entry pressure for liquids in hydrophobic membranes. J Colloid Interf Sci 115:1910–1916
87. Stanojević M, Lazarević B, Radić D (2003) Review of membrane contactors designs and applications of different modules in industry. FME Transactions 31:91–98
88. Dindore VY, Brilman DWF, Versteg GF (2005) Hollow fiber membrane contactor as a gas-liquid model contactor. Chem Eng J 60:467–479
89. Simons K, Nijmeijer K, Wessling M (2009) Gas-liquid membrane contactors for CO_2 removal. J Membr Sci 340:214–220
90. Li JL, Chen BH (2005) Review of CO_2 absorption using chemical solvents in hollow fiber membrane contactors. Sep Purif Technol 41:109–122
91. Rongwong W, Jiraratananon R, Atchariyawut S (2009) Experimental study of membrane wetting in gas-liquid membrane contacting process for absorption by single and mixed absorbents. Sep Purif Technol 69:118–125
92. Rajabzadeh S, Yoshimoto S, Teramoto M, Al-Marzouqi M, Matsuyama H (2009) CO_2 absorption by using PVDF hollow fiber membranes contactors with various membrane structure. Sep Purif Technol 69:210–220
93. Rajabzadeh S, Teramoto M, Al-Marzouqi M, Kamio E, Ohmukai Y, Maruyama T, Matsuyama H (2010) Experimental and theoretical study on propylene absorption by using PVDF hollow fiber membrane contactors with various membrane structures. J Membr Sci 346:86–97
94. Kartohardjono S, Nata PA, Prasetio E, Yuliusman (2009) Performance of hollow fiber membrane gas-liquid contactors to absorb CO_2 using diethanolamine (DEA) as a solvent. Makara Teknolgi 13:86–90
95. Faiz R, Fallanza M, Ortiz I, Li K (2013) Separation of olefin/paraffin gas mixtures using ceramic hollow fiber membrane contactors. Ind Eng Chem Res 52:7918–7929
96. Marjani A, Shirazian S (2011) CFD simulation of SO_2 removal from gas mixtures using ceramic membranes. World Acad Sci Eng Technol 58:868–871
97. Al-Saffar HB, Ozturk B, Hughes R (1997) A comparison of porous and non porous gas-liquid membrane contactors for gas separation. Chem Eng Res Des 75:685–692
98. Qi Z, Clussler EL (1985) Microporous hollow fibers for gas absorption: 1. Mass transfer in the liquid. J Membr Sci 23:321–332
99. Karoor S, Sirkar KK (1993) Gas absorption studies in microporous hollow fiber membrane modules. Ind Eng Chem Res 32:674–684
100. Zhikang X., Jianli W, Wei C, Youyi X (2001) Separation and fixation of carbon dioxide using polymeric membrane contactor. First national conference on carbon sequestration, U.S. Department of Energy, NETL, 14–17
101. Keshavarz P, Faithikalajahi J, Ayatollahi S (2008) Analysis of CO_2 separation of a partially wetted hollow fiber membrane contactor. J Hazard Mater 152:1237–1247

102. Ghasem N, Al-Marzouqi M (2012) Effect of nonsolvent additive on the effectiveness of polyvinylidene fluoride membrane fabricated with thermal induced phase separation method for carbon dioxide absorption. J Chem Eng Process Technol 3:1, http://dx.doi.org/4172/2157-7048.1000125
103. Usachov V, Laguntsov N, Okunev A, Teplyakov V, Glukhov S (2003) Experimental study of the membrane contactor for gas dehumidification. Ars Separatoria Acta 2:36–46
104. Bottino A, Capannelli G, Comite A, Firpo R, Felice RD, Pinacci P (2006) Separation of carbon dioxide from the flue gases using membrane contactors. Desalination 200:609–611
105. Nymeijer DC, Visser T, Assen R, Wessling M (2004) Composite hollow fiber gas-liquid membrane contactors for olefin/paraffin separation. Sep Purif Technol 37:209–220
106. Yan SP, Fang MX, Zhang WF, Wang SY, Xu ZK, Luo ZY, Cen KF (2007) Experimental study on the separation of CO_2 from flue gas using hollow fiber membrane contactors without wetting. Fuel Process Technol 88:501–511
107. Zhu B, Chen W, Wang J, Xu Y, Xu Z (2003) Separation of carbon dioxide from gas mixture by membrane contactor. Huan Jing Ke Xue 24:34–38
108. Wang R, Zhang HY, Feron PHM, Liang DT (2005) Influence of membrane wetting on CO_2 capture in microporous hollow fiber membrane contactors. Sep Purif Technol 46:33–40

Chapter 7
Characterization of Membranes

7.1 Introduction

Gas transport through polymers is an area of growing interest as materials with unique transport properties continue to find uses in new, specialized applications ranging from extended life tennis balls [1] to natural gas systems. Membrane users (manufacturers and membrane scientists) require knowledge of membrane characteristics in order to choose an appropriate one for application in different processes. Understanding these characteristics will help to determine membrane casting conditions, control membrane quality, and develop membrane transport. Membrane mechanisms and characteristics include surface morphology, and various chemical and physical properties. An ideal characterization method should be non-destructive, accurate, repeatable, and fast and should maximize data. Many methods of characterization have been devised, which can be classified according to the physical mechanisms they exploit.

Detailed and systematic studies have significantly enhanced our understanding of membranes and have allowed for control of transport phenomena in polymer membranes of various types. Knowledge of the structural characteristics of membranes is necessary to control their quality, characteristics, and transport mechanisms. Nowadays, many researchers are working on membrane development for gas separation, and all of them attempt to find the cause and effect relationship between membrane fabrication, membrane morphology, and membrane performance. The ultimate goal of the research is to be able to choose fabrication conditions to achieve membrane morphology that are designed to enable the desired separation performance. In this regard, membrane morphology characterization is one of the indispensable components of membrane research. Different approaches can be used to characterize the membranes, and they can be classified according to structure and permeation parameters. There are various well-established methods to characterize membranes, which are discussed in this chapter.

A.F. Ismail et al., *Gas Separation Membranes*,
DOI 10.1007/978-3-319-01095-3_7

7.2 Mass Transport

The ability to relate polymer molecular structures to gas transport properties is crucial in any attempt to rationally design materials for specific permeability applications such as gas barriers. Historically, the availability of experimental permeability data has been limited mostly to common/commercial polymers and this information has demonstrated that gas transport rates of polymers vary by many orders of magnitude [2]. Gas permeability is the ability of a barrier material to allow gases (O_2, N_2, CO_2, etc.) to permeate through at a specific time. Gas permeability may vary with temperature, humidity and pressure. A variety of techniques have been used to measure gas permeation through polymer membranes.

Characterization of gas separation membranes involves evaluation of three fundamental transport parameters—(1) permeability, (2) diffusion, and (3) solubility coefficient. In general, there are two methods that are employed for membrane characterization:

1. Constant pressure (CP): In this system the permeate side of the membrane is open to atmosphere, or swept with a inert/carrier gas at atmospheric pressure, and thus, gas permeation tests are performed at a constant transmembrane pressure.
2. Constant volume (CV): In this system the permeate side of the membrane is initially in a vacuum; as the gas permeates through the membrane, the pressure at the permeate side increases. Consequently, the gas permeation tests in the CV system are executed at a variable transmembrane pressure.

The membrane tested in a CP system is exposed to atmosphere, unlike constant volume systems (CV), where permeate flow rate is measured via a flow meter device. Figure 7.1 presents a diagram of a typical CP system used for measuring the permeation gas A through a membrane. The gas permeation rate is evaluated based on gas

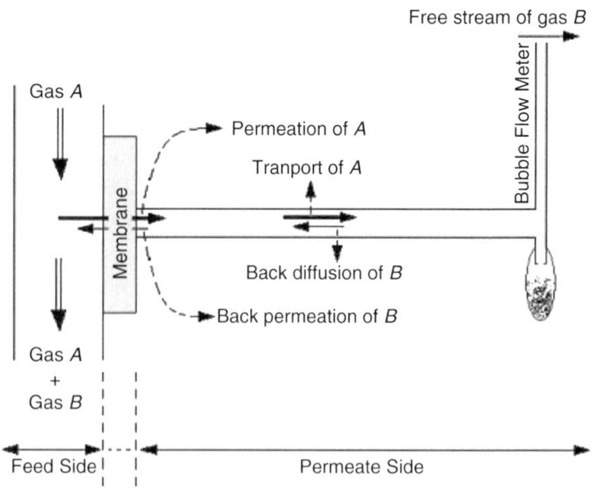

Fig. 7.1 Measurement of the permeation rate of gas through a membrane in a constant pressure system

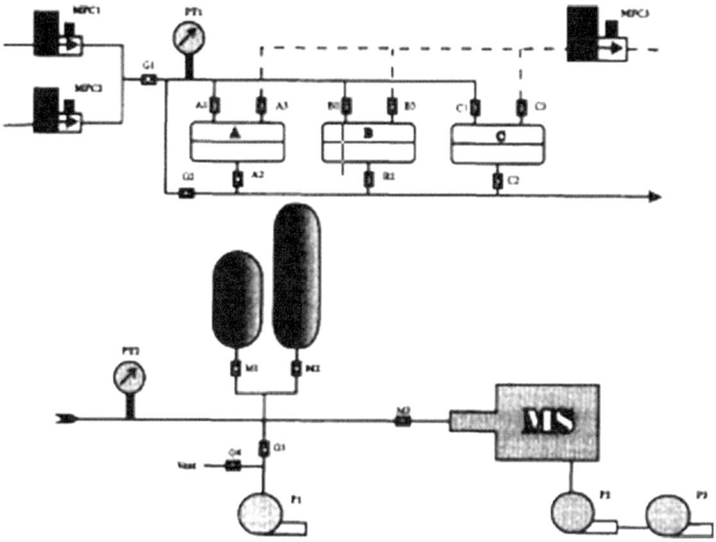

Fig. 7.2 Schematic diagram of the constant volume separation system for measurement of low permeation rates with high accuracy
Components of the system; MFC1, MFC2, MFC3—mass flow controllers; PT1, PT2—pressure transducers; P1, P2, P3; pumps; MS—mass spectrometer; A, B, C—separation cells; A1, A2, B1, B2, C1, C2, G1, G2, G3, G4—pneumatic valves; A3, B3, C3—check valves; M1, M2, M3—manual valves.

flow rate measured by a soap bubble flow meter, which is attached to the permeate side of the membrane. The soap bubble is introduced at the entrance of a calibrated column and is driven upwards by the flowing gas. Because of the negligible weight of the soap bubble, the speed of the bubble through the calibrated column can be correlated to the volumetric flow rate of the permeate gas. Lashkari et al. [3] evaluated the errors involved in a CV system by detailed analysis. In the system depicted in Fig. 7.1, the membrane is open to atmosphere of gas B which in this case is air (in a CP system with sweep gas, B can be any gas). If A and B are different gases, B may diffuse towards the membrane and then permeate through the membrane. The direction of the diffusion and permeation of B is opposite to the direction of the permeation and flow of A, and as a result may affect the evaluation of the permeation of gas A. The authors presented a mathematical model that allows estimation of an error arising from back diffusion and back permeation CP systems.

Tabe Mohammadi et al. [4] designed a fully automated gas permeation system for the measurement of low permeation rates and permeate compositions in CV systems (Fig. 7.2). This system allows for the measurement of permeation rates of 1×10^{-8} cm^3 s^{-1}, or lower, with extreme accuracy. The composition of the permeate gas is determined simultaneously using a mass spectrometer connected to the system. The gas(es) was fed into the separation cell through one or both mass flow controllers. The permeate was collected in the downstream tubing and cylinders of known volume.

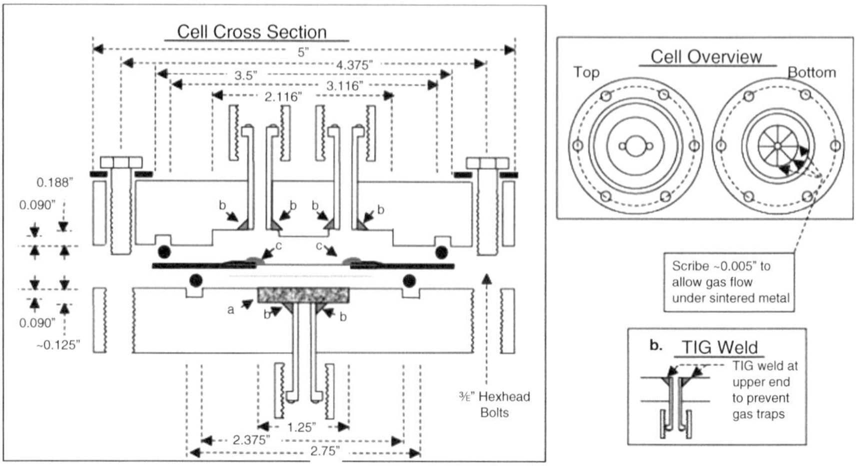

Fig. 7.3 Schematic of the permeation cell

O'Brien et al. [5] described a system based on a combination of manometric and a gas chromatography technique to measure mixed gas permeation through polymer films. This technique allowed straightforward determination of film permeabilities and selectivities over a wide range of feed pressures and compositions. This technique eliminated the need to use a downstream sweep gas and could be implemented by simple modification of pure gas permeation cells.

The isochoric (constant volume, variable pressure) technique for measuring permeation through membranes has been used extensively for a long time. In principle, this technique should be well suited for measuring the permeability of any gas separation membranes. In practice, accurate determination of the permeability becomes difficult as the membrane permeability decreases, due to outgoing and finite leakage of atmospheric gases into the permeate reservoir. To solve this problem, Moore at al. [6] designed an isochoric system incorporating a novel membrane cell to minimize leakage through the cell into the system.

A drawing of the recommended cell, including relevant dimensions, is included in Fig. 7.3. Seventy durometer o-rings of size 229 (lower cell) and 236 (upper cell) have proven to be optimal for general purpose use. (The durometer is a measure of the o-ring hardness and the numbers for the given sizes are all industry standards.) The smaller o-ring is used in the vacuum face of the cell to minimize rubber on the metal downstream sealing area, which is the biggest source of leaks in the cell. The o-ring grooves are slightly wider than the o-rings to allow for expansion and contraction of the o-ring when operated beyond ambient temperature. The depth of this groove was chosen to provide sufficient compression (greater than 25 % of the diameter) and to achieve a good seal. Selecting an o-ring material of relatively low permeability that is compatible with the fluid in the system is obviously recommended. Viton® (DuPont) is a reasonable choice for many applications. Actual o-ring toler-

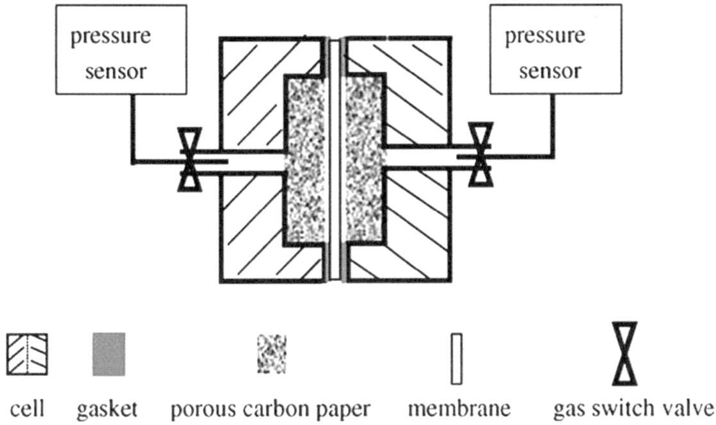

Fig. 7.4 Cell for gas permeability measurements [7]

ances vary from manufacturer to manufacturer, so it is important to verify that o-rings with near nominal dimensions are used in order to obtain proper compression. Finally, to achieve a good seal, it is necessary to polish the metal in direct contact with the o-ring. The gas permeability of the membranes also can be measured by means of a two chamber cell as shown in Fig. 7.4. [7].

A sample membrane is fixed by porous carbon plates with the membrane edges sealed by gaskets. During the measurement, a pressure difference of up to 1×10^6 Pa is applied using pressure sensors. The temperature is controlled by immersing the system in a temperature-controlled bath. One chamber of the cell is filled with pressurized gases while the other is always kept under vacuum. The mole number of the gas, n, passed through the membrane can be calculated from the decreased pressure P_d on one side of the membrane (Pa) with a certain time (t in second) by using the equation:

$$n = \{(P_d.V)/(RT)\}(\mathrm{mol}) \tag{7.1}$$

where R is the gas constant, T the temperature (K), and V is the volume of the gas chamber (m^3). The gas permeability coefficient, P, can be calculated:

$$P = \{(n.L)/(A.t.P_a)\}\left(\mathrm{mol\,m\,m^{2}s^{-1}Pa^{-1}}\right) \tag{7.2}$$

where L (m) is the membrane thickness, A (m^2) is the area of the membrane for gas diffusion, and P_a is the pressure difference (Pa) across the membrane. Before the measurement, the membrane should be dried under vacuum at appropriate temperature for more than 1 h.

7.3 Membrane Morphology

There are newly developed methods for the characterization of membranes, especially for surface morphology. The methods for morphology characterization are summarized in this section.

7.3.1 Microscopic Method

An important trend in membrane characterization places emphasis on the characterization of a membrane's surface morphology. This is to be expected since membrane separation is fundamentally the reflection of surface phenomena, as manifested by the structure of integrally skinned asymmetric membranes and thin-film composite membranes. There are three main microscopic techniques to study the morphology of membrane surfaces:

1. Atomic force microscopy (AFM).
2. Scanning electron microscopy (SEM).
3. Transmission electron microscopy (TEM).

7.3.1.1 Atomic Force Microscopy

Atomic force microscopy (AFM) was invented by Binnig et al. [8]. It is an important tool for imaging surfaces down to the atomic scale even under ambient atmosphere. The surface of a membrane plays a main role in its performance for separation purposes. The AFM consists of a cantilever with a sharp tip (probe) at its end that is used to scan the specimen surface. The cantilever is typically silicon or silicon nitride with a tip radius of curvature on the order of nanometers. When the tip is brought into proximity of a sample surface, forces between the tip and the sample lead to a deflection of the cantilever according to Hooke's law [9]. This force is kept small and at a constant level with the feedback mechanism. When the tip is moved sideways it will follow the surface contours such as the trace B in Fig. 7.5. The deflection is measured using a laser spot reflected from the top surface of the cantilever into an array of photodiodes. Other methods that are used include optical interferometry, capacitive sensing, or piezoresistive AFM cantilevers. These cantilevers are fabricated with piezoresistive elements that act as a strain gauge. Using a Wheatstone bridge, strain in the AFM cantilever due to deflection can be measured, but this method is not as sensitive as laser deflection or interferometry. Imaging of soft samples, for example polymeric membranes, sometimes suffers from plastic deformations introduced by the tip. When imaging under ambient atmosphere, a thin water layer present on the surface causes capillary forces which are adhesive. Different strategies have been described in order to minimize the interaction between

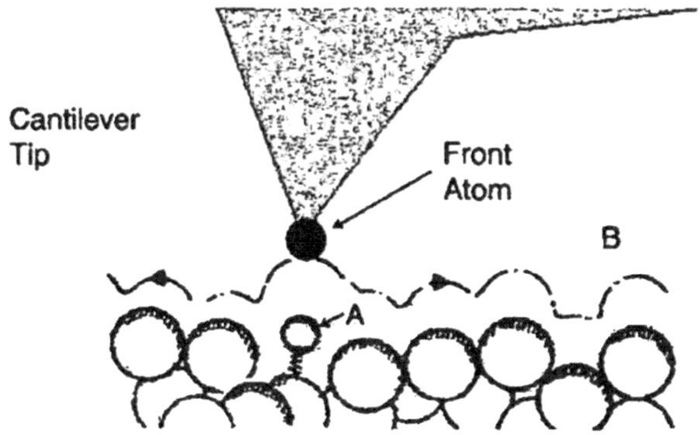

Fig. 7.5 Mechanism of AFM

the tip and the surface [10]. Among other techniques, such as immersing the system into a liquid [11], the idea of a tip not being in contact with surface has been suggested (non-contact AFM) by Martin et al. [12]. When the tip is vibrated within nanometers of the surface, van der Waals or electrostatic forces modify the spring constant and thus the resonance frequency of the spring. This modification can be sensed in order to obtain a topographic image of the surface. A principal drawback of this method is that the lateral resolution is limited by the tip–sample separation, which cannot be lower than the vibrational amplitude. By reducing this amplitude too much, the resulting low vibrational energies involved increase the chance of the tip being caught in the layer consisting of adsorbed gases (condensed water vapor and other contaminants) or sticking to the surface.

Tapping Mode AFM (TM-AFM) has been used successfully in the interpretation of the morphology of synthetic polymeric membranes [9]. The basic objective of the operation of the AFM is to measure the forces (at the atomic level) between a sharp probing tip and a sample surface (Fig. 7.5). Scanning the sample relative to the probing tip and measuring the deflection of the cantilever as the sample is scanned under the stylus on x, y, and z directions, the piezoelectric translator produces images. Typical spring constants (amount of force required to bend a cantilever some given amount) are in the range between 0.001 and 100 N/m and motions from microns to ~0.1 Å are measured by the deflection sensor. Typical forces between tip and sample range from 10^{-11} to 10^{-6} N. For comparison, the interaction between two covalently bonded atoms is in the order of 10^{-9} N at separations of ~1 Å. Therefore, non-destructive imaging is possible with these small forces. Binnig et al. [8] proposed the scanning tunnelling microscope as a method to measure forces as small as 10^{-18} N. A flexible cantilever with a very low spring constant could be produced. With a cantilever that induces forces smaller than inter-atomic forces, the topography of the sample could be measured without replacing the atom.

There are different modes to get surface images by AFM. The main ones include contact mode, non-contact mode and tapping mode (TM.).

Contact mode: The contact mode, where the tip scans the sample in close contact with the surface, is the most common mode used in the force microscope. The force on the tip is repulsive with a mean value of 10^{-9} N. This force is set by pushing the cantilever against the sample surface with a piezoelectric positioning element. In contact mode AFM, the deflection of the cantilever is sensed and compared in a DC feedback amplifier to some desired value of deflection. Problems with contact mode are caused by excessive tracking forces applied by the probe to the sample. The effects can be reduced by minimizing the tracking force of the probe on the sample, but there are practical limits to the magnitude of the force that can be controlled by the user during operation in ambient environments.

Non-contact Mode: In this mode the tip hovers 50–150 angstroms above the sample surface. Attractive van der Waals forces acting between the tip and the sample are detected, and topographic images are constructed by scanning the tip above the surface. Unfortunately the attractive forces from the sample are substantially weaker than the forces used by the contact mode; therefore, the tip must be given a small oscillation so that AC detection methods can be used to detect the small forces between the tip and the sample by measuring the change in amplitude, phase, or frequency of the oscillating cantilever in response to force gradients from the sample.

Tapping mode: This technique allows high resolution topographic imaging of sample surfaces that are easily damaged, are loosely held to their substrate, or difficult to image by other AFM techniques. Tapping mode overcomes problems associated with friction, adhesion, electrostatic forces, and other difficulties that plague conventional AFM scanning methods, by alternately placing the tip in contact with the surface to provide high resolution and then lifting the tip to avoid dragging it across the surface. Tapping mode imaging is implemented in ambient air by oscillating the cantilever assembly at or near the cantilever's resonant frequency using a piezoelectric crystal. The piezo motion causes the cantilever to oscillate with a high amplitude (typically greater than 20 nm) when the tip is not in contact with the surface. The oscillating tip is then moved toward the surface until it begins to lightly touch, or tap the surface. During scanning, the vertically oscillating tip alternately contacts the surface and lifts off, generally at a frequency of 50,000–500,000 cycles per second. As the oscillating cantilever begins to intermittently contact the surface, the cantilever oscillation is necessarily reduced due to energy loss caused by the tip contacting the surface. The reduction in oscillation amplitude is used to identify and measure surface features.

Using materials sensing modes such as lateral force and phase contrast, it is possible to differentiate the types of materials at a polymer/membrane surface. AFM was first applied to polymer membrane surfaces by Albrecht et al. [13] in 1988 shortly after its invention. AFM opened a new avenue for the study of membrane surfaces. Nodule observation, measurement of roughness parameters, pore size, and pore size distribution are discussed below in detail. AFM can also be used for:

1. True three-dimensional surface topographic imaging;
2. Complete image analysis of all surface or irregularities;

3. Surface elasticity or compressibility measurements;
4. Surface adhesion measurements; and
5. Quantitative summary statistics.

7.3.2 *Observation of Nodules*

Nodules are structural units observable at the polymer surface in general, and at the membrane surface in particular. To obtain the pore sizes and nodule sizes, cross-sectional line profiles are selected to traverse micron scan surface areas of the TM-AFM images. The size of nodule is determined from the cross-sectional profiles of the data along a reference line. An example of the measurements of nodule diameters is shown in Fig. 7.6. The bright sites are nodules and the dark sites are interstitial domains. For each pair of cursor (pointers), horizontal and the vertical distances are given in the right window. The diameter of nodules (bright sites), i.e., maximum width of the cross section of the bright site, can be measured by the help of a pair of cursors (also indicated in Fig. 7.6). By measuring the diameters of a large number of bright sites (at least 25), the average size of nodules, nodule aggregates and supernodular aggregates are obtained, depending on the size of the bright sites.

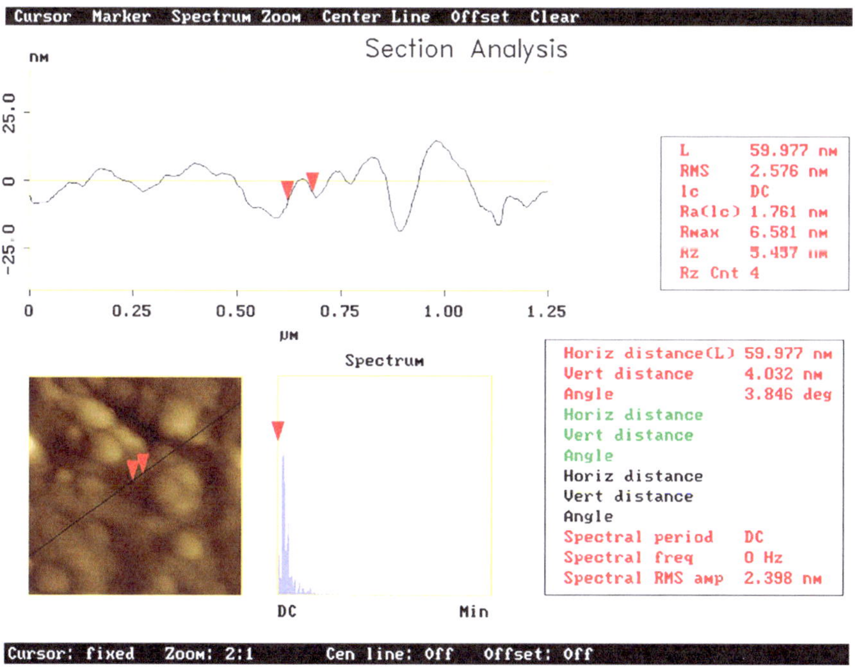

Fig. 7.6 Section analysis of a TM-AFM image: a vertical displacement of the top surface of the dense PPO-TCE membrane (membrane was prepared by casting PPO solution in trichloroethylene)

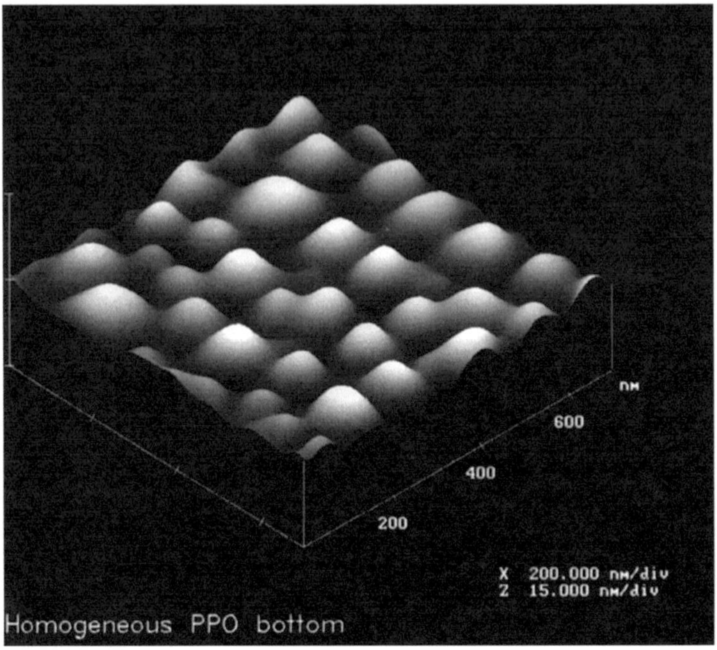

Fig. 7.7 TM-AFM image of the homogeneous membrane's bottom surface

Similarly, the width of the dark sites, which could be the openings of pores in porous membranes, can be measured.

A, B, C show the pair of cursors of each measurement. (Distance between the cursors shows the pore size diameter.)

A homogeneous membrane was prepared from polyphenylene oxide (PPO)–trichloroethylene (TCE) solution by casting the solution on a glass plate and removing the solvent by evaporation at room temperature. The surface morphology was studied by tapping mode AFM [14].

Quite unexpectedly, a significant difference between the top and bottom surface was observed. Figures 7.7 and 7.8 illustrate the 3D images of the bottom and the top surfaces of the homogeneous membranes, respectively. Both surfaces show a relatively uniform nodular structure; however, nodules of the bottom surface (d = 137.5 nm) are twice as large as those on the top surface (d = 63.8 nm).

Surface roughness of the membrane affects the permeate flux [15] as well as membrane fouling due to particle deposition on the surface [16]. Hence, the measurement of surface roughness is important.

Three roughness parameters—i.e., the mean roughness (R_a), the mean square of the Z data (R_q), and the mean difference in height between the five highest peaks and the five lowest values (R_z)—are used to represent the surface roughness. The roughness parameters depend on the curvature and the size of the TM-AFM tip, as well as on the treatment of the captured surface data (plane-fitting, flattening,

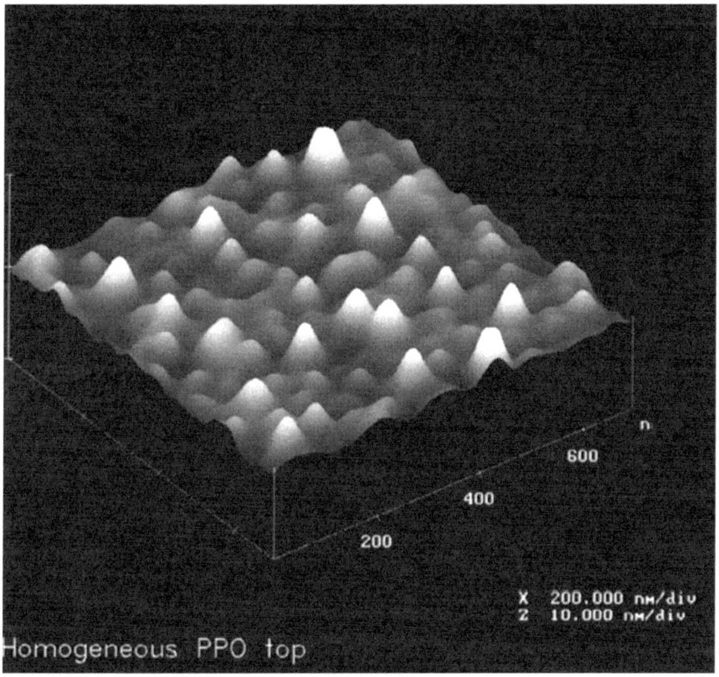

Fig. 7.8 TM-AFM image of the homogeneous membrane's top surface

filtering, etc.). Therefore, the roughness parameters should not be considered absolute roughness values.

The mean roughness is the mean value of surface relative to the center plane, where the volume enclosed by the image above and below this plane are equal, calculated as,

$$R_a = \frac{1}{L_x L_y} \int_0^{L_x} \int_0^{L_y} \left| f(x,y) \right| dxdy \tag{7.3}$$

where $f(x, y)$ is the surface relative to the center plane. L_x and L_y are the dimensions of the surfaces.

The root mean square of the Z values (R_q) is the standard variation of the Z values within the given area and is calculated as

$$R_q = \sqrt{\frac{\sum \left(Z_i - Z_{\text{avg}} \right)^2}{N}} \tag{7.4}$$

where, Z_i means the height of ith pixel, Z_{avg} is the average of Z_i in the given area, and N is the number of points within the given area.

It has been reported that the mean roughness of the membrane (measured by AFM) is directly proportional to the permeability of gases [17]. Kesting reported that the surface roughness could distinguish between the amorphous and crystalline regions, and irregularities on the surface could affect the film's physicochemical properties [18].

7.3.2.1 Electron Spectroscopy for Chemical Analysis (ESCA) and Scanning Electron Microscopy (SEM)

A modern light microscope has a maximum magnification of about 1,000×. The resolving power of such a microscope is not only limited by the number and quality of the lenses but also by the wavelength of the light used for illumination. White light has wavelengths from 400 to 700 nanometers. The average wavelength is 550 nm, which results in a theoretical limit of resolution (not visibility) of the light microscope in white light of about 200–250 nm. Figure 7.9 shows two points at the limits of detection and the two individual spots can still be distinguished. The right image shows the two points so close together that the central spots overlap.

Electron spectroscopy is an analytical technique to study the electronic structures and dynamics in atoms and molecules. The electron microscope was developed when wavelength became the limiting factor in light microscopes. Electron microscopy is widely used to study the morphological structure of surfaces. Electron microscopy (EM) is one of the techniques that can be used for membrane characterization. .Two basic techniques can be distinguished: scanning electron microscopy (SEM) and transmission electron microscopy (TEM). These methods are well documented and well known in membranes field [19–21].

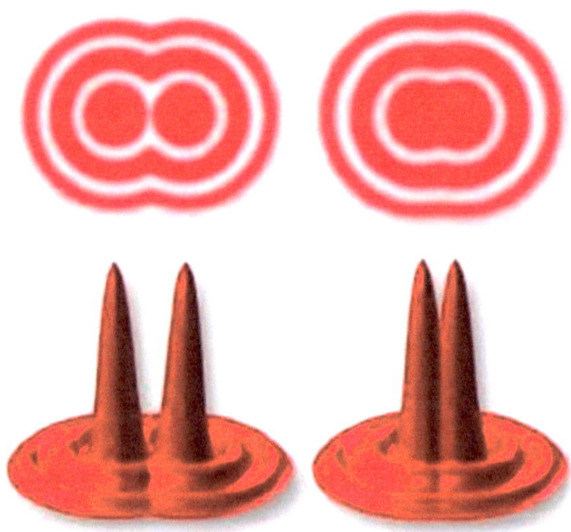

Fig. 7.9 Two points showing the limits of detection

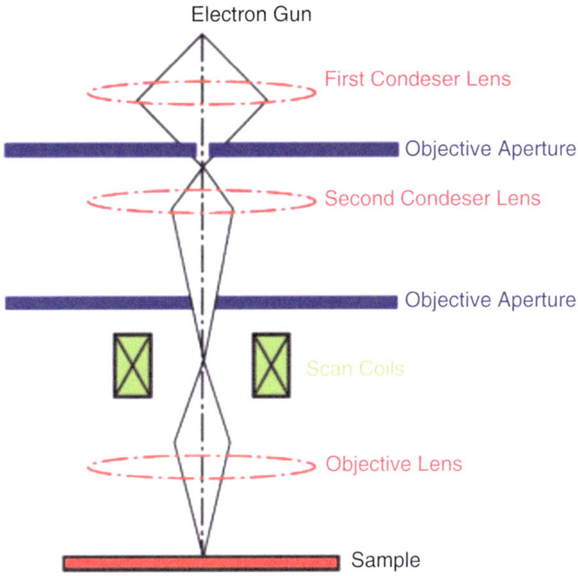

Fig. 7.10 Simple schematic of the scanning electron microscope

Electrons have much shorter wavelengths, enabling a much better resolution. The main components of the SEM include:

1. Source of electrons;
2. Column down which electrons travel with electromagnetic lenses;
3. Electron detector;
4. Sample chamber; and
5. Computer and display to view the images.

Figure 7.10 shows the schematic diagram of the SEM.

The position of the electron beam on the sample is controlled by scan coils, which are situated above the objective lens. These coils allow the beam to be scanned over the surface of the sample. This beam rastering, or scanning, as the name of the microscope suggests, enables information about a defined area on the sample to be collected. As a result of the interaction of the electron beam with the sample, a number of signals are produced. This signal can then be detected by appropriate detectors.

A Scanning electron microscope (SEM) scans a focused electron beam over a surface to create an image. The electrons in the beam interact with the sample, producing various signals that can be used to obtain information about the surface topography and composition.

TEM works in transmission geometry and requires (microtomed) thin samples, whereas SEM measures low energy secondary electrons emitted from the specimen surface due to excitations in the specimen itself, produced by the primary electron beam. TEM is not a surface reflection method but can image surfaces with excellent resolution. TEM can provide images that have higher magnification and greater resolution than images produce by SEM. SEM produces 3D images, whereas TEM provides only flat (2D) images. The 3D images provide more information about the shape of features and also about the location of features relative to each other. In some cases this information is very useful and more important than the higher resolution and magnification that is offered by TEM.

The first electron micrographs of a polymeric membrane were published by Riley et al. in 1964 [22] using a TEM-replica technique. In 1980, Merin and Cheryan [23] succeeded to visualize membrane surface pores by using replica techniques and TEM. By using field emission scanning electron microscopy (FESEM), Koutake et al. [24] first observed pores on UF membrane surfaces. Zeman and Denault [25] and Zeman [26] developed a computerized quantitative image analysis method of SEM photographs to characterize the structure of MF membranes.

The maximum resolution obtained in an SEM depends on multiple factors, like the electron spot size and interaction volume of the electron beam with the sample. While they cannot provide atomic resolution, many SEMs have been known to achieve resolution below 1 nm. Typically, modern full-sized SEMs provide resolution between 1 and 20 nm, whereas desktop systems can provide a resolution of 30 nm or more. SEM can be applied to study a membrane's problems including swelling, asymmetry, and void size, pore size, rugosity, deswelling. [20].

In SEM, samples are coated with metals. There are different ways of coating to prepare the samples for characterization. Conductive coating can enhance the image contrast due to higher secondary electron yield. The resolution of SEM with field emission may go down to a range of 0.6–3 nm. However, in order to obtain such high resolution with polymer samples, preparation artifacts have to be minimized. One of the reasons for the artificial changes to the observed surface is the effect of impact energy induced during the conductive coating process. Usually, metals such as gold, palladium, chromium, platinum and carbon, and their mixtures, are used for the conductive coating [27] by the following methods.

1. *Magnetron sputter coating*: This is the most popular way of applying a conductive layer on a non-conductive specimen. For this method, the sputter source (metal) and the sample are located in a common vacuum chamber. The pressure in the chamber is kept at about 10 Pa and a noble gas such as argon or xenon is introduced in to the chamber.
2. *Ion beam sputtering coating*: In this procedure the sample is placed under a much higher vacuum (8×10^{-3} Pa). An ion source generates a directed ion beam, which hits the target material, ejecting the atoms from the solid surface in towards the specimen surface.
3. *Penning sputter coating* (*PSC*): This procedure combines plasma generation and ejection of target material in one piece of equipment. From the source, a directed beam of neutral particles is emitted with energy comparable to the kinetic energy of an ion beam sputter coater. In addition, the emitted beam is directed through an electrical field, which filters out charged particles. The sample is maintained under high vacuum and must be kept in motion for a continuous coating. Using neutral particle sources, thin coatings are possible, even at room temperature.
4. *Electron beam evaporation*: Evaporation through heating of the target material in a high vacuum is a reliable (especially in the TEM) and widely used procedure.

The directed beam of uncharged particles hits the sample with low kinetic energy. The same conditions concerning the geometry of the coating are valid in the ion beam and penning sputter procedures.

Due to rapid development of computer hardware, image processing is now a useful automated technique to measure the morphological parameters quantitatively. How to prepare membrane sample without any artifacts is a problem for scanning electron microscopy (SEM) and transmission electron microscopy (TEM) studies. Careful drying is the first step to prepare a membrane sample, and in order to avoid collapse of the original structure, the freeze dry technique using liquid nitrogen or critical-point drying method with carbon dioxide is usually employed.

In order to observe cross sections by SEM, the dried membrane is first fractured at liquid nitrogen temperature, and fixed perpendicularly to the sample holder. For a TEM study, the dried sample is first embedded, if necessary, and then cut by a microtome. An example of the artifacts introduced in the process of conductive coating was shown by Schossig-Tiedemann and Paul [27].

When imaging sputter-coated polymer membranes, nodular structures can be observed on the membrane surface [18, 28]. Figure 7.11 shows the surface of an

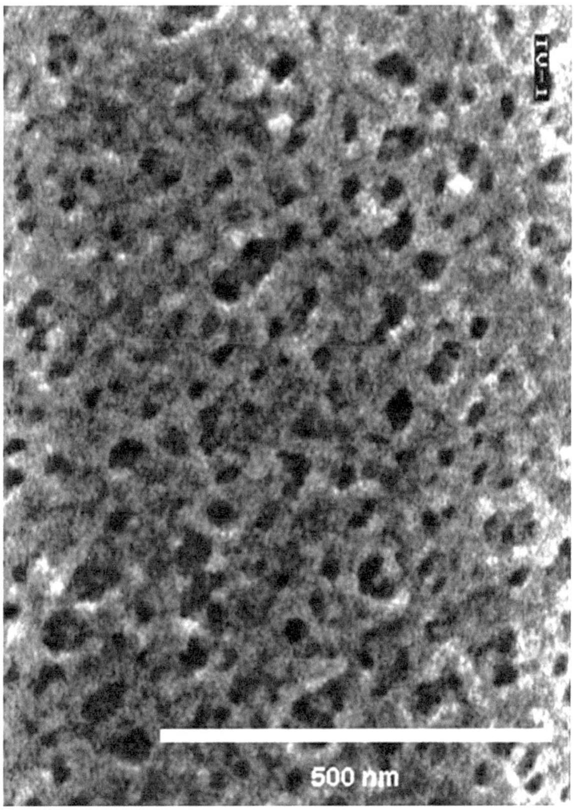

Fig. 7.11 PEI-membrane, native specimen-surface

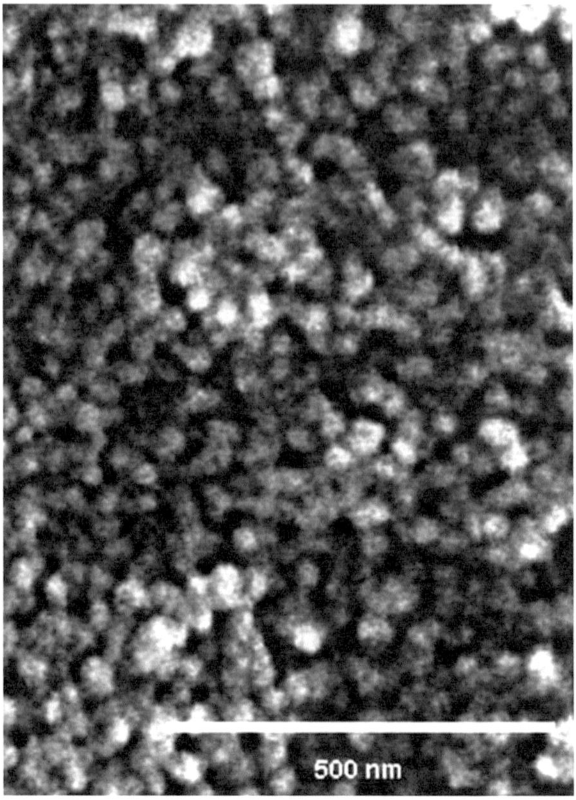

Fig. 7.12 PEI-membrane, magnetron sputtered at room temperature

uncoated porous polyetherimide membrane. The uncoated sample was examined by using a DSM 682 Gemini microscope with $E = E_2$ at 1 kV [27]. The magnetron sputtered surfaces, represented in Figs.7.12 and 7.13 were prepared at room temperature, and at liquid nitrogen temperature, respectively. The nodule size reduced with a decrease in the preparation temperature. It was concluded that for polymer samples with structural elements on a submicron scale, the surface structure could be significantly altered by energy impacts resulting from the preparation method. Recent developments in the area of conductive coating should suppress surface charging, minimize radiation damage, and increase electron emission from the surface. The selection of a suitable coating procedure could lead to significantly improved results in SEM images.

Electron spectroscopy is an analytical technique to study the electronic structure and its dynamics in atoms and molecules. In general, an excitation source such as X-rays, electrons or synchrotron radiation will eject an electron from an inner-shell orbital of an atom. Detecting photoelectrons that are ejected by X-rays is called X-ray photoelectron spectroscopy (XPS) or electron spectroscopy for chemical analysis (ESCA). Detecting electrons that are ejected from higher orbitals

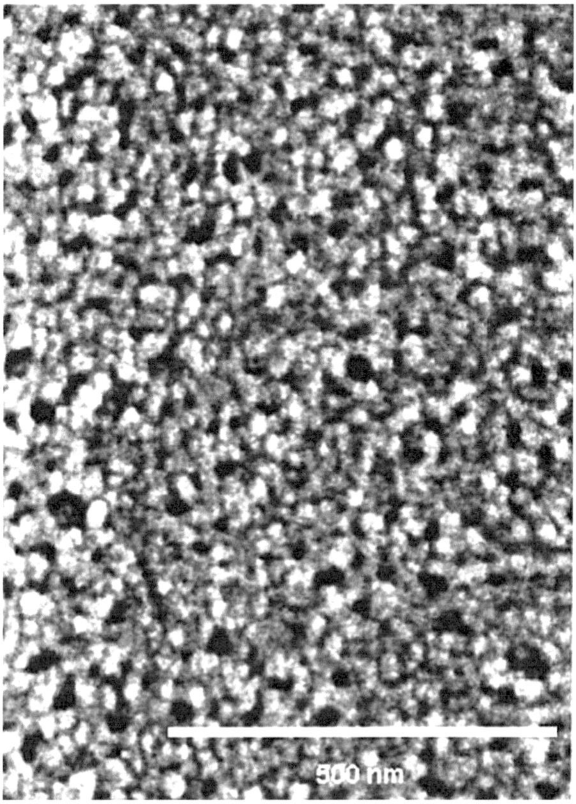

Fig. 7.13 PEI-membrane magnetron sputtered at 143 K

Table 7.1 ESCA results for PSf films reacted with PFPA at 25 °C

	Reaction time			
Atomic ratio	No reaction[a]	10 min	20 min	30 min
O/C	0.15	0.335	0.313	0.278
S/C	0.037	0.027	0.0264	0.0265
F/C	–	0.004	0.006	0.013

[a]Theoretical ratio

to conserve energy during electron transitions is called Auger electron spectroscopy (AES).

Rhim et al. [29] modified polysulfone (PSf) membranes for gas separation using the fluorine chemicals, PFPA (pentafluoropropionic anhydride) with varying reaction times. The resulting membrane surfaces were analyzed by ESCA. Table 7.1 ESCA results for PSf films reacted with PFPA at 25 °C.

As the reaction time passed, PFPA was reacted with PSf. As expected, the number of C atoms was greater and as a result, O/C and S/C ratios decreased with increasing reaction times. F/C ratio increased since the number of fluorine atoms was greater than that of carbon atoms in PFPA.

Quantitative angle-dependent ESCA coupled with IR spectroscopy and DSC measurements can be used to investigate the miscibility of polymers in each other. By using this technique Clark et al. [30] studied miscibility of poly(ε-caprolactone) (PCL) with poly(vinyl chloride) (PVC) blends films.

7.3.3 Spectroscopic Method

7.3.3.1 Infrared (IR) and Fourier Transform Infrared (FTIR) Spectroscopy

Infrared spectroscopy (IR spectroscopy) is the spectroscopy that deals with the infrared region of the electromagnetic spectrum, a region of longer wavelength and lower frequency than visible light. It covers a range of techniques, mostly based on absorption spectroscopy. As with all spectroscopic techniques, it can be used to identify and study chemicals. For a given sample—solid, liquid, or gas—the method or technique of infrared spectroscopy uses an instrument called an infrared spectrometer (or spectrophotometer) to produce an infrared spectrum. A basic IR spectrum is essentially a graph of infrared light absorbance (or maybe transmittance) on the vertical axis versus frequency or wavelength on the horizontal axis. Typical units of frequency used in IR spectra are reciprocal centimeters (sometimes called wave numbers), abbreviated as cm^{-1}. Units of IR wavelength are commonly given in microns, abbreviated as μm, which are related to wave numbers in a reciprocal way. A common laboratory instrument that uses this technique is an infrared spectrophotometer.

A beam of infrared light is produced and split into two separate beams. One passes through the sample, the other passes through a reference, which is often the substance the sample is dissolved in. Both obtained beams are reflected back towards a detector; however, first they pass through a splitter that quickly alternates which of the two beams enters the detector. The two signals are then compared and a printout is obtained. A reference is used for two reasons: (1) this prevents fluctuations in the output of the source affecting the data; (2) this allows the effects of the solvent to be cancelled out. (The reference is usually a pure form of the solvent the sample is in.) The wavelength of light absorbed is characteristic of the chemical bond, as seen in the attached Infrared Spectroscopy Correlation in Table 7.2.

The modern instruments that have been using for qualitative and quantitative analysis are FTIR instruments. Fourier transform infrared (FTIR) spectroscopy is a

Table 7.2 Infrared spectroscopy correlation

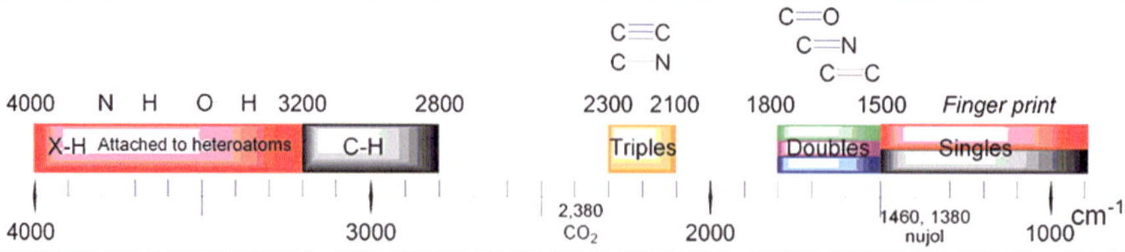

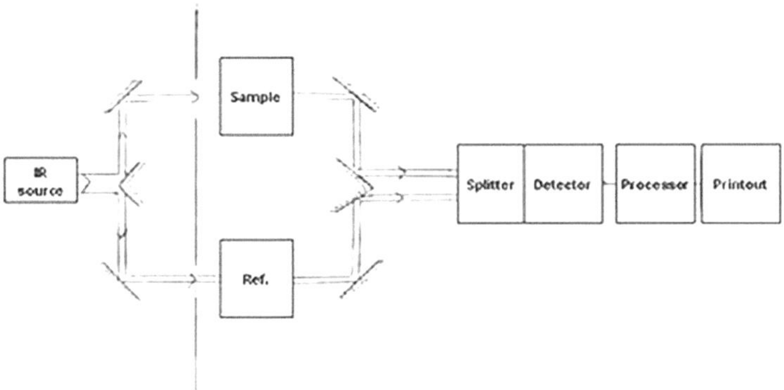

Fig. 7.14 Typical FTIR apparatus

powerful tool for identifying types of chemical bonds in a molecule by producing an IR absorption spectrum that is like a molecular *fingerprint*; thus, FTIR spectroscopy is a measurement technique for collecting infrared spectra. Instead of recording the amount of energy absorbed when the frequency of the infra-red light is varied (monochromator), the IR light is guided through an interferometer. After passing through the sample, the measured signal is the interferogram. Performing a Fourier transform on this signal data results in a spectrum identical to that from conventional (dispersive) infrared spectroscopy. FTIR is perhaps the most powerful tool for identifying types of chemical bonds (functional groups). Today's FTIR instruments are computerized, which makes them faster and more sensitive than the older dispersive instruments. Figure 7.14 shows a schematic diagram of FTIR.

FTIR can be utilized to quantitate some components of an unknown mixture. It can be applied to the analysis of solids, liquids, and gases. By interpreting the infrared absorption spectrum, the chemical bonds in a molecule can be determined. For most common materials, the spectrum of an unknown compound can be identified by comparison to a library of known compounds. There are several infrared spectral libraries including on-line computer libraries. To identify less common materials, IR will need to be combined with nuclear magnetic resonance, mass spectroscopy, X-ray diffraction, and/or other techniques.

Nowadays IR is used together with many other techniques to characterize membranes more precisely. For example, angle-dependent electron spectroscopy for chemical analysis (ESCA) or X-ray photoelectron spectroscopy (XPS) coupled with IR spectroscopy and differential scanning calorimetry (DSC) can be used to investigate the miscibility of blended polymers [30].

7.3.3.2 Positron Annihilation Spectroscopy (PALS)

The positron annihilation technique is quite promising for studying free-volume holes (FVH) in polymer structures. Information on free-volume holes is useful for understanding many physical and mechanical properties of polymeric materials.

However, some problems in rationalizing the annihilation characteristics obtained by measuring positron lifetimes still remain unsolved.

The study of free volume (FV) in polymer systems is of great interest because the size and concentration of its elements (holes) affect numerous transport, mechanical, and other physiochemical properties. Positron annihilation lifetime (PAL) spectroscopy is one of the most efficient approaches for investigation of free volume (FV) distribution of membrane materials. The foundations of this technique for probing polymers are based in particular on Walker–Brandt–Berko's free volume model [31]. According to Walker–Brandt–Berko's model, positronium Ps (a bound atomic system which consists of an electron and a positron) tends to be localized or trapped before its annihilation in FV or, in other words, in areas with reduced electron density. Accordingly, annihilation characteristics (life-times and intensities of longer lifetime components of annihilation radiation) give information on concentration sizes and distribution [32–35]. A computer program PATFIT, which represents annihilation lifetime distribution in a discrete manner, i.e., as a sum of several exponents, is employed for this purpose. Gregory et al. [36, 37] and Deng and Jean [38] proposed the use of continuous lifetime analysis, which is based on the Laplace inversion program CONTIN to obtain the continuous probability density function of annihilation with a given lifetime from annihilation lifetime spectra. In this way one can obtain the size distribution of FV in polymers. The CONTIN analysis of positron annihilation lifetime distribution of positrons was accomplished successfully in a number of polymers: epoxy polymers, polystyrene, and polycarbonates with different structures.

Positron annihilation life time spectra can be analyzed in terms of three lifetime components [39] viz: *para*-positronium (*p*-Ps) annihilation, τ_1; free positron and positron-molecular species annihilation, τ_2; and *ortho*-positronium (*o*-Ps) annihilation, τ_3. While τ_1 and τ_2 are in the order of few 100 ps, τ_3 is in the order of nanoseconds. Each lifetime has intensity, l, corresponding to the fraction of annihilations taking place with the respective lifetimes. The parameters τ_3 and l_3, corresponding to decay of *o*-Ps, provide the size-specific for free volumes and pores.

By using the results of *o*-Ps lifetime, the mean free-volume hole radius can be calculated by the following equation [39, 40].

$$\tau_3 = \frac{1}{2}\left[1 - \frac{R}{R_0} + \frac{1}{2\pi}\sin\left(\frac{2\pi R}{R_0}\right)\right]^{-1} \tag{7.5}$$

Where τ_3 (*o*-Ps life-time) and R (hole radius) are expressed in ns (nanoseconds) and $\times 10^{-10}$ m (Å), respectively, R_o is equal to $R + \Delta R$, where ΔR is a fitted empirical electron layer thickness ($= 1.66 \times 10^{-10}$ m (1.66 Å)). The cavity volumes can be calculated from $V_h = 4\pi R^3/3$.

Further, the fractional free volume f may be estimated from the following empirical relation [41].

$$f = CV_F l_3 \tag{7.6}$$

Where, V_F is free volume and C is the scaling factor; C can be obtained from the variation of free volume with temperature. However, in absence of such data, it may be typically assigned a value of 1.0 [42] in which case the values of f obtained will be proportional to the actual free volume fraction.

Because Ps has a relatively small size (~1.59 Å) and small scale of probe lifetime (~ns), positron annihilation lifetime spectroscopy (PALS) is very sensitive in measuring small holes and free volume in a size (1 Å ~ 20 Å) and at a time of molecular motion from 10^{-10} s and longer. The positrons and Ps are localized in preexisting holes and free-volumes in polymers; therefore, the measurements depend on a function of the temperature, pressure, degree of crystallinity, and time of aging [43].

Chen et al. [44] studied free-volume depth profiles of polymeric membrane systems prepared by interfacial polymerization by using PALS. The obtained layer structures of asymmetric polymeric membranes by PALS were supported by the data obtained by AFM. Tung et al. [45] used PALS coupled with a slow positron beam to characterize in situ the layer structure and depth profile of the cavity size in thin film composite (TFC) polyamide nanofiltration (NF) membranes prepared by the interfacial polymerization method. The membranes have a composite structure containing three layers: a selective polyamide layer, a transition layer, and a porous support prepared by the phase inversion technique. It was noticed that the cavity size distribution in the selective top layers plays an important role in determining the performance of NF membranes. Shantarovich et al. [32, 46] applied PALS to measure free volume size distribution in polymer samples with unusually long lifetimes: dense films of poly(trimethylsilylpropyne) (PTMSP) and in porous membranes prepared from poly(phenylene oxide) (PPO). In the case of PTMSP, the longer lifetime corresponds to continuous free-volume size distribution in the range R=4.5–5.5 Å. These radii were much larger than those characteristic for "normal" glassy polymers. A good agreement was obtained between the parameters of PALS distributions obtained by means of finite-term lifetime analysis (PATFIT program) and continuous lifetime analysis (CONTIN program) in PTMSP and PPO dense films. Shantarovich et al. [46] provided the first direct evidence for nonuniformity of free-volume distribution in glassy polymers by using a positron annihilation technique.

The PALS has been used in polymeric membranes to control the gas flow, for the application of field-assisted positron moderation, to alter the structural conformation of a polymeric chain, to observe hydrophilic/hydrophobic characteristics, to study vacancy profiles associated with salt selectivity, and to measure the relationship between gas permeability and free volume [43].

Huang et al. [47] studied the structures of polyamide thin-film composite membranes in regard to the variation in the free volume in the active polyamide layer of a thin-film composite membrane. From PALS, the variation in the free volume in the ethylenediamine (EDA) and trimesoyal chloride (TMC) active layer (on the surface of a modified polyacrylonitrile (mPAN)) showed a good correlation with the pervaporation performance.

Marques et al. [48] obtain free-volume parameters in various urethane/urea membranes by PALS and Doppler broadening measurements technique. On bi-soft segment membranes, a correlation was found between the composition of membranes, the normalized free volume, the radii of the holes and gas permeability. On the other hand, the correlation was not clear when PU (polyurethane) data was considered. It was concluded that other features such as chain flexibility, chemistry at the free volume walls, and diffusion barrier, beside free volume, radii of holes, must also play an important role in the permeation mechanism. Kobayashi et al. [49] applied PALS to a series of cardopolyimide and polysulfone membranes. The authors reported that a favorable Ps formation was observed for polyimide membranes with 6FDA moiety with lower electron affinity. However, no Ps formed in most polyimide membranes with PMDA and BTDA moieties with higher electron affinity. This observation supports the assertion that the acid anhydride has an important impact on Ps formation in polyimide. The relationship between gas diffusivities and polymer free volume is not so simple. A number of factors should be taken into consideration in optimal design of gas separation membranes.

The data on positronium annihilation in glassy polymers, in particular in candidate membrane materials, provide the first direct evidence for nonuniformity of free volume distribution in these systems. A general correlation between the obtained values of specific free volume and the mobility, permeability, and solubility of oxygen in some glassy membrane polymers was reported [46].

7.3.3.3 X-ray Analysis

Knowledge of crystalline morphology is essential in understanding the permeability and permselectivity of polymer membranes such as dense membranes, dialysis membranes and the membrane separation of gases. X-ray is very common to study the crystallinity in polymeric membranes. Crystalline structure of a polymer membrane includes dimensions of unit cell, percentage of crystallinity, crystallite size, and orientation. The most generally applicable technique that provides information is the X-ray diffraction method. Permeation sites may be either amorphous material or interstices between crystallites. Most polymer membranes used for gas separation are of low or no crystallinity.

An X-ray diffraction technique is often used to determine the crystalline structure of polymers. This technique specifically refers to Bragg Peaks analysis scattered to wide angles, which (by Bragg's law) implies that they are caused by sub-nanometer sized structures. Wide-angle X-ray scattering (WAXS) is the same technique as small-angle X-ray scattering (SAXS). The only difference is that the distance from the sample to the detector is shorter for WAXS, and thus, diffraction maxima are observed at larger angles.

A sample is scanned in a wide angle X-ray goniometer, and the scattering intensity is plotted as a function of the 2θ angle. When X-rays are directed into solids they will scatter in predictable patterns based upon the internal structure of the solid. A crystalline solid consists of regularly spaced atoms (electrons) that can be described by imaginary planes. The distance between these planes is called the d-spacing. The intensity of the d-space pattern is directly proportional to the number of electrons (atoms) that are found in the imaginary planes. Every crystalline solid will have a unique pattern of d-spacings (known as the powder pattern), which is a "finger print" for that solid. In fact, solids with the same chemical composition but different phases can be identified by their pattern of d-spacing.

In general, the crystalline phase may be regarded as impermeable, so that gas permeability in a semicrystalline polymer membrane is substantially lower than in the more amorphous membrane because of the reduced space available for diffusion and the winding path around the crystallites. Permeation sites may be of either amorphous material or interstices between crystallites. Most polymer membranes used for gas separation are of low or no crystallinity. The results obtained from X-ray diffraction of dense poly(2,6-dimethyl-1,4-phenylene oxide) membranes, prepared from different solvents, showed that the physical properties of solvents have a significant effect on the conformation of the membrane [50]. The diffraction intensities obtained by X-ray analysis were correlated with the selectivity (CO_2/CH_4) and the permeability of CO_2. In view of the results obtained by Khulbe et al. [50], when crystallization is low, the state of the amorphous polymer may have an equally important role in the permeability of membranes for gas separation.

Six polyimides were synthesized by combining two dianhydrides, pyromellitic dianhydride (PMDA) and 5,5'-[2,2,2-trifluoro-1-(trifluoromethyl)ethylidene]bis-1,3-isobenzofurandione (6FDA), and three dianilines—oxydianiline (ODA), methylene dianiline (MDA), and isopropylidenedioxianiline (IPDA) [51]. Their chemical structures were studied by wide angle X-ray diffraction (WAXD) technique. It was observed that by introducing bulky pendant groups to the backbone chain of polyimide, d-spacing was increased. This led to an increase in gas permeability due to the increase in intersegmental distance while maintaining selectivity due to the increase in chain stiffness.

Interpretation of d-spacing determined by WAXS is of great interest in the study of gas diffusivity or permeability in glassy polymers. Various authors report [52–54] that the appearance of large Van der Waals peaks on X-ray scattering profiles in many polymers seem to be the norm for the amorphous state of polymers, and a marked large Van der Waals peak represents a "polymer" property corresponding to intermolecular distances for polymer chains. However, the d-spacing does not always correspond to the intermolecular distance that dominates gas diffusivity or permeability [52]. It is difficult to interpret the broad amorphous peaks of many amorphous polymers in X-ray scattering profiles. In order to obtain a clearer interpretation of WAXS profiles, characterization of the scattering profile on the atomic level is needed. Shimazu et al. used a new approach for WAXS analysis and of molecular modeling of polyimide synthesized from 2,2-bis(3,4-carboxyphenyl) hexafluoropropane-dianhydride (6FDA). The corresponding theoretical X-ray scattering functions were calculated from the atomic coordinates using Debye's equation and then compared with the experimental data. The amorphous polymer model was found to be superior to the tetramer conformation model in predicting experimental X-ray scattering function. It was noticed that the d-spacing for 6FDA-BAAF polyimide was clearly affected by intramolecular distances containing F atoms.

X-ray diffraction techniques resolved that spinning does not induce crystallization in hollow fibers [55], while it does happen in flat sheet membranes.

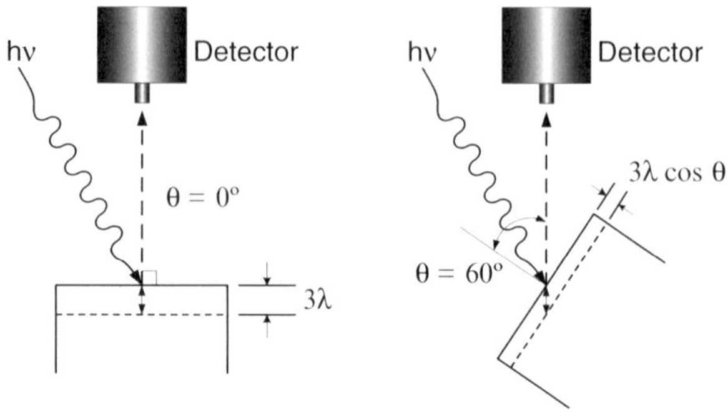

Fig. 7.15 A schematic diagram showing the setup of AR-XPS used at different take off angles [56]

7.3.3.4 X-ray Photoelectron Spectroscopy (XPS)

When irradiated by a monoenergetic beam of x-photons, solid atoms emit electrons. By measuring the energy of emitted electrons, E_K, the binding energy E_B is obtained by subtracting the E_K from the energy of the incident x-photon. The binding energy of the core electron is specific to an atom. Therefore, the identification and quantification of atoms become possible.

A schematic illustration of the angle resolved-XPS (AR-XPS) is shown in Fig. 7.15. The angle (θ) between the normal to the sample surface and the electron trajectory into the detector is defined as the take-off angle (TOA). The effective sampling depth, z, can be calculated by

$$z = 3\lambda \cos\theta \tag{7.7}$$

where λ is the effective mean path for electrons to escape the surface. Using $\lambda = 2.1$ nm, $z = 6.3$ nm at $\theta = 0°$, and 3.15 nm at $\theta = 60°$.

According to this principle, identification and quantification of atoms by nm depth from the top surface are possible by XPS [56]. This method is widely used to provide quantitative and qualitative chemical information of the top 1–20 nm of a surface. Generally, the samples are analyzed at a series of take-off angles (measured from the surface sample to the X-ray lens) to determine whether a compositional gradient exists near the surface.

Since membrane surface modification is currently a popular research topic, XPS is frequently used to determine the surface composition of the membrane. For example, migration of fluorine containing surface modifying macromolecules (SMMs) to the top surface of the polyethersulfone membrane during membrane casting, was confirmed by determining the surface concentration of fluorine by XPS [57, 58]. The XPS results further showed that SMM migrated to the surface, where it dominated the surface properties of the membrane and also that the orientation of the SMM was such that the fluorine tails were present at the surface [56]. Chitosan membranes after oxygen plasma treatment were characterized by an angle–resolved X-ray photoelectron spectrometer (ARXPS) to study the spatial orientation of surface chemical group [59].

7.3.3.5 Small Angle Neutron Scattering (SANS)

Neutron diffraction or elastic neutron scattering is the application of neutron scattering to the determination of the atomic and/or magnetic structure of a material. A sample to be examined is placed in a beam of thermal or cold neutrons to obtain a diffraction pattern that provides information on the structure of the material. The technique is similar to X-ray diffraction but due to their different scattering properties, neutrons and X-rays provide complementary information.

Small angle neutron scattering is in many respects very similar to small-angle X-ray scattering (SAXS); both techniques are jointly referred to as small-angle scattering (SAS). Advantages of SANS over SAXS are its sensitivity to light elements, the possibility of isotope labelling, and the strong scattering by magnetic moments. There are numerous SANS instruments available worldwide at neutron facilities such as research reactors or spallation sources. Neutron scattering is routinely used in modern science to understand material properties on the atomic scale.

The technique of small angle neutron scattering (SANS) is used for studying the structure of a material on length scales of 10–1,000 Å. In particular, SANS is used to determine the shapes and sizes of the particles dispersed in homogenous medium.

Neutron scattering (especially in situ studies) provides unique information about both bulk and surface properties. It is a tool ideally suited to interrogate lateral heterogeneity in model membranes, primarily due to its unique spatial resolution (i.e., 5–100 nm) and its ability to resolve structure with minimal perturbation to the membrane [60]. He et al. described a technique of neutron in-plane scattering for studying the structures of peptide pores in membranes [61].

Knowledge of microstructural parameters is essential for membrane optimization. Non-destructive characterisation of the pore microstructure was carried out by a small-angle neutron scattering technique by Strunz et al. [62]. The combined results from pinhole and double-crystal facilities enabled the authors to determine microstructural parameters of the nanoporous membrane (pore-to-pore distance, raft thickness, pore volume fraction, specific interface). Ye et al. [63] studied the microstructure of plasticized PVC membranes in the dry state and, during the process of soaking in heavy water, by small-angle neutron scattering. In the dry membrane homogeneities were found. It was concluded that the particles consisted of unplasticized PVC, probably in the crystalline state.

7.3.3.6 Raman Spectroscopy (RS)

The main advantage of Raman spectroscopy in polymer characterization is that a good spectrum can be obtained with little or no sample handling. This process is extremely valuable in many different cases for polymer membrane characterization. It can be used to determine the functional groups, and group structure, conformation and orientation of chains and to follow changes in the structural parameters as the polymers are exposed to environmental or mechanical stresses. Hydrogen bonding, crystal field splitting, and chain packing are all examples of modification to the surrounding field of the molecular unit, which can be studied by RS. It is also sensitive to the thermal history of polymers. Raman polarization studies can reveal the information about the distribution and orientation of structural units for both crystalline and non crystalline regions of the polymer. However, optical clarity and fluorescence can cause problems, especially with impure and chemically complex samples. Another problem specific to polarization studies is the scrambling caused by multiple scattering in heterogeneous systems. Nevertheless, Raman polarization measurements on polymers can provide extremely valuable structural information. An important area of research in polymeric membrane studies involves how the physical and mechanical properties of a polymer are influenced by molecular orientation induced by drawing.

The infrared and Raman spectra of a molecule compliment each other and information on the complete vibrational spectra of a molecule often requires both infrared and Raman vibration. The increased energy may be at the level of the electronic, vibrational or rotational energy of the molecule. Interaction between molecular units and their surroundings can be readily detected by perturbations in the Raman spectra.

In RS, light scattered by the molecules contains frequencies other than that of incident monochromatic light. It is the differences between the frequencies of scattered light and the frequency of the incident light that correspond to the normal vibrational frequencies of the molecules. When a molecule absorbs radiations its energy increases in proportion to the photon.

$$E = h\nu = \frac{hc}{\lambda} \tag{7.8}$$

where c is the velocity of light, h the Planck's constant, λ the wave length of the radiation, and ν the frequency. The increased energy may be at the level of the electronic vibrational or rotational energy of the molecule. Interaction between molecular units and their surroundings can be readily detected by perturbation in the Raman spectra.

An important area of research in polymer studies involves how the physical and mechanical properties of a polymer are influenced by molecular orientation induced by drawing. Polymers are varied between amorphous and crystalline states. The crystallinity content depends on the molecular weight of the polymer. Low molecular weight polymer will have a high content of crystallites in comparison with high molecular weight polymer. Various morphologies are possible between a completely crystalline and completely amorphous conformation. The formation of a crystalline region depends on the time allowed for the polymer to crystallize from solution. Crystalline polymers have a number of morphological features that can be studied by RS.

There are three different Raman spectroscopic methods, which can be used for surface studies: microprobe, internal reflection and surface enhanced Raman spectroscopy (SERS). Analyses of a cross section of a specimen by SERS can be used for depth profiling with resolution in the order of 1 μm. Structures of surface can be studied also for laminated polymers and embedded materials. RS can be obtained ranging from a few angstroms to several micrometers of thickness of polymer coated on a surface.

Polymeric membranes are processed in a variety of ways (casting of solutions by doctor blade or spinning etc.), and thus, it is fairly common that some orientation, either in plane or out of plane, will be induced. The morphology of the membrane, including the orientation of polymer chain and the degree of crystallinity of the polymer in the membrane, can be studied by RS, which also enables monitoring of the morphology change during membrane formation. The morphology–membrane transport relationship will help to reveal the mechanism of mass transport of the membrane. RS may also contribute to establish the transport of permeating molecules, either in gas or liquid form. However, so far no work has been done to correlate RS and permeation properties of membranes.

Chemical modification, such as that occurring during ion implantation in polymers, has been monitored by RS [64]. Khulbe et al. [17, 65, 66] studied poly(phenylene oxide) (PPO) powder and dense homogeneous membranes prepared by casting PPO using different solvents—i.e., carbon disulfide, benzene, 1,1,2-trichloroethylene (TCE), toluene, chlorobenzene and bromobenzene—and studied both surfaces (top and bottom layers) of the membranes by RS. These membranes were also subjected to gas separation experiments. It was revealed that the structural changes noticed by RS were correlated to the membrane performance. In another RS study, Khulbe et al. [17, 66] found that the top layer (active layer) was more amorphous than the bottom layer of the dense membrane and the crystalline phase in the dense membrane depended on the boiling point (BP) of solvent used for the preparation of membrane.

RS studies on polymeric membranes appear to hold a great potential in the future for many different applications. Study of crystalline structure of polymers will add to the existing knowledge of intramolecular forces in crystals and their effect on stable polymer structure. The use of vibration spectroscopy will help to understand the relationship between structure and transport properties, which could have an important impact on designing membranes for specific separation problems. RS study of isotropic polymer membranes can be related to permeability by using free volume as an intermediary property; however, no such work has been carried out so far.

7.3.3.7 Electron Spinning Resonance (ESR)

ESR is based on the same principal as NMR except that microwave (rather than radio wave) frequencies are employed, and spin transitions of unpaired electrons rather than nuclei are recorded. Species that contain unpaired electrons (namely, free radicals, odd-electron molecules, transition metal complexes, rare earth ions, etc.) can therefore be detected by ESR.

When an atomic or molecular system with unpaired electrons is subjected to a magnetic field, the electronic energy levels of the atom or molecule will split into different levels. The magnitude of the splitting is dependent on the strength of the applied magnetic field. The atom or molecule can be excited from one split level to another in the presence of an external radiation of frequency, corresponding to the frequency obtained from the difference in energy between the split levels. Such an excitation is called a magnetic resonance absorption. The atom or molecule under investigation may be in different environments in an actual sample. The magnetic resonance frequency will, hence, be influenced by the local environment of the atom or molecule. The electron spin resonance technique is, therefore, a probe for detailed identification of the various atomic and molecular systems, their environments, and all associated parameters. Unlike NMR spectra, where absorption is recorded directly, ESR spectrometers plot the first derivatives of the absorption curve [14].

1. Electron spin resonance, ESR, is a powerful non-destructive and non-intrusive analytical method. It can be used to study the following:

   (a) Molecular structure.
   (b) Crystal structure.
   (c) Reaction kinetics.
   (d) Valence electron wave functions.
   (e) Molecular motion.
   (f) Relaxation properties.
   (g) Electron transport.
   (h) Crystal/ligand fields.
   (i) Reaction mechanisms.

Any substance that has unpaired electrons will give ESR. Some examples of substances that exhibit these characteristics are given below:

1. Atoms or ions having partially filled inner electron shells that are all of the transition elements of the iron series, rare earth's platinum series.
2. Molecules having an odd number of electrons in their outer shells (e.g., NO or ClO_2).
3. Molecules with an even number of electrons in their outer shells but with resultant magnetic moments (e.g., O_2).
4. Free radicals, which are naturally or artificially produced.
5. Conduction electrons in metals and acceptors and donors in semiconductors.
6. Modified crystal structure and defects in crystals, e.g., color centers.

Quantum mechanics reveals that every electron acts as a magnetic dipole and, as such, if it is placed in a static magnetic field, H, can have only two possible orientations, with or against the field. In thermal equilibrium the greater number of electrons occupy the lower energy levels according to the Boltzmann statistics. In the presence of an electromagnetic variation field electrons in the lower energy levels can absorb photons of energy, $h\nu$ (h is the Planck's constant and ν the frequency) and thereby are excited to a higher energy level, which corresponds to the energy absorbed. The absorption energy can be observed as a function of external applied field value by means of high frequency technique. The relation between field value and measuring frequency can be described quantitatively as follows:

$$h\nu = g\beta H \tag{7.9}$$

where g is the spectroscopic splitting factor or Lande's constant and β the Bohr constant. The g value of a free electron is 2.0035. The magnetic interactions between the electron spins and nuclear spins cause the ESR spectrum to consist of a number of lines rather than a single line. The arrangement of the resulting group of lines in the ESR spectrum is called the hyperfine structure of the spectrum.

Polymers themselves contain paramagnetic free radicals. It is possible that these radicals may take part in the transportation of gases through the membrane. It has been observed that these radicals are affected reversibly with gases. Khulbe et al. observed that the PPO radicals in PPO powder, and in membranes prepared from it, contain free radicals that are affected by the conditions of the environment [67]. Froyer et al. [68] prepared poly(phenylene) (PP) by using two different polymerization procedures. Although the two products were very similar from a structural point of view, their magnetic properties as seen by ESR were very different. They interpreted this by correlating the number of spins with the number of structural defects that were probably formed during the polymerization process. Lou et al. [69] suggested that the ESR has a significant potential of a multifrequency ESR approach to the complex dynamics of membranes.

A stable radical can also be introduced into polymeric material. The radical, so introduced, is often called a spin label or a spin probe. It is invariably a nitroxide radical, which exhibits a three-line hyperfine structure. The peak shape and splitting depend on the radical's environments. The nitroxide label is a monitor of motion. The shape of the ESR signal depends also on the orientation of the magnetic field relative to the axis of the radical. Thus, the spin label method is useful to study the environment of radicals at a molecular level [14]. Probably Stone et al. [70] are the first who introduced the paramagnetic nitroxide radical into the polymer (synthetic polypeptides).

Spin labelling in synthetic polymers has been discussed by Miller [71]. The spin labelling method was applied by Khulbe et al. to investigate synthetic polymeric membranes for the first time [72]. Figure 7.16 shows the ESR spectra of the membrane-casting solution (polyethersulfone + polyvinylpyrrolidone + *N*-methyl pyrrolidone) containing TEMPO. (2,2,6,6-tetramethylpiperridinoxy free radical) [73]. The isotropic spectrum is due to the rapid and randomly tumbling nitroxides.

Figure 7.17 shows the ESR spectra of TEMPO in the membrane. Unlike Fig. 7.16, the spectrum is no longer isotropic but similar to the one reported by Griffith and Waggnor [74] when the TEMPO was in glass or in a highly viscous liquid. These spectra suggest that the radical is very immobile in the polymer. On keeping the membrane in water for 24 h, no appreciable change either in shape or in peak intensity was noticed. It seems the radicals are diffused in the polymer matrix where water cannot affect them.

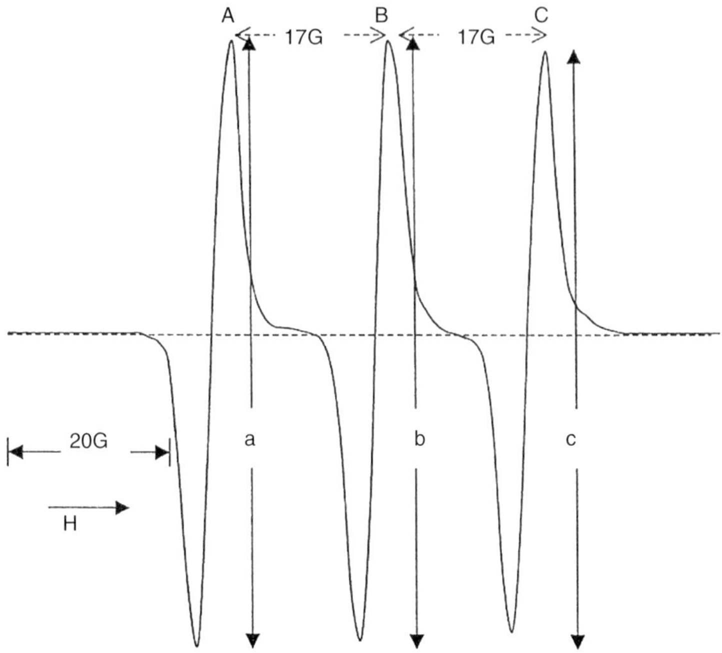

Fig. 7.16 ESR spectra of the membrane-casting solution containing TEMPO

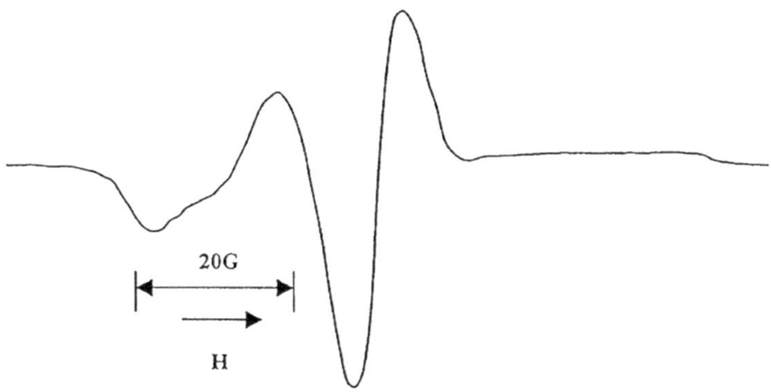

Fig. 7.17 ESR spectra of membrane prepared (evaporation period 1 min) from a casting solution incorporated with TEMPO

Khulbe et al. [72] prepared dense homogeneous polyphenyleneoxide (PPO) membranes by casting a solution, which consists of PPO, TEMPO (spin probe), and tetrachloroethylene (TCE) solvent. The solvent was evaporated at 22, 4, and −10 °C. Membranes were subjected to ESR spectroscopy as well as to permeation experiments, thus correlating ESR signals to the membrane performance data.

The ESR technique offers an investigator or a manufacturer several advantages over other methods of membrane characterization. In particular, ESR provides information without disturbing the membrane. Although the application of the ESR method to the synthetic polymeric membrane is not as popular as to the biological membrane, it is possible to obtain information, at a molecular level, on the environment of the permeating molecules.

7.3.3.8 Nuclear Magnetic Resonance

Nuclear Magnetic Resonance (NMR) spectroscopy is an analytical chemistry technique used in quality control and research for determining the content and purity of a sample as well as its molecular structure. The process works through analyzing absorption of electromagnetic energy (typically radio waves) by the nuclei of atoms placed in a strong magnetic field. The nuclei of different atoms absorb unique frequencies of radiation depending on their environment; thus, by observing which frequencies are absorbed by a sample placed in a strong magnetic field (and later emitted again, when the magnetic field is removed), it is possible to learn much about the sample's makeup and structure.

Solid-state NMR (SSNMR) spectroscopy is a kind of NMR spectroscopy, characterized by the presence of anisotropic (directionally dependent) interactions. Solid-state NMR spectroscopy is widely applicable to the investigation of non-crystalline or amorphous materials, e.g., polymers, glasses, protein precipitates, and membrane proteins [75–77]. One area of solid-state NMR spectroscopy that has proven fruitful with regard to the investigation of membranes is ^{2}H NMR spectroscopy. Solid-state NMR spectroscopy is a powerful tool for the investigation of crystalline and amorphous solids. It can provide a large amount of information about local structure around selected atoms/nuclei and is extensively employed in the studies of new inorganic, organic, and hybrid materials [78].

7.4 Other Techniques

7.4.1 Optical Technique

Optical microscopy was historically the first technique used to observe what could not be seen with the naked eye. Development of the magnifying glass and then of the optical microscope occurred some four hundred years ago. The first generation of (simple) optical microscopes used one focusing lens. Microscopes have evolved to using multiple (compound) lenses in order to enhance magnification.

One of many objective lenses–ocular lens pieces can usually be used for a specific magnification. Magnification factors have improved over the years to a range between 4 and 1,000. Actually, the best present-day optical microscopes can magnify up to 2,000 times. The optical microscope's best resolution is around 0.2 μm. Optical microscopy uses white (visible) light [79, 80].

The thickness of a membrane film can be measured by the optical technique, which is nondestructive and relatively inexpensive. Interferometry relies on the interference of two or more beams of light, e.g., from the air/film surface and the film–substrate interface, where the optical path difference is related to film thickness. An optical film thickness sensor is usually used to measure the thickness of membranes. It is based on either interferometric or reflectometric technology, which enables the tool to non-destructively measure films of a few millimeters down to just 10 nm in thickness. These tools are available at market [81].

7.5 Thermal Properties

7.5.1 Differential Scanning Calorimeter (DSC) and Differential Thermal Analysis (DTA)

Differential scanning calorimetry (DSC) measures the amount of heat absorbed or emitted by a system as well as by a "reference" sample undergoing heating and/or cooling cycles. The difference in the amount of heat required to increase the temperature of both the measured and reference samples is measured as a function of temperature. Both samples are maintained at nearly the same temperature throughout the measurement [82].

DSC is a thermodynamical tool for direct assessment of the heat energy uptake, which occurs in a sample within a regulated increase or decrease in temperature. The calorimetry is particularly applied to monitor the changes of phase transitions. A differential scanning calorimeter (DSC) and differential thermal analysis (DTA) are in fact identical techniques used to measure transitions or chemical reactions in a polymer sample. DSC determines the energy (dQ/dt) necessary to counteract any temperature difference between the sample and the reference, whereas DTA determines the temperature difference (ΔT) between the sample and the reference upon heating or cooling. DSC is a thermal analysis apparatus measuring how physical properties of a sample change along with temperature against time. In other words, the device is a thermal analysis instrument that determines the temperature and heat flow associated with material transitions as a function of time and temperature. During a change in temperature, DSC measures a heat quantity, which is radiated or absorbed excessively by the sample on the basis of a temperature difference between the sample and the reference material.

Various types of calorimeters have been developed, and their applications could improve the thermoanalysis of a wide range of materials. On the other hand, in the last decades, considerable progress has been made in applications of DSCs in microfluidics, drug discoveries, pharmaceuticals, molecular biology, and nanoscience [83].

There are other microscopic methods to characterize the membranes such as static secondary ion mass spectrometry (SSIMS), energy dispersive X-ray spectroscopy

(EDXS), laser confocal scanning microscopy (LCSM), environmental scanning electron microscopy (ESEM), and contact angle measurement [84].

Scanning electron microscopy (SEM) combined with energy dispersive X-ray (EDX or EDS) analysis has proved to be valuable tools in determining the chemical composition of membrane and the distribution of elements.. Energy-dispersive X-ray spectroscopy (EDS, EDX, or XEDS), sometimes called energy dispersive X-ray analysis (EDXA) or energy dispersive X-ray microanalysis (EDXMA), is an analytical technique used for the elemental analysis or chemical characterization of a sample. It relies on an interaction of some source of X-ray excitation and a sample. Its characterization capabilities are due in large part to the fundamental principle that each element has a unique atomic structure allowing unique set of peaks on its X-ray spectrum [85]. Electron images in the SEM display compositional contrast results from different atomic number elements and their distribution. Energy dispersive spectroscopy (EDS) allows one to identify what those particular elements are and their relative proportions (Atomic % for example). EDS analysis usually involves in generation of an X-ray spectrum from the entire area of SEM.

7.6 Mechanical Properties

7.6.1 Tensile Strength

Tensile strength is a measure of the mechanical properties of a material. The ability to resist breaking under tensile stress is one of the most important and widely measured properties of materials used in structural applications. The force per unit area (MPa or psi) required to break a material is the ultimate tensile strength, or tensile strength to break. In general, tensile strength quantifies how much stress the material will hold before suffering permanent damages [86].

Nghiem and Schäfer [87] used tensile strength as a parameter for membrane autopsy studies. To measure the tensile strength of membrane is to understand what level the membrane can be subjected to (stress and strain) during its application (gas separation), and to quantify how much the material has altered as compared to a fresh polymer membrane. The membrane after long years of operation is expected to become more brittle and more susceptible to integrity problems.

The tensile test is a destructive test method. It can be carried out using any standard testing equipment and using the standard test method for the tensile properties of plastics, which are specified in ASTM D638-10. When a material is subjected to a tensile pull force, it undergoes elongation until it breaks. The results can be seen in the form of a curve showing how the material reacted to the forces being applied. The point of interest is the percent elongation of the material and the stress at the point of ultimate failure where the materials (membrane) break or snap. ASTM D882 is commonly used for testing the tensile strength of polymer films. Figure 7.18 gives a diagram for measuring tensile strength.

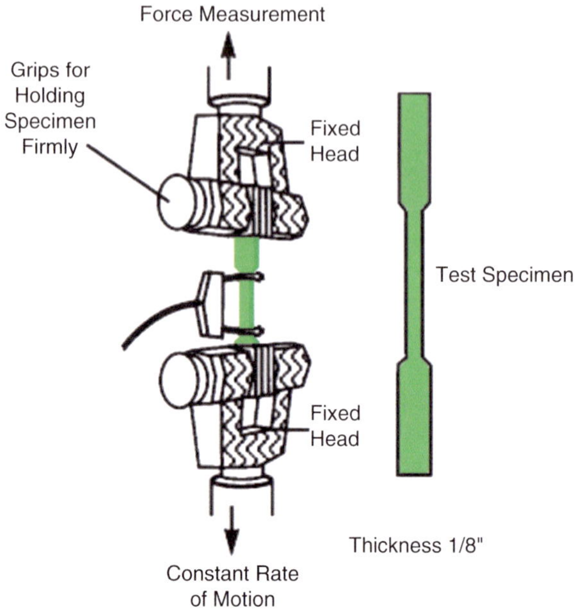

Fig. 7.18 Schematic (main) for measuring the tensile strength of films

Ginga and Sitaraman [88] demonstrated a new fracture testing technique that can be used to determine the tensile strength of low-strength thin films. The technique avoids the issues associated with testing thin films by utilizing simple photolithography-based sample preparation, not requiring fixturing of thin films, and is driven by the high intrinsic stress of the support layer film. This technique uses finite element analysis to extract the tensile strength from the experimental data.

7.6.2 Young's Modulus or Tensile Modulus of Elasticity

When rigid materials are subject to particular stresses or forces, deformation (compression, twisting, stretching, etc.) may occur. For many materials, when suffering from force or stress, the resisting or restoring force that tends to return the material to its original shape is proportional to the deformation. Young's Modulus, E, is a constant that describes a material's mechanical property of stiffness and is expressed as the ratio of stress to strain for a material experiencing tensile or compressive stress.

$$\lambda = \text{Stress} / \text{strain} \tag{7.10}$$

where lambda (λ) is the elastic modulus; stress is the restoring force caused due to the deformation divided by the area to which the force is applied; and strain is the ratio of the change caused by the stress to the original state of the object. If stress is measured in pascals, since strain is a dimensionless quantity, then the units of λ are pascals as well [89].

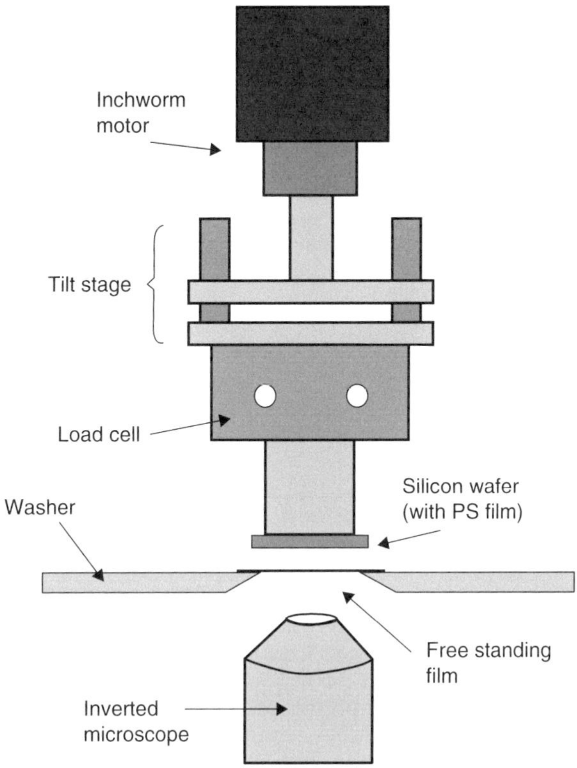

Fig. 7.19 Schematic of the experimental apparatus

In other words an elastic modulus, or modulus of elasticity, is the mathematical description of an object's or substance's tendency to be deformed elastically (i.e., non-permanently) when a force is applied to it. The elastic modulus of an object is defined as the slope of its stress–strain curve in the elastic deformation region [90]. As such, a stiffer material will have a higher elastic modulus.

Raegen et al. presented a novel method of probing adhesion energies and Young's modulus for polymers. This technique uses the axisymmetric deformation of thin spincast polymer membranes brought into contact with a flat substrate to probe the work of adhesion [91]. Figure 7.19 shows a schematic of the experimental set-up. Flat substrates (films) mounted on a Transducer Techniques SGO Series 50 g Load Cell, is attached via a Newport MM-1 tilt stage to a Burleigh Inchworm linear stepper motor. This module is mounted axisymmetrically above the free-standing film, atop an Olympus IX70 inverted microscope equipped with a Pixelink CCD camera. The entire apparatus is placed on a Halcyonics MOD-1 anti-vibration table. A bandpass filter of 630, 530, or 460 nm was used to observe interference fringes set up between the two films, and to enhance the visibility of the contact patch as needed (more on the use of the filters below). A lab view program enables control of the motor, automation of image capture, and data acquisition.

The use of a thin membrane minimizes uncertainty in the radius of contact, while the use of spincast films provides very smooth surfaces by means of a simple method. The experimental set up enables the measurement of Young's modulus for thin films.

Rafiq et al. [92] reported that with the increase in the PI content in the polysulfone–polyimide (PSF-PI) blended polymeric membranes, the mechanical properties of the membranes, like Young's modulus, tensile strength and elongation at break, increased. Permeance of CO_2 and CH_4 through the membrane increased with the increase in PI content.

Metal nanoparticle-containing film can show enhanced mechanical properties as well as impart multifunctionality like catalysis and gas separation capability. Similar results were reported by Yang and Brown [93] with polyetherketone hollow-fiber membrane for gas separation.

References

1. Schmidhauser JC, Longley KL (1990) Gas transport through bisphenol-containing polymers. ACS Symp Ser 423:159–176
2. George SC, Thomas T (2001) Transport phenomena through polymeric systems. Prog Polym Sci 26:985–1017
3. Lashkari S, Tran A, Kruczek B (2008) Effect of back diffusion and back permeation of air membrane characterization in a constant pressure system. J Membr Sci 324:162–172
4. Tabe Mohammadi A, Matsuura T, Sourirajan S (1995) Design and construction of a gas permeation system for the measurement of low permeation rates and permeate compositions. J Membr Sci 98:281–286
5. O'Brien KC, Koros WJ, Barabari TA, Sanders ES (1986) A new technique for the measurement of multicomponent gas transport through polymeric films. J Membr Sci 29:229–238
6. Moore TT, Damle S, Williams PJ, Koros WJ (2004) Characterization of low permeability gas separation and barrier materials; design and operation conditions. J Membr Sci 245:227–231
7. Peighambardoust SJ, Rowshanzamir S, Amjadi M (2010) Review of the proton exchange membranes for fuel cell preparation. Int J Hydrogen Energy 35:9349–9384
8. Binnig G, Quate CF, Gerber C (1986) Atomic force microscope. Phys Rev Lett 56:930–933
9. Khulbe KC, Feng CY, Matsuura T (2008) Synthetic polymeric membranes, characterization by atomic forces microscopy. Springer, Heidelberg
10. Schmitz I, Schreiner M, Friedbacer G, Grasserbaur M (1997) Phase imaging as an extension to tapping mode AFM for the identification of material properties on humidity-sensitive surfaces. Appl Surf Sci 115:190–198
11. Drake B, Prater CB, Weisenhorn AL, Gould SAC, Albrecht TR, Quate CF, Cannel PS, Hansma HG, Hansma PK (1989) Imaging crystals, polymers, and processes in water with the atomic force microscope. Science 243:1586–1589
12. Martin Y, Williams CC, Wickramasinghe GH (1987) Atomic force microscope-force mapping and profiling on A sub 100-Å scale. J Appl Phys 61:4723–4729
13. Albrecht TR, Dovek MM, Lang CA, Grutter P, Quate CF, Kuan SNJ, Frank CW, Pease RFW (1988) Imaging and modification of polymers by scanning tunneling and atomic force microscopy. J Appl Phys 64:1178–1184
14. Khulbe KC, Matsuura T (2000) Characterization of synthetic membranes by Raman spectroscopy, electron spin resonance, and atomic force microscopy: a review. Polymer 41: 1917–1935
15. Hirose M, Itoh H, Minamizaki Y, Kamiyama Y (1996) Ultra-low-pressure reverse osmosis membranes ES 10. In: Proceedings of the international congress on membranes and membrane process, Yokohama, Japan, 18–23 August 1996

16. Elimelech M, Zhu X, Childress AE, Hong S (1997) Role of membrane surface morphology in colloidal fouling of cellulose acetate and composite aromatic polyamide reverse osmosis membranes. J Member Sci 127:101–109
17. Khulbe KC, Matsuura T, Lamarche G, Kim HJ (1997) The morphology characterisation and performance of dense PPO membranes for gas separation. J Membr Sci 135:211–223
18. Kesting RE (1990) The four tiers of structure in integrally skinned phase inversion membranes and their relevance to the various separation regimes. J Appl Polym Sci 53:2739–2752
19. Mulder M (1992) Basic principles of membrane technology. Kluwer Academic Publishers Group, The Netherlands
20. Kesting RE, Engdahl M, Stone W Jr (1969) The application of scanning electron microscopy to membrane morphology. J Macromol Sci A—Chemistry 3(1):157–167
21. Patterson DA, Havill A, Costello S, See-Toh YH, Livingston AG, Turner A (2009) Membrane characterisation by SEM, TEM and ESEM: the implications of dry and wetted microstructure on mass transfer through integrally skinned polyimide nanofiltration membranes. Sep Puri Technol 66:90–97
22. Riley RL, Gardner JO, Merten U (1964) Cellulose acetate membranes: electron microscopy of structure. Science 143:801–803
23. Merin U, Cheryan M (1980) Ultrastructure of the surface of a polysulfone ultrafiltration membrane. J Appl Polym Sci 25:2139–2142
24. Koutake M, Uchida Y, Kimura T, Sagara A, Watanabe A, Nakao S (1985) Observation of UF membrane pores through a scanning electron microscope and their pure water fluxes. Maku 10:310–312
25. Zeman L, Denault L (1992) Characterization of microfiltration membranes by image analysis of electron micrographs: part I. Method development. J Membr Sci 71:221–231
26. Zeman L (1992) Characterization of microfiltration membranes by image analysis of electron micrographs: part II. Functional and morphological parameters. J Membr Sci 71:233–246
27. Schossig M, Paul D (2001) Improved preparation of membrane surfaces for field-emission scanning electron microscopy. J Membr Sci 187:85–91
28. Wienk IM, Boomgaard TV, Smolders CA (1994) The formation of nodular structures in the top layer of ultrafiltration membranes. J Appl Polym Sci 53:1011–1023
29. Rhim JW, Kim JR, Park YI, Lee KH (2001) Modification of polysulfone membranes and their applications to gas separations. J Ind Eng Chem 7:299–304
30. Clark MB, Burkhardt CA, Gardella JA (1991) Surface of polymer blends. 4. An ESCA, IR, and DSC study of the effect of homopolymer molecular weight on crystallinity and miscibility of poly(ε-caprolactone)/poly(vinyl chloride) homopolymer blends. Macromolecules 24:799–805
31. Brandt W, Berko S, Walker WW (1960) Positronium decay in molecular substances. Phys Rev 120:1289–1295
32. Shantarovich VP, Azamatova ZK, Novikov YA, Yampolskii YP (1998) Free-volume distribution of high permeability membrane materials probed by positron annihilation. Macromolecules 31:3963–3966
33. Tao SJ (1972) Positronium annihilation in molecular substances. J Chem Phys 56:5499–5510
34. Eldrup M, Lightbody D, Sherwood JN (1981) The temperature dependence of positron lifetime in solid pivalic acid. Chem Phys 63:51–58
35. Shantarovich VP, Yampolskii YP, Kevdina IB (1994) Free volume and time of a life of positronium in polymeric systems. Chem High Energ 28:53–59
36. Gregory RB, Yongkang Z (1991) Positron and positron chemistry (Jean YC, Ed.). World Scientific, Singapore
37. Gregory RB (1991) Free-volume and pore-size distributions determined by PALS. J Appl Phys 70:4665–4670
38. Deng Q, Jean YC (1993) Free-volume distribution of an epoxy polymer probed by positron annihilation spectroscopy: pressure dependence. Macromolecules 26:30–34
39. Satyanarayana SV, Subrahmanyam VS, Verma HC, Sharma A, Bhattacharya PK (2006) Application of positron annihilation: study of pervaporation dense membranes. Polymer 47:1300–1307

40. Jean YC, Malton PE, Schrader DM (2003) Principles and applications of positron and positronium chemistry. World Scientific Publishers, New Jersey
41. Jean YC, Sandreczki TC, Ames DP (1986) Positronium annihilation in amine-cured epoxy polymers. J Polym Sci B Polym Phys 24:1247–1258
42. Hill AJ, Weinhold S, Stack GM, Tant MR (1996) Effect of copolymer composition on free volume and gas permeability in poly(ethylene terephthalate)-poly(1,4 cyclohexylenedimethylene terephthalate) copolyesters. Eur Polym J 32:843–849
43. Yin HH, Yin Z, Ma W, Zhu D (2005) A review of studies of polymeric membranes by positron annihilation lifetime spectroscopy. Plasma Sci Tech 7:3062–3064
44. Chen H, Hung WS, Lo CH, Huang SH, Cheng ML, Guang L, Lee KR, Lai JY, Sun YM, Hu CC, Suzuki R, Ohdaira T, Oshima N, Jean YC (2007) Free-volume depth profile of polymeric membranes studied by positron annihilation spectroscopy: layer structure from interfacial polymerization. Macromolecules 40:7542–7557
45. Tung KL, Jean YC, Nanda D, Lee KR, Hung WS, Lo CH, Lai JY (2009) Characterization of multilayer nanofiltration membranes using positron annihilation spectroscopy. J Membr Sci 343:147–156
46. Shantarovich VP, Kevdina IB, Yampol'skii YP (2000) Evaluation of nonuniformity of polymeric membrane materials by positron annihilation technique. High Energy Chem 34:265–272
47. Huang SH, Hung WS, Liaw DJ, Li CL, Kao ST, Wang DM, Guzman MD, Hu CC, Jean YC, Lee KR, Lai JY (2008) Investigation of multilayer pervaporation membrane by positron annihilation spectroscopy. Macromolecules 41:6438–6443
48. Marques MFF, Gil CL, Gordo PM, Kajcsos Z, Lima AP, Queiroz DP, Pinho MN (2003) Free-volume studies in polyurethane membranes by positron annihilation spectroscopy. Radiat Phys Chem 68:573–576
49. Kobayashi Y, Kazama S, Inoue K, Toyama T, Nagai Y, Haraya K, Mohamed HFM, O'Rouke BE, Oshima N, Kinomura A, Suzuki R (2014) Positron annihilation in cardo-based polymer membranes. J Phys Chem B 118:6007–6014
50. Khulbe KC, Matsuura T, Lamarche G, Lamarche AM (2000) X-ray diffraction analysis of dense PPO membranes. J Membr Sci 170:81–89
51. Kim TH, Koros WJ, Husk GR, O'Brien KC (1988) Relationship between gas separation properties and chemical structure in a series of aromatic polyimdes. J Membr Sci 37:45–620
52. Shimazu A, Miyazaki T, Ikeda K (2000) Interpretation of d-spacing determined by wide angle x-rays scattering in 6FDA-based polyimide by molecular modeling. J Membr Sci 166: 113–118
53. Miller RL, Boyer RF (1984) Regularities in x-ray scattering patterns from amorphous polymers. J Polym Sci Phys Ed 22:2043–2050
54. Boyer RF (1987) Order in the amorphous state of polymers. Plenum Press, New York
55. Khayet M, García-Payo MC (2009) X-ray diffraction study of polyethersulfone polymer, flat sheet and hollow fibers prepared from the same under different gas-gaps. Desalination 246:121–127
56. Suk D (2005) Development of surface modifying macromolecule blended polyethersulfone membranes for vacuum membrane distillation. Ph.D. thesis, Department of Chemical Engineering, University of Ottawa
57. Fang Y, Pham VA, Matsuura T, Santerre PJ, Narbaitz R (1994) Effect of surface-modifying macromolecules and solvent evaporation time on the performance of polyethersulfone membranes for the separation of chloroform/water mixtures by pervaporation. J Appl Polym Sci 54:1937–1943
58. Ho JY, Matsuura T, Santerre JP (2000) The effect of fluorinated surface modifying macromolecule on the surface morphology of polyethersulfone membranes. J Biomater Sci Polym Ed 11:1085–1104
59. Wang Y, Yin S, Ren L, Zhao L (2009) Surface characterization of the chitosan membrane after oxygen plasma treatment and its aging effect. Biomed Mater 4:035003
60. Pan J, Heberle FA, Petruzielo RS, Katsaas J (2013) Using small-angle neutron scattering to detect nanoscopic lipid domains. Chem Phys Lipids 170–171:19–32

61. He K, Ludtke SJ, Worcester DL, Huang HW (1996) Neutron scattering in the plane of membranes: structure of alamethicin pores. Biophys J 70:2659–2666
62. Strunz P, Mukherji D, Saroun J, Keiderling U, Rösler J (2010) Pore structure characterization and in-situ diffusion test in nanoporous membrane using SANS. J Phys Conf Ser 247:012023
63. Ye Q, Borbély S, Horvai G (1999) Microstructure of ion-selective plasticized PVC membranes studied by small-angle neutron scattering. Anal Chem 71:4313–4320
64. Gall MJ, Hendra PJ, Peacock CJ, Cudby MEA, Willis HA (1972) Laser-Raman spectrum of polyethylene: Part 1. Structure and analysis of the polymer. Polymer 13:104–108
65. Khulbe KC, Kruczek B, Chowdhury G, Gagne S, Matsuura T, Verma S (1996) Characterization of membranes prepared from PPO by Raman scattering and atomic force microscopy. J Membr Sci 111:57–70
66. Khulbe KC, Matsuura T, Kim HJ (2000) Raman scattering of PPO membranes. J Appl Polym Sci 77:2558–2560
67. Khulbe KC, Chowdhury G, Matsuura T, Lamarche G (1997) Characterization of PPO [poly(phenylene oxide)] powder and membranes from it by ESR technique. J Membr Sci 123:9–15
68. Froyer G, Maurice F, Bernier P, Mc Andrew P (1982) EPR studies on polyphenylene)s synthesized by two different routes. Polymer 23:1103–1105
69. Lou Y, Ge M, Freed JH (2001) A multifrequency ESR study of the complex dynamics of membranes. J Phys Chem B 105:11053–11056
70. Stone TJ, Buckman T, Nordio PC, McConnel HM (1965) Spin-labeled biomolecules. Proc Natl Acad Sci U S A 54:1010–1017
71. Miller WG (1976) Spin labeled synthetic polymers molecular biology spin labeling II: Theory and application. Academic Press, New York
72. Khulbe KC, Chowdhury G, Kruczek B, Vujosevic R, Matsuura T, Lamarche G (1997) Characterization of the PPO dense membrane prepared at different temperatures by ESR, atomic force microscope and gas permeation. J Membr Sci 126:115–122
73. Khulbe KC, Matsuura T, Feng CY, Lamarcdhe G, Lamarche AM (2002) Characterization of ultrafiltration membrane prepared from poly ethersulfone by using electron spin resonance technique. Sep Purif Technol 29:15–22
74. Griffith OH, Waggnor AS (1969) Nitroxide free radicals: Spin labels for probing biomolecular structure. Acct Chem Res 2:17–24
75. Porbeni FE, Shin ID, Shuai XT, Wang XW, White JL, Jia X, Tonelli AE (2005) Morphology and dynamics of the poly(ϵ-caprolactone)-b-poly(L-lactide) diblock copolymer and its inclusion compound with α-cyclodextrin: A solid state ^{13}C NMR study. J Polym Sci Part B: Polym Phys 43:2086–2096
76. Wang LY, Fang PF, Ye CH, Feng JW (2006) Solid-state NMR characterizations on phase structures and molecular dynamics of poly(ethylene-co-vinyl acetate). J Polym Sci Part B: Polym Phys 44:2864–2879
77. Kurosu H, Yamamoto Y, Fujikawa A, Kawabata E, Sone M, Naga N (2009) Structure and dynamics of poly(ethylene-co-1,5-hexadiene) as studied by solid state 13C NMR and quantum chemical calculations. J Mol Struct 921:208–214
78. Laws DD, Bitter H-ML, Jerschow A (2002) Solid-state NMR spectroscopic methods in chemistry. Angew Chem Int Ed 41:3096–3129
79. Pawley JB (ed) (2006) Handbook of biological confocal microscopy. Springer, Berlin
80. Ho DL, Hammouda B, Kline S, Chen WR (2006) Unusual phase behavior in mixtures of poly(ethylene oxide) and ethyl alcohol. J Polym Sci Part B: Polym Phys 44:557–564
81. The MicroSpy® FT, FRT, Fries Research & Technology GmbH. http://www.frt-gmbh.com
82. Dean JA (1995) The analytical chemistry handbook. McGraw Hill, New York
83. Chiu MH, Prenner EJ (2011) Differential scanning calorimetry: an invaluable tool for a detailed thermodynamic characterization of macromolecules and their interactions. J Pharm Bioallied Sci 3:39–59
84. Xu ZK, Huang XJ, Wan LS (2009) Surface engineering of polymer membranes (advanced topics in science and technology in China). Springer, New York

85. Joseph IG, Dale EN, Patrick E, David CJ, Charles EL, Eric L, Linda S, Joseph RM (2003) Scanning electron microscopy and x-ray microanalysis. Springer, New York
86. Ashby MF, Jones DRH (1996) Energy materials I, 2nd edn. Butterworth Heinemann, Oxford
87. Nghiem LD, Schäfer AI (2006) Fouling autopsy of hollow-fiber MF membranes in wastewater reclamation. Desalination 188:113–121
88. Ginga NJ, Sitarama SK (2011) New method to measure tensile strength of low modulus thin films. Int J Fract 170:199–206
89. Beer F, Johnston ER, John D, Mazurek D (2009) Mechanics of materials. McGraw Hill, Houston, TX
90. Askeland DR, Phulé PP (2006) The science and engineering of materials. Cengage Learning, United Kingdom
91. Raegen AN, Dalnoki-Veress K, Wan KT, Jones RA (2006) Measurement of adhesion energies and Young's modulus in thin polymer films using a novel axisymmetric peel test geometry. Eur Phys J E Soft Matter 19:453–459
92. Rafiq S, Man Z, Maitra S, Maulud A, Ahmad F, Muhammad N (2011) Preparation of asymmetric polysulfone/polyimide blended membranes for CO_2 separation. Korean J Chem Eng 28:2050–2056
93. Yang J, Brown P (2007) Highly gas permselective polyetherketone hollow fibre membranes using aqueous sulfuric acid solution as coagulant. e-Polymers 7:884–895

Index

A.F. Ismail et al., *Gas Separation Membranes*,
DOI 10.1007/978-3-319-01095-3

MIX
Papier aus verantwortungsvollen Quellen
Paper from responsible sources
FSC® C105338

If you have any concerns about our products,
you can contact us on
ProductSafety@springernature.com

In case Publisher is established outside the EU,
the EU authorized representative is:
Springer Nature Customer Service Center GmbH
Europaplatz 3, 69115 Heidelberg, Germany

Printed by Libri Plureos GmbH
in Hamburg, Germany